Monitoring the Comprehensive Nuclear-Test-Ban Treaty: Data Processing and Infrasound

Edited by
Zoltan A. Der
Robert H. Shumway
Eugene T. Herrin

Springer Basel AG

Reprint from Pure and Applied Geophysics
(PAGEOPH), Volume 159 (2002), No. 5

Editors:

Zoltan A. Der
ENSCO Inc.,
5400 Port Royal Road
Springfield, VA 22151
USA
e-mail: der@ensco.com

Eugene T. Herrin
Department of Geological Sciences
Southern Methodist University
PO Box 750395
Dallas, Texas 75275-0395
USA
e-mail: herrin@passion.isem.smu.edu

Robert H. Shumway
Division of Statistics
380 Kerr Hall
One Shields Ave.
University of California
Davis, CA 95616
USA
e-mail: shumway@wald.ucdavis.edu

A CIP catalogue record for this book is available from the Library of Congress,
Washington D.C., USA

Deutsche Bibliothek Cataloging-in-Publication Data

Monitoring the comprehensive nuclear-test-ban treaty. - Basel ; Boston ; Berlin : Birkhäuser
(Pageoph topical volumes)

Data processing and infrasound / ed. by Zoltan A. Der ... - 2002
    ISBN 978-3-7643-6676-6   ISBN 978-3-0348-8144-9 (eBook)
    DOI 10.1007/978-3-0348-8144-9

© 2002 Springer Basel AG
Originally published by Birkhäuser Verlag, Basel, Switzerland in 2002
Printed on acid-free paper produced from chlorine-free pulp

9 8 7 6 5 4 3 2 1

Contents

Pure appl. geophys. 159 (2002) 905–906
0033–4553/02/050905–02 $ 1.50 + 0.20/0

❙ **Pure and Applied Geophysics**

Monitoring the Comprehensive Nuclear-Test-Ban Treaty

Preface

The first nuclear bomb was detonated in 1945, thus ushering in the nuclear age. A few political leaders quickly saw a need to limit nuclear weapons through international cooperation and the first proposals to do so were made later that same year. The issue of nuclear testing, however, was not formally addressed until 1958 when the United States, the United Kingdom, and the Soviet Union, initiated talks intended to establish a total ban on that testing (a Comprehensive Test-Ban Treaty or CTBT). Those talks ended unsuccessfully, ostensibly because the participants could not agree on the issue of on-site verification.

Less comprehensive treaties did, however, place some constraints on nuclear testing. The United States, the Unites Kingdom, and the Soviet Union, in 1963, negotiated the Limited Test-Ban treaty (LTBT) which prohibited nuclear explosions in the atmosphere, outer space and under water. The Threshold Test-Ban Treaty (TTBT), signed by the United States and the Soviet Union in 1974, limited the size, or yield, of explosions permitted in nuclear tests to 150 kilotons.

Seismological observations played an important role in monitoring compliance with those treaties. Many of the world's seismologists set aside other research projects and contributed to that effort. They devised new techniques and made important discoveries about the Earth's properties that affect our ability to detect nuclear events, to determine their yield, and to distinguish them from earthquakes. Seismologists are rightfully proud of their success in developing methods for monitoring compliance with the LTBT and TTBT.

Although seismologists have also worked for many years on research related to CTBT monitoring, events of recent years have caused them to redouble their efforts in that area. Between 1992 and 1996 Russia, France and the United States all placed moratoria on their nuclear testing, though France did carry out a few tests at the end of that period. In addition, the United States decided to use means other than testing to ensure the safety and reliability of its nuclear arsenal, and all three counties, joined by the United Kingdom, agreed to continue moratoria as long as no other country tested. Those developments, as well as diplomatic efforts by many nations, led to the renewal of multilateral talks on a CTBT that began in January 1994. The talks led to the Comprehensive Nuclear-Test-Ban Treaty. It was adopted by the United Nations

General Assembly on 10 September, 1996 and has since been signed by 161 nations. Entry of the treaty into force, however, is still uncertain since it requires ratification by all 44 nations that have some nuclear capability and, as of 1 November 2001, only 31 of those nations have done so.

Although entry of the CTBT into force is still uncertain, seismologists and scientists in related fields, such as radionuclides, have proceeded with new research on issues relevant to monitoring compliance with it. Results of much of that research may be used by the International Monitoring System, headquartered in Vienna, and by several national centers and individual institutions, to monitor compliance with the CTBT. New issues associated with CTBT monitoring in the 21st century have presented scientists with many new challenges. They must be able to effectively monitor compliance by several countries that have not previously been nuclear powers. Effective monitoring requires that we be able to detect and locate considerably smaller nuclear events than ever before and to distinguish them from small earthquakes and other types of explosions. We must have those capabilities in regions that are seismically active and geologically complex, and where seismic waves might not propagate efficiently.

Major research issues that have emerged for monitoring a CTBT are the precise location of events, and discrimination between nuclear explosions, earthquakes, and chemical explosions, even when those events are relatively small. These issues further require that we understand how seismic waves propagate in the solid Earth, the oceans and atmosphere, especially in regions that are structurally complex, where waves undergo scattering and, perhaps, a high degree of absorption. In addition, we must understand how processes occurring at explosion and earthquake sources manifest themselves in recordings of ground motion.

Monitoring a CTBT has required and will continue to require, the best efforts of some of the world's best seismologists. They, with few exceptions, believe that methods and facilities that are currently in place will provide an effective means for monitoring a CTBT. Moreover, they expect that continuing improvements in those methods and facilities will make verification even more effective in the future. This topical series on several aspects of CTBT monitoring is intended to inform readers of the breadth of the CTBT research program, and of the significant progress that has been made toward effectively monitoring compliance with the CTBT.

The following sets of papers, edited by Drs. Zoltan Der, Robert Shumway, and Eugene Herrin, present research results on Data Processing and Infrasound Signal Detection that are applicable to monitoring a CTBT. These are the seventh and eighth, as well as the last two topics, addressed by this important series on *Monitoring the Comprehensive Nuclear-Test-Ban Treaty*. Previously published topics are Source Location, Hydroacoustics, Regional Wave Propagation and Crustal Structure, Surface Waves, Source Processes and Explosion Yield Estimation, and Source Discrimination.

Brian J. Mitchell
Saint Louis University
Series Editor

Pure appl. geophys. 159 (2002) 907–908
0033–4553/02/050907–02 $ 1.50 + 0.20/0

| Pure and Applied Geophysics

Introduction

ZOLTAN A. DER,[1] ROBERT H. SHUMWAY[2] and EUGENE T. HERRIN[3]

This issue covers two important aspects of CTBT monitoring: Data processing and infrasound measurements. Data processing is important because a successful nuclear monitoring program will require the handling and rapid processing of large volumes of data. Infrasound detection is of primary importance in detecting the detonation of explosions in the atmosphere.

The topic 'data processing' is defined here as a collection of various approaches to data handling which needed nuclear monitoring. It includes more than signal processing in the conventional sense. Consequently, the six data processing papers in this volume are varied in topic and methodology and provide glimpses into a few aspects of the total task of data processing.

The paper by M. Bahavar and R. North describes the procedures in characterizing seismic background noise characteristics on the seismic stations used in the International Monitoring System (IMS). It discusses various factors that determine the spatial and time variations in background noise at the primary seismic stations used in the IMS.

The paper by G. Leonard, Z. Somer, Y. Bartal, Y. Ben Horin, M. Villagran, and M. Joswig provides an outline of the data processing system used at the NDC in Israel from the point of view of a computer scientist. The description emphasizes the object-oriented design of a user interface, the processing flow and the integration of the system with a geographical information system (GIS).

Two papers by T. Kvaerna, F. Ringdal, J. Schweitzer, and L. Taylor discuss a methodology for estimating the lower magnitude threshold for an event at a given location to be detectable by considering the expected signal amplitudes and background noise levels in a network of seismic stations. The use of this methodology is demonstrated on regional signals in the Novaya Zemlya region and teleseismic signals from the Indian and Pakistani nuclear test sites.

[1] ENSCO Inc., 5400 Port Royal Road, Springfield, VA 22151, U.S.A.
E-mail: der@ensco.com
[2] Division of Statistics, 380 Kerr Hall, One Shields Ave., University of California, Davis, CA 95616, U.S.A. E-mail: shumway@wald.ucdavis.edu
[3] Department of Geological Sciences, Southern Methodist University, PO Box 750395, Dallas, Texas 75275-0395, U.S.A. E-mail: herrin@passion.isem.smu.edu

A paper by J. Wang describes procedures for designing the initial beam deployment at arrays in the IMS when scant specific prior information regarding the signal and noise correlation characteristics is available.

Another paper by J. Wang discusses an adaptive method for training neural networks for automatic discrimination between teleseismic, regional P and S arrivals from background noise used in the IMS.

Atmospheric nuclear explosions are not likely to be recorded by conventional seismic stations. Neither may they be recorded by hydroacoustic sensors if they are high above the surface of the ocean. For those reasons it is important for CTBT monitoring to have the capability to detect and locate atmospheric nuclear explosions with a global array of infrasound sensors. Five papers cover various aspects of infrasound recording within the context of CTBT monitoring.

The first paper is by J.L. Stevens, I.I. Divnov, D.A. Adams, J.R. Murphy, and V.N. Bourchik. They use infrasound recordings from the data archives of the Institute for Dynamics of the Geopheres in Moscow and compare them with previously developed scaling and attenuation relations. They use scaling relations to define an infrasound magnitude and to estimate the detection capability of the International Monitoring System (IMS).

M.T. Hagerty and W.-Y. Kim describe infrasonic signals at an observatory in Kazakstan that were caused by large mining blasts in Kazakstan and Siberia. Using ray-tracing for various atmosphere models they predict times of enhanced reception produced by favorable wind directions and show that the first arrival travels through the troposphere and the second through the stratosphere.

D.J. Brown, C.N. Katz, R.L. Bras, M.P. Flanagan, and A.K. Gault describe an automatic and interactive data processing system for locating impulsive atmospheric sources with a yield of at least a kiloton. It can utilize all available data, seismic, hydroacoustic or infrasound, to locate a single source. The infrasonic subsystem they use will ultimately process data from the 60-station global infrasound network that is under development.

In order to find sites with relatively low infrasonic noise M.A.H. Hedlin, J. Berger, and F.L. Vernon conducted a survey of noise on oceanic islands. They present results of that survey for two islands, Sao Miguel (the main island of the Azores) and Maio (Cape Verde) where they used a microbarometer to measure noise levels. The surveys illustrate difficulties in site selection and allow the authors to recommend array dimensions and numbers of instruments to use.

In the final paper of this issue, G. Sorrells, J. Bonner, and E. Herrin employ infrasound observations in southwest Texas to detect "shuttle-quakes," precursors to space shuttle re-entry shock fronts. They develop a theoretical model to account for the origin of "shuttle-quakes" and describe the phases that they produce.

Although the papers in this volume do not cover all aspects of data handling and processing, or of infrasound recording, required in monitoring a CTBT, they do provide enlightenment of the wide variety of methodologies needed in this endeavor.

Data Processing

Pure appl. geophys. 159 (2002) 911–944
0033–4553/02/050911–34 $ 1.50 + 0.20/0

Pure and Applied Geophysics

Estimation of Background Noise for International Monitoring System Seismic Stations

MANOUCHEHR BAHAVAR[1] and ROBERT NORTH[1]

Abstract — The prototype International Data Centre (IDC) in Arlington, Virginia has been acquiring data from seismic stations at locations designated in the Comprehensive Test-Ban Treaty for the International Monitoring System (IMS) since the start of 1995. A key characteristic of these stations is their background noise levels and their seasonal and diurnal variability. Since June 1997 an automated sample selection effort has collected over 700,000 individual noise sample spectra from 39 primary and 57 auxiliary stations. Monthly median and 5 and 95 percentile estimates have been calculated for each channel of every station. Compatibility of median spectra obtained for the same station and channel in the same month for two different years confirms the consistency of the noise-sampling algorithm used. A preliminary analysis of the results shows strong (more than a factor of two) seasonal variation at a quarter of all stations. Strong diurnal variations at half of the sites indicate that many of the selected sites are poorly located with respect to cultural noise sources. The results of this study are already being used to evaluate station quality, improve those processes that require background noise values, such as automatic association and requesting auxiliary station data, and to improve the estimation of station and network detection and location thresholds.

Key words: Seismic, noise, IMS network, CTBT.

Introduction

Since January 1995 a prototype International Data Centre (PIDC) in Arlington, Virginia has been acquiring, processing and analyzing data from a global network of seismic stations. These stations were initially those contributed on a voluntary basis by those countries participating in the Group of Scientific Experts Third Technical Test (GSETT-3). When the Comprehensive Nuclear-Test-Ban Treaty (CTBT) was tabled for signature in September 1996, its text included an Annex listing the stations that would comprise the International Monitoring System (IMS). Many of the IMS seismic stations were at the same locations as those then providing data to the PIDC, which at the end of 1996 discontinued the use of data from stations not listed as being part of the IMS. The requirements that a station must satisfy for certification as an IMS station are quite stringent, and no stations have yet (July, 1999) been so certified. In all cases upgrades are required before existing stations can be certified,

---

[1] Center for Monitoring Research, 1300 North 17th Street, Suite 1450, Arlington VA 22209, U.S.A.

although many are close to meeting the requirements. The seismic component of the IMS consists of "primary" stations which provide data continuously in near real time, and "auxiliary" stations which provide data for a specific time interval upon request.

A key characteristic of seismic stations which was taken into account in the selection of sites for the IMS was actual or estimated background noise. GSETT-3 had provided preliminary noise information in its third volume of documentation (1995) and this, supplemented by data from other studies such as the Federation of Digital Seismograph Networks (FDSN) Station Book (1994), was used in network capability estimation codes to help design the seismic network. The primary stations that define the detection capability of the network were preferentially located at sites of known or predicted low background noise. The auxiliary stations that are intended primarily to improve location capability were of necessity sometimes situated at positions such as islands that are known to have inherently high background noise levels. A number of the automatic processes developed for the PIDC use background noise estimates in conjunction with other factors to calculate the probability of detection of a particular event at a given station – examples include the automatic association of signal detections to form event hypotheses, and the selection of auxiliary stations from which data are requested for a given event.

The FDSN Station Book (1994) contains plots of seismic background noise prepared by the Incorporated Research Institutions for Seismology (IRIS) from data available in their archive for 1992. The stations studied include one of the IMS primary stations (HIA) and instrumentation located at the site of six further primary stations, plus 27 auxiliary stations. Diurnal and seasonal variations are assessed in the Station Book.

In June 1997 a daily process that automatically selected noise samples from the data provided by all the stations contributing to the PIDC was initiated at the Center for Monitoring Research. Spectra are calculated from each sample and the results stored in a noise database. The initial data selection involved short-period and/or broadband channels of three-component stations and one representative short-period vertical channel of each array. Some of the three-component stations provided long-period channels, rather than broadband, and samples of these were added starting in March 1998. In May 1998 noise sampling for arrays was expanded to include infinite-velocity beams, calculated as the average of all short-period vertical channels in a given time window. This paper describes the procedures employed and summarizes the results obtained through the end of 1998.

Noise Sample Selection and Processing

The PIDC background noise database is continuously updated through a set of automated processes which are launched once the PIDC's Reviewed Event Bulletin

(REB) is published, indicating that data acquisition, signal and event processing, and subsequent analysis for a given day have been completed. Figure 1 shows the flow of processes in the noise sampling and processing system. The processes involved may be divided into Pre-processing, Processing and Post-Processing groups.

Pre-processing

Pre-processing includes both administrative and triggering actions. Administrative actions include updating the noise station database in accordance with the

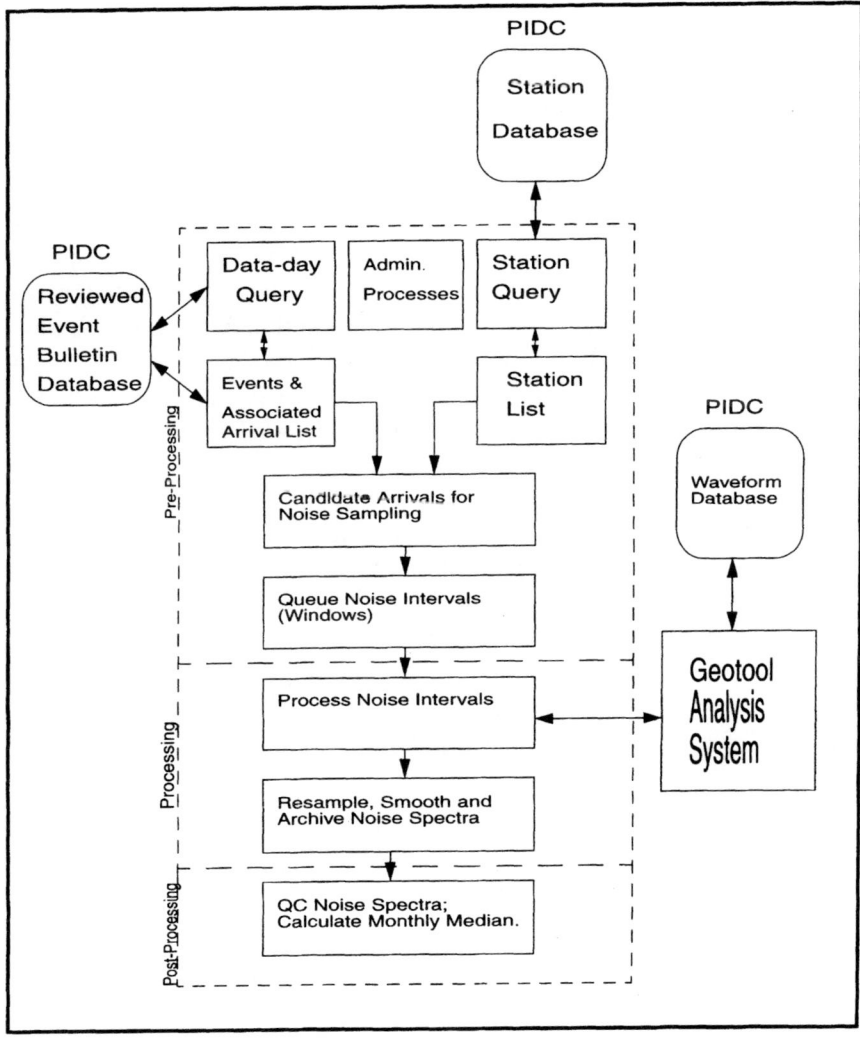

Figure 1
Flow diagram for the noise sampling and processing system and its interaction with the PIDC database and the Geotool Analysis System.

PIDC's operations database and creating necessary disk directories for new stations and data days. Administrative processes also send e-mail notifications to the supervisor if a new station is added which requires noise parameter setting, and if any of the noise processes is in state "running" for more than a pre-specified time limit. Finally, a daily state-of-health report is sent to the supervisor. Noise sampling and processing is triggered through querying the REB database and detecting new bulletin(s). Once a trigger is activated, event information is extracted from the REB database and noise intervals are created for stations contributing to those events. A spectrum is calculated for each interval if:

- continuous waveforms exist for the interval;
- the length of the interval allows accurate estimation of low frequency noise, and
- the interval contains no undesired seismic arrivals (signals)

To meet the above criteria, window length and position must be selected judiciously.

Window length: A good representation of noise levels at various frequencies requires selecting a window which is at least five to ten times longer than the longest period of interest. For the broadband seismometers utilized in the PIDC network, periods of at least 100 seconds should be considered, and thus a time window of about 500 seconds or about 10 minutes is needed. Such a time window is usually readily available from the waveform database for the primary stations. For auxiliary stations waveform segments start about 90 seconds prior to the expected arrival time, therefore the noise window length for the auxiliary stations is set to 60 seconds. As a result, low frequencies (less than about 0.1 Hz) are not well represented in the auxiliary station spectra.

Window position: Proper positioning of time windows is an important factor in avoiding possible contamination by seismic arrivals. A good criterion is to position the time window before the station's first observed arrival for a given event, on the premise that such a time window is less likely to be contaminated by seismic arrivals. Presence of any signal detection in the database for that station, within the selected time window, disqualifies it as a candidate noise sample. The number of windows selected for each station in a given day depends on the number of events to which the station contributed, but not more than one per hour.

Processing

Once a time window is selected, its spectrum is obtained using the Geotool Data Analysis Package (COYNE and HENSON, 1995). The scheme interface of Geotool allows automatic processing of the selected time windows based on pre-defined parameters. The following parameters are used in calculation of the noise spectrum for a given time window:

- the end time of the window is positioned 10 seconds before the arrival to avoid possible contamination by signals;
- time windows for which the waveforms contain gaps are flagged and rejected;

- for beams, channels with abnormal amplitudes are rejected, and the average (sum of all accepted traces divided by the number of such traces) trace is calculated
- spectral estimates for each window are obtained by averaging spectra computed from 40-second and 10-second Hanning windows (before January 20, 1998) and 100-second and 20-second Hanning windows (starting January 20, 1998) with 10% tapering and 67% overlap, for the primary and auxiliary stations respectively; these correspond to between 5 and 42 overlapping windows used for each estimate.
- the spectrum is corrected for instrument response; and
- the spectral values are converted to decibels of displacement power (dB relative to 1 nm**2/Hz).

Post-Processing

Before archiving the calculated spectrum, it is first converted to acceleration power (nm**2/s**4/Hz) and then smoothed and resampled to 27 equally spaced samples per decade of frequency. The resulting spectrum is saved in a file named according to the station, channel and arrival identifier, organized into directories by data day. Database entries are made to facilitate future references and extractions via database queries. Tables 1 and 2 list all the stations, channels and time periods covered in this study.

Quality control: Each month, all individual spectra for each station and channel are visually inspected and outliers are removed. Outliers are defined as those spectra that appear to be contaminated by signals or are otherwise significantly different from adjacent samples. The rejected outlier spectra are flagged but are not removed from the database. After visual inspection of data, monthly median and 5 and 95 percentiles of the accepted spectra are calculated and saved. Tables 3 and 4 list the monthly number of samples available per channel for each station in the noise database. The repeatability of the results is high. Figure 2 shows median monthly spectra for the Canadian array YKA over several years – the substantial seasonal variation at this site is consistent from year to year.

Results

Background noise characteristics over the entire PIDC network and for some representative stations are discussed in the following sections. Emphasis is placed on three frequencies −0.2, 1 and 5 Hz. The lowest frequency of 0.2 Hz is chosen to demonstrate the influence of natural sources, primarily microseisms, on the background noise, while the frequencies of 1 and 5 Hz are within the frequency range used by automated detection and subsequent processes. The effects of cultural noise are expected to be most evident at higher frequencies.

Table 1

*Primary stations in the noise database. Stations currently not sampled are denoted by a * next to the station code. Start and end dates are in year/day-of-year*

Code	Name	Type	Channels	Start	End
ABKT	Alibek	Single	be bn bz	1997216	1998140
ARCES	ARCESS	Array	le ln lz sz beam	1997247	1999143
ASAR	Alice Springs	Array	le ln lz sz beam	1997171	1999143
BDFB	Brasilia	Single	be bn bz sz	1997171	1999143
BGCA	Bogoin	Single	be bn bz sz	1997171	1999143
BJT	Baijiatuan	Single	be bn bz se sn sz	1997171	1999142
BOSA	Boshof	Single	be bn bz sz	1997171	1999142
BRAR	Belbasi	Array	be bn bz sz beam	1998143	1999141
CMAR	Chiang Mai	Array	be bn bz le ln lz sz beam	1997172	1999141
CPUP	Villa Florida	Single	be bn bz sz	1997171	1998311
DBIC	Dimbokro	Single	be bn bz sz	1997171	1999143
ESDC	Sonseca	Array	be bn bz le ln lz sz beam	1997212	1999143
FINES	FINESS	Array	sz beam	1997171	1999143
GERES	GERESS	Array	sz beam	1997171	1999143
HFS	Hagfors	Array	sz beam	1997171	1999138
HIA	Hailar	Single	be bn bz se sn sz	1997171	1999142
ILAR	Eielson	Array	be bn bz sz beam	1997211	1999141
KBZ	Khabaz	Single	se sn sz	1997171	1997239
KSAR	Wonju	Array	be bn bz sz beam	1997212	1999143
LPAZ	La Paz	Single	be bn bz sz	1997171	1999143
MAW	Mawson	Single	be bn bz	1997172	1999143
MJAR	Matsushiro	Array	ez	1998016	1999143
MNV*	Mina	Single	be bn bz	1997171	1999080
NOA	NORSAR	Array	be bn bz sz	1998151	1999143
NORES*	NORESS	Array	sz	1997171	1998006
NRIS	Norilsk	Single	lz se sn sz	1998125	1999130
NVAR	Mina Array	Array	be bn bz sz beam	1999090	1999143
PDAR	Pinedale	Array	be bn bz sz beam	1997212	1999143
PDY	Peleduy	Single	lz se sn sz	1997171	1999143
PLCA	Paso Flores	Single	be bn bz sz	1997171	1999143
ROSC	El Rosal	Single	be bn bz	1997200	1998292
SCHQ	Schefferville	Single	be bn bz	1997171	1999143
SPITS	Spitsbergen	Array	sz beam	1997190	1999142
STKA	Stephens Creek	Single	be bn bz	1997171	1999143
TXAR	TXAR	Array	be bn bz sz beam	1997212	1999113
ULM	Lac du Bonnet	Single	be bn bz	1997171	1999143
VNDA	Vanda	Single	be bn bz sz	1997171	1999099
WRA	Warramunga	Array	sz beam	1997171	1999143
YKA	Yellowknife	Array	be bn bz sz beam	1997171	1999143
ZAL	Zalesovo	Single	lz se sn sz	1997171	1999143

Overview

Figures 3, 4 and 5 show average noise levels over the PIDC network during a one-year period. The height of the bars depends upon the average of monthly median

Table 2

Auxiliary stations in the noise database. Stations currently not sampled are denoted by a * next to the station code. Start and end dates are in year/day-of-year

Code	Name	Type	Channels	Start	End
AAE*	Addis Ababa	Single	se sn sz	1997192	1997229
AFI*	Afiamalu	Single	be bn bz	1997171	1998336
ALQ	Albuquerque	Single	be bn bz	1997175	1998310
AQU*	L'Aquila	Single	be bn bz	1998068	1998326
ARU	Arti	Single	be bn bz	1997171	1999143
ATTU	Attu	Single	be bn bz sz	1998075	1999143
BBB	Bella Bella	Single	be bn bz	1997175	1999142
BORG	Borgarfjordur	Single	be bn bz se sn sz	1997190	1999122
CTA	Charters Towers	Single	be bn bz	1997171	1999141
DAV*	Davao	Single	be bn bz	1997171	1998335
DAVOS	Davos	Single	se sn sz	1997172	1999143
DLBC	Dease Lake	Single	be bn bz	1997171	1999143
EIL	Eilat	Single	be bn bz	1997200	1999141
EKA	Eskdalemuir	Array	sz beam	1997212	1999143
ELK	Elko	Single	be bn bz	1997177	1999142
FITZ	Fitzroy Crossing	Single	be bn bz	1997290	1999143
FRB	Iqaluit	Single	be bn bz	1997174	1999137
FURI*	Mt. Furi	Single	be bn bz se sn sz	1998068	1998164
GNI*	Garni	Single	be bn bz se sn sz	1997225	1997308
HFS	Hagfors	Array	sz beam	1997171	1999138
HNR*	Honiara	Single	be bn bz	1997173	1998335
INK	Inuvik	Single	be bn bz	1997171	1999141
JCJ	Chichijima	Single	be bn bz	1997275	1999143
JHJ	Hachijojima	Single	be bn bz	1997275	1999143
JKA	Kamikawa-asahi	Single	be bn bz	1997275	1999143
JNU	Ohita	Single	be bn bz	1998037	1999143
JOW	Kunigami	Single	be bn bz	1997275	1999143
KDAK	Kodiak Island	Single	be bn bz se sn sz	1997248	1998355
KIEV*	Kiev	Single	be bn bz	1997175	1998314
KVAR	Kislovodsk	Array	sz beam	1998015	1999143
LBTB	Lobatse	Single	be bn bz sz	1997324	1999143
LSZ*	Lusaka	Single	be bn bz	1998131	1998335
MA2*	Magadan	Single	be bn bz se sn sz	1997305	1998335
MLR	Muntele Rosu	Single	se sn sz	1997200	1999133
MRNI	Meron	Single	be bn bz	1999071	1999142
MSEY	Mahe	Single	be bn bz se sn sz	1998336	1999024
NEW	Newport	Single	be bn bz	1997181	1999143
NIL	Nilore	Single	be bn bz se sn sz	1997199	1999143
NNA	Nana	Single	be bn bz	1997177	1999142
NWAO*	Narrogin	Single	be bn bz	1997172	1998335
OBN	Obninsk	Single	be bn bz	1997171	1999131
PARD*	Parod	Single	be bn bz	1997290	1997329
PFO	Pinon Flat	Single	be bn bz	1997171	1999142
PMG*	Port Moresby	Single	be bn bz	1997171	1998335
RAR*	Rarotonga	Single	be bn bz	1997238	1998324
RPN	Rapanui	Single	be bn bz	1997177	1999025
SADO	Sadowa	Single	be bn bz	1997172	1999143
SFJ*	Sondre Stromfjord	Single	be bn bz	1997171	1998335

Table 2

Continued

Code	Name	Type	Channels	Start	End
SNZO*	South Karori	Single	b1 b2 be bn bz	1997171	1998335
SPITS	Spitsbergen	Array	sz beam	1997190	1999142
SUR	Sutherland	Single	be bn bz se sn sz	1997175	1999142
TKL	Tuckaleechee Caverns	Single	se sn sz	1997171	1999143
TSUM*	Tsumeb	Single	be bn bz	1997185	1998335
ULN*	Ulaanbaatar	Single	be bn bz	1997171	1998336
VRAC	Vranov	Single	be bn bz	1997200	1999142
YAK*	Yakutsk	Single	be bn bz se sn sz	1997244	1998210
YSS*	Yuzhno Sakhalinsk	Single	be bn bz se sn sz	1997324	1998335

spectra between November of 1997 and October of 1998 at frequencies of 0.2, 1 and 5 Hz, respectively. The values are obtained from the broadband vertical channels of broadband stations and the short-period vertical channel of arrays and short-period stations. For display and comparison purposes, the resulting values for each station are normalized by the Low Noise Model (LNM) value (PETERSON, 1993) at the corresponding frequency. The maximum vertical scale is set to 3000 and values of the stations with extremely high noise levels have been truncated accordingly. For stations with noise levels more than three orders of magnitude above the LNM, station names and corresponding values are given on the map.

Background noise with frequencies around 0.2 Hz is mainly due to oceanic noise caused by wave action (microseisms, see e.g., HEDLIN and ORCUTT, 1989 and LAY and WALLACE, 1995). DARBYSHIRE (1990) analyzed twenty microseism storms during the winter of 1987–1988 and observed a quantitative relationship between the spectra of 7–10 second period microseisms and the sea-wave spectra. As shown in Figure 3, in the present study the highest noise levels at frequencies of about 0.2 Hz are observed for the island stations or stations in the coastal regions. Stations SPITS (Spitsbergen, Arctic) and BORG (Iceland) at high northern latitudes show average noise levels more than four orders of magnitude above the LNM.

Figure 4 is a similar display for a frequency of about one Hertz. Here the concentration of stations with high noise levels in the coastal regions also suggests that background noise levels at this frequency are still largely dominated by microseisms, but there is also some indication of higher noise levels at many inland stations.

The map showing average noise levels at 5 Hertz (Fig. 5) differs markedly from the previous two. The stations in the interior of Asia and North America that exhibited the lowest noise levels at the lower frequencies are now generally those with the highest levels. This would appear to indicate that many of these stations are located close to artificial noise sources.

As a more specific example and based on the above observations, median spectra for 12 stations during August of 1998 are selected and displayed in Figure 6. At the low-frequency end many stations show noise levels less than an order of magnitude

Table 3

Number of noise samples per channel, per month for the primary stations

Sta	Chan	97/06	07	08	09	10	11	12	98/01	02	03	04	05	06	07	08	09	10	11	12	99/01	02	03	04	05
ABKT	bb	0	0	23	0	0	0	0	0	0	0	12	13	0	0	0	0	0	0	0	0	0	0	0	0
ARCES	sz	0	0	0	155	198	252	308	303	235	266	200	246	260	251	260	252	119	264	207	214	221	317	295	305
	beam	0	0	0	0	0	0	0	0	0	0	0	49	241	235	250	247	114	236	23	207	221	311	294	302
	lp	0	0	0	0	0	0	0	0	0	94	188	224	226	220	234	232	116	220	20	199	202	282	265	277
ASAR	sz	73	160	174	193	349	365	419	365	273	449	219	415	347	345	376	342	191	378	118	261	391	434	436	350
	beam	0	0	0	0	0	0	0	0	0	0	0	82	347	345	376	329	188	383	0	261	391	434	436	350
	lp	0	0	0	0	0	0	0	0	0	0	0	0	60	345	376	340	190	263	0	162	0	13	434	350
BDFB	bb sz	47	109	125	116	143	88	138	125	49	131	85	113	103	111	103	87	67	94	95	71	0	53	114	80
BGCA	bb sz	60	160	91	2	7	44	54	57	57	153	120	192	159	215	162	197	106	213	191	187	246	127	169	104
BJT	bb sp	22	88	127	93	49	47	78	107	104	69	100	124	93	98	127	69	0	75	78	141	163	159	156	153
BOSA	bb sz	42	110	78	112	111	65	131	101	95	106	49	97	69	82	105	81	50	109	101	106	114	123	123	117
BRAR	sz	0	0	0	0	0	0	0	0	0	0	0	0	79	114	144	153	93	135	14	161	186	190	181	182
	bb	0	0	0	0	0	0	0	0	0	0	0	35	154	116	144	153	93	135	14	183	186	190	181	182
	beam	0	0	0	0	0	0	0	0	0	0	0	0	80	116	144	153	92	133	14	175	186	189	181	182
CMAR	sz	78	217	200	223	316	277	316	333	269	281	144	127	252	158	254	256	143	298	288	264	342	328	320	311
	bb lp	0	0	0	0	0	0	0	0	0	57	72	63	126	79	127	128	71	132	14	150	136	152	160	155
	beam	0	0	0	0	0	0	0	0	0	0	0	0	252	158	254	256	138	244	28	290	342	328	320	299
CPUP	bb sz	42	132	111	116	12	5	17	20	42	104	9	12	92	108	72	46	19	27	0	0	0	0	0	0
DBIC	bb sz	44	129	120	120	142	110	172	184	122	190	98	132	99	99	112	92	65	90	118	145	167	138	166	134
ESDC	sz	0	5	119	134	166	156	187	127	122	144	88	128	131	120	140	127	68	118	11	132	121	153	166	118
	beam	0	0	0	0	0	0	0	0	0	0	0	16	131	120	140	123	66	108	11	125	121	153	166	118
	bb lp sz	0	0	0	0	0	0	0	0	0	33	37	54	56	51	60	55	29	50	4	53	51	65	71	50
FINES	sz	68	186	192	229	291	267	352	270	256	399	225	269	269	246	299	293	150	281	149	283	321	369	337	311
	beam	0	0	0	0	0	0	0	0	0	0	0	66	269	246	299	271	145	249	30	301	321	369	337	311
GERES	sz	75	169	160	218	251	241	298	228	205	296	185	262	220	200	252	188	130	171	103	241	243	293	273	230
	beam	0	0	0	0	0	0	0	0	0	0	0	42	221	201	252	179	120	158	23	247	243	293	273	230
HFS	beam	0	0	0	0	0	0	0	0	0	0	0	4	35	45	46	78	55	71	5	53	56	62	68	43
	sz	27	63	49	58	88	112	74	46	40	70	53	33	35	45	46	78	56	76	27	53	56	62	68	43
HIA	bb sp	31	104	113	138	86	41	183	141	142	162	119	186	146	105	31	67	0	126	75	191	215	228	179	154
ILAR	sz	0	9	174	200	333	301	351	336	318	374	229	260	295	305	290	313	180	300	28	283	371	378	389	360
	bb	0	0	0	0	0	0	0	0	0	146	229	273	285	305	290	313	184	286	28	319	371	378	388	360

Table 3

Continued

Sta	Chan	97/06	07	08	09	10	11	12	98/01	02	03	04	05	06	07	08	09	10	11	12	99/01	02	03	04	05
KBZ	beam sz	0	0	0	0	0	0	0	0	0	0	0	0	147	152	145	156	90	145	14	153	185	189	194	180
	sp	32	93	65	0	0	0	0	0	0	0	0	0	0	0	0	0	0	0	0	0	0	0	0	0
KSAR	sz	0	5	156	179	196	197	227	225	206	245	138	222	161	175	166	186	117	200	17	190	230	226	239	239
	bb	0	0	0	0	0	0	0	0	0	91	138	222	161	175	166	180	71	195	17	175	230	226	239	239
	beam sz	0	0	0	0	0	0	0	0	0	0	0	23	80	87	83	90	54	107	8	92	115	112	119	120
LPAZ	bb sz	42	158	116	109	95	97	61	111	84	140	113	187	118	153	127	88	68	135	7	0	0	5	146	103
MAW	bb	45	35	0	132	177	138	115	33	21	34	65	144	110	123	133	153	104	123	49	84	56	69	137	106
MJAR	ez	0	0	0	0	0	0	0	111	183	220	135	228	213	193	213	183	119	174	90	180	186	200	181	171
MNV*	bb	44	76	96	118	49	122	86	82	83	140	96	161	107	94	136	89	84	99	71	112	130	112	0	0
NOA	sz	0	0	0	0	0	0	0	0	0	0	0	3	239	216	231	272	125	230	103	214	195	261	251	240
	bb	0	0	0	0	0	0	0	0	0	0	0	3	243	216	231	261	134	220	22	203	195	261	251	240
NORES*	sz	76	179	156	214	265	205	281	48	0	0	0	0	0	0	0	0	0	0	0	0	0	0	0	0
NRIS	lz sp	0	0	0	0	0	0	0	0	0	0	0	131	131	113	133	124	74	147	55	112	131	150	140	24
NVAR	sz	0	0	0	0	0	0	0	0	0	0	0	0	0	0	0	0	0	0	0	0	0	9	262	266
	bb	0	0	0	0	0	0	0	0	0	0	0	0	0	0	0	0	0	0	0	0	0	9	262	266
	beam	0	0	0	0	0	0	0	0	0	0	0	0	0	0	0	0	0	0	0	0	0	9	262	266
PDAR	sz	0	6	155	163	176	225	266	257	241	250	162	178	142	121	0	62	156	149	27	222	257	256	265	240
	bb	0	0	0	0	0	0	0	0	0	120	162	190	166	150	57	60	106	135	27	204	143	136	268	206
	beam sz	0	0	0	0	0	0	0	0	0	0	0	15	83	75	27	31	73	75	13	118	128	128	134	126
PDY	lz sp	45	131	103	129	151	140	165	185	129	214	13	161	209	100	219	203	99	193	112	188	223	237	223	135

Station	Type																								
PLCA	bb sz	56	145	117	145	159	125	140	158	72	173	106	134	149	111	114	119	71	129	67	120	102	165	169	144
ROSC	bb	0	40	2	6	0	0	0	0	0	0	0	33	22	0	0	31	43	0	0	0	0	0	0	0
SCHQ	bb	32	103	88	82	27	0	0	27	133	173	92	124	100	76	131	96	39	87	59	123	159	181	150	138
SPITS	sz	0	27	26	29	9	50	53	34	28	49	36	30	42	25	37	43	34	44	59	59	81	90	71	37
	beam	0	0	0	0	0	0	0	0	0	0	0	11	42	25	37	44	34	45	3	59	81	90	71	37
STKA	bb	75	198	197	202	268	239	305	243	229	338	154	325	276	269	307	317	159	297	152	294	324	328	354	300
TXAR	bb	0	0	0	0	0	0	0	0	0	72	110	203	112	157	192	137	53	0	0	0	0	0	0	0
	sz	0	8	162	129	157	179	292	311	250	262	110	203	113	157	192	137	51	7	0	6	17	12	0	0
	sz	0	0	0	0	0	0	0	0	0	0	0	0	0	0	0	0	0	0	0	0	0	0	3	2
	bb	0	0	0	0	0	0	0	0	0	0	0	0	0	0	0	0	0	0	0	0	0	0	3	2
	beam sz	0	0	0	0	0	0	0	0	0	0	0	8	56	79	99	69	35	4	0	7	11	9	7	0
ULM	bb	37	146	78	90	143	136	198	139	161	204	108	156	155	99	139	125	74	119	78	161	168	210	209	149
VNDA	bb sz	50	124	94	160	4	19	154	152	77	217	125	176	161	6	0	0	0	136	75	108	95	136	58	0
WRA	sz	61	178	116	79	335	325	375	361	245	490	242	421	364	303	392	355	254	313	166	329	368	378	373	335
	beam	0	0	0	0	0	0	0	0	0	0	0	83	364	303	392	355	249	310	30	329	368	378	373	335
YKA	beam	0	0	0	0	0	0	0	0	0	0	0	49	241	138	161	190	66	135	32	320	341	377	350	323
	sz	63	181	139	100	156	203	290	292	312	358	223	256	241	138	154	189	62	140	107	338	332	370	339	316
ZAL	bb	0	0	0	0	0	0	0	0	0	131	224	243	232	137	175	187	66	137	33	325	341	377	344	323
	lz sp	48	128	95	118	100	142	151	143	117	114	60	134	196	141	209	203	114	217	89	174	217	240	223	129

Table 4

Number of noise samples per channel, per month for the auxiliary stations

Sta	Chan	06	07	08	09	10	11	12	01	02	03	04	05	06	07	08	09	10	11	12	01	02	03	04	05
AAE*	sp	0	2	1	0	0	0	0	0	0	0	0	0	0	0	0	0	0	0	0	0	0	0	0	0
AFI*	bb	24	29	28	32	44	33	35	24	26	55	32	33	24	43	10	38	34	59	8	18	15	22	20	19
ALQ	bb	1	8	0	0	2	47	38	28	62	84	79	64	26	38	83	52	46	6	0	0	0	0	0	0
AQU*	bb	0	0	0	0	0	0	0	0	0	7	14	11	14	4	4	0	0	1	0	0	0	0	0	0
ARU	bb	28	75	59	38	60	92	126	76	3	0	36	52	33	20	69	26	0	1	0	46	71	107	68	64
ATTU	bb sz	0	0	0	0	0	0	0	0	0	1	0	15	16	8	21	16	16	12	13	18	15	22	20	19
BBB	bb	1	3	1	0	0	2	5	6	4	2	1	2	8	8	9	8	3	1	2	4	3	3	5	10
BORG	bb sp	0	1	0	2	0	1	1	0	1	1	0	1	2	1	0	0	1	2	2	0	1	1	0	1
CTA	bb	67	130	115	141	163	168	155	140	101	142	83	117	106	76	120	75	152	144	70	81	115	144	142	109
DAV*	bb	10	8	5	10	20	21	23	32	13	27	25	15	19	25	25	53	31	40	2	0	0	0	0	0
DAVOS	sp	11	38	32	40	65	78	64	46	43	39	43	34	34	30	43	41	37	40	36	43	37	33	59	35
DLBC	bb	20	44	17	33	44	64	74	51	35	51	35	43	44	42	51	59	36	59	63	67	64	56	57	63
EIL	bb	0	5	11	10	29	35	17	10	9	9	10	18	14	19	13	11	22	10	7	25	21	22	15	26
EKA	beam sz	0	0	0	0	0	0	0	0	0	0	0	0	0	0	0	0	0	0	0	0	0	0	0	0
	sz	0	2	48	68	102	138	99	99	68	84	74	66	33	21	30	47	40	34	1	37	46	44	50	40
ELK	bb	3	16	8	7	14	18	13	10	14	8	11	10	66	43	60	94	81	69	2	75	93	89	100	80
FITZ	bb	0	0	0	0	67	58	192	138	73	166	41	54	13	16	13	13	25	19	8	27	14	26	11	25
FRB	bb	2	9	6	6	0	0	0	0	0	0	0	6	51	99	129	201	207	193	6	135	135	248	236	172
FURI*	bb sp	0	0	0	6	0	0	0	0	0	1	0	1	20	4	15	5	9	6	17	31	26	37	15	13
GNI*	bb sp	0	0	0	6	9	4	0	0	0	0	0	0	0	0	0	0	0	0	0	0	0	0	0	0
HFS	beam	0	0	0	0	0	0	0	0	0	0	0	0	0	0	0	0	0	0	0	0	0	0	0	0
	sz	27	63	49	58	88	112	74	46	40	70	53	33	35	45	46	78	55	71	5	53	56	62	68	43
HNR*	bb	5	3	6	12	20	20	16	8	10	8	4	10	35	45	46	78	56	76	27	53	56	62	68	43
INK	bb	31	97	81	72	80	162	234	146	141	130	162	134	1	4	8	19	20	19	1	0	0	0	0	0
JCJ	bb	0	0	0	0	56	70	46	33	38	37	46	74	82	77	143	126	112	138	83	191	189	193	190	164
JHJ	bb	0	0	0	0	49	65	34	34	28	22	46	61	36	42	54	50	62	61	30	43	56	57	47	64
JKA	bb	0	0	0	0	50	90	159	77	67	75	50	98	72	65	88	66	92	80	46	109	81	101	67	64
JNU	bb	0	0	0	0	0	0	0	0	0	0	0	0	72	65	88	66	92	80	46	109	81	101	67	64
JOW	bb	0	0	0	0	49	65	67	32	38	67	35	83	50	59	80	58	66	75	49	74	67	71	63	76
KDAK	bb sp	0	0	0	15	18	34	27	33	22	13	14	12	44	56	73	54	71	72	39	66	52	78	61	0
KIEV*	bb	1	5	3	10	11	20	16	11	5	12	9	9	5	12	20	22	15	20	5	0	0	0	0	0

Station	Type																						
KVAR	sz	0	0	0	0	0	0	0	12	15	59	44	24	34	0	0	0	8	58	17	109	63	76
	beam	0	0	0	0	0	0	0	12	15	15	0	0	0	0	0	0	0	58	17	110	63	76
LBTB	bb sz	0	0	0	0	0	6	12	23	20	20	12	17	18	12	6	21	15	31	28	18	0	35
LSZ*	bb	0	0	0	0	0	0	0	0	1	1	0	0	0	2	0	3	1	0	0	0	0	0
MA2*	bb sp	0	0	0	0	0	14	45	10	9	20	25	26	23	45	14	2	0	0	0	0	0	0
MLR	sp	7	22	14	38	90	35	35	17	26	32	20	20	28	35	90	25	15	44	41	43	24	18
MRNI	bb	0	0	0	0	0	0	0	0	0	0	0	0	0	0	0	0	0	0	0	3	20	16
MSEY	bb sp	0	0	0	0	0	0	0	0	0	0	0	0	0	0	0	0	0	0	0	0	0	0
NEW	bb	25	14	16	24	26	26	44	26	23	5	17	5	16	26	5	32	5	36	33	35	19	30
NIL	bb sp	9	34	15	32	38	38	37	39	28	24	39	33	39	40	15	44	15	35	28	52	32	33
NNA	bb	56	26	29	15	28	25	23	24	37	20	32	41	32	30	40	35	32	0	21	1	25	36
NWAO*	bb	0	15	33	50	25	40	41	31	12	37	31	44	34	40	1	29	1	0	0	0	0	0
OBN	bb	9	10	9	30	8	6	23	12	12	12	12	10	26	9	0	12	0	15	14	12	18	4
PARD*	bb	0	0	0	8	67	41	54	0	0	0	0	0	0	6	0	0	0	0	0	0	0	0
PFO	bb	35	74	43	67	76	41	54	42	40	44	42	28	41	54	41	42	5	0	0	32	38	38
PMG*	bb	33	42	31	76	98	85	63	62	67	65	89	63	78	85	98	54	3	0	0	0	0	0
RAR*	bb	0	0	2	4	3	1	0	2	7	2	1	0	5	1	3	2	0	0	0	0	0	0
RPN	bb	1	4	2	10	3	7	3	8	4	1	2	3	8	7	3	3	5	3	0	0	0	0
SADO	bb	9	41	20	34	20	32	35	27	37	34	32	35	18	32	20	34	43	35	33	43	35	40
SFJ*	bb	4	14	3	8	5	5	4	5	16	5	0	4	6	21	5	0	1	0	0	0	0	0
SNZO*	bb	2	4	9	8	5	0	0	1	0	0	20	0	0	0	5	21	0	0	0	0	0	0
SPITS	sz	27	27	29	9	50	53	28	25	36	30	37	28	34	53	59	44	59	59	81	90	71	37
	beam	0	0	0	0	0	0	0	25	0	11	37	0	0	0	3	45	3	59	81	90	71	37
SUR	bb sp	6	18	15	12	6	24	11	9	14	13	13	11	13	24	6	9	6	9	4	5	42	10
TKL	sp	10	42	33	20	22	37	23	29	38	47	36	23	34	37	33	36	33	41	42	50	42	36
TSUM*	bb	0	34	15	21	23	27	23	19	21	26	27	23	27	27	22	27	2	0	0	0	15	0
ULN*	bb	29	44	33	18	61	38	27	20	52	29	0	23	10	38	61	0	2	16	19	17	0	0
VRAC	bb	0	10	15	33	23	19	11	12	22	16	15	11	12	19	23	15	7	16	19	0	0	18
YAK*	bb sp	0	0	1	5	6	19	3	4	5	4	4	3	5	19	6	0	0	0	0	0	0	0
YSS*	bb sp	0	0	0	0	9	43	24	21	20	24	34	24	29	43	9	28	1	0	0	0	0	0

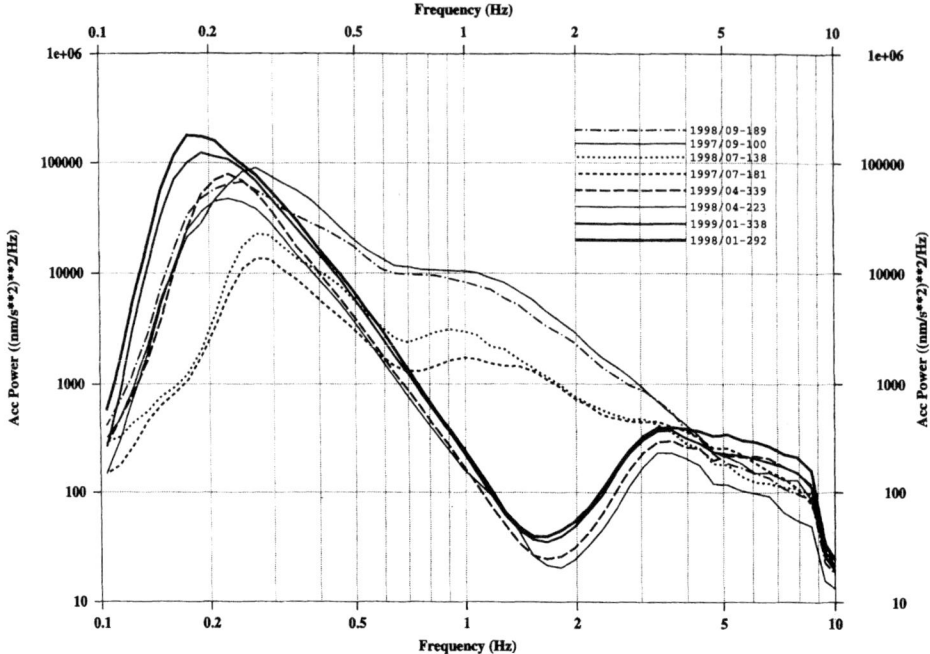

Figure 2
Comparison of monthly median noise spectra of YKR8 (Yellowknife, Canada) for Jan., Apr., July and Oct. for two different years.

above the LNM, while noise levels at SPITS and RPN (Easter Island) are four orders of magnitude above the LNM. As the frequency increases and approaches those frequently associated with cultural noise, the stations divide into two groups. The first group includes stations like ARCES (Norway), ILAR (Alaska), SPITS and BBB (Canada) with relatively low noise levels. The second group includes stations like BRAR (Turkey), CMAR (Thailand), KSAR (Korea), NIL (Pakistan), ARU (Russia), RPN and HIA (China) showing higher noise levels.

Of the three frequencies considered above, noise levels at one and five Hertz have the most impact on the automated detection algorithms. In the following sections their long-term (seasonal) and short-term (diurnal) variations are discussed, and the well-known ability of arrays to suppress noise is illustrated.

Seasonal Variations

Both environmental and cultural conditions may vary from one season to the next. Stormy seas, freezing and unfreezing of large nearby masses of water, and decreases in cultural activities during extreme climatic conditions are all factors affecting the level of the background noise on a seasonal scale. Two time periods are selected to show the extent of these changes for the PIDC network. The first interval

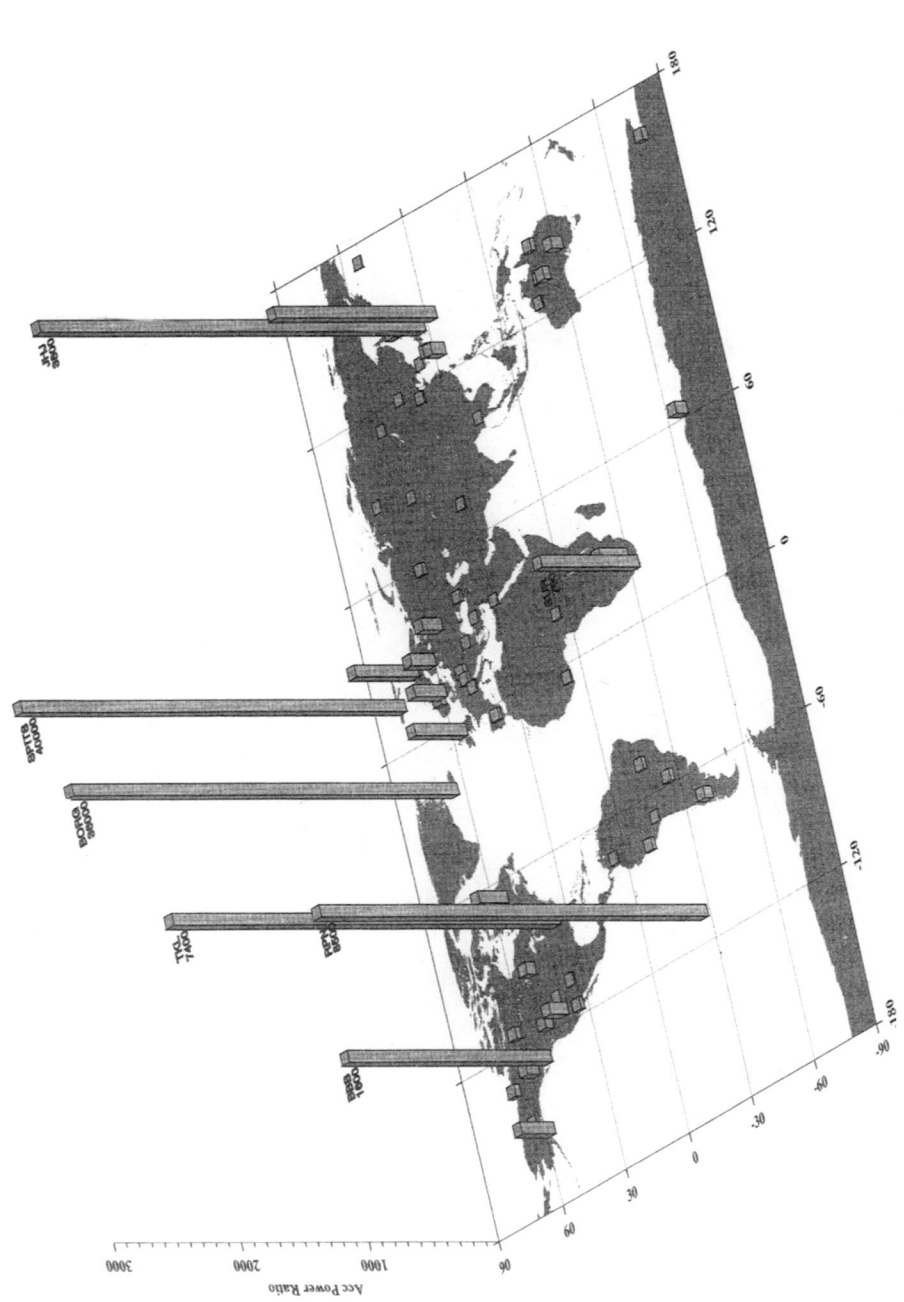

Figure 3

Ratio of average monthly median noise spectra between November 1997 and October 1998 to that of the Low Noise Model (LNM) at 0.2 Hz.

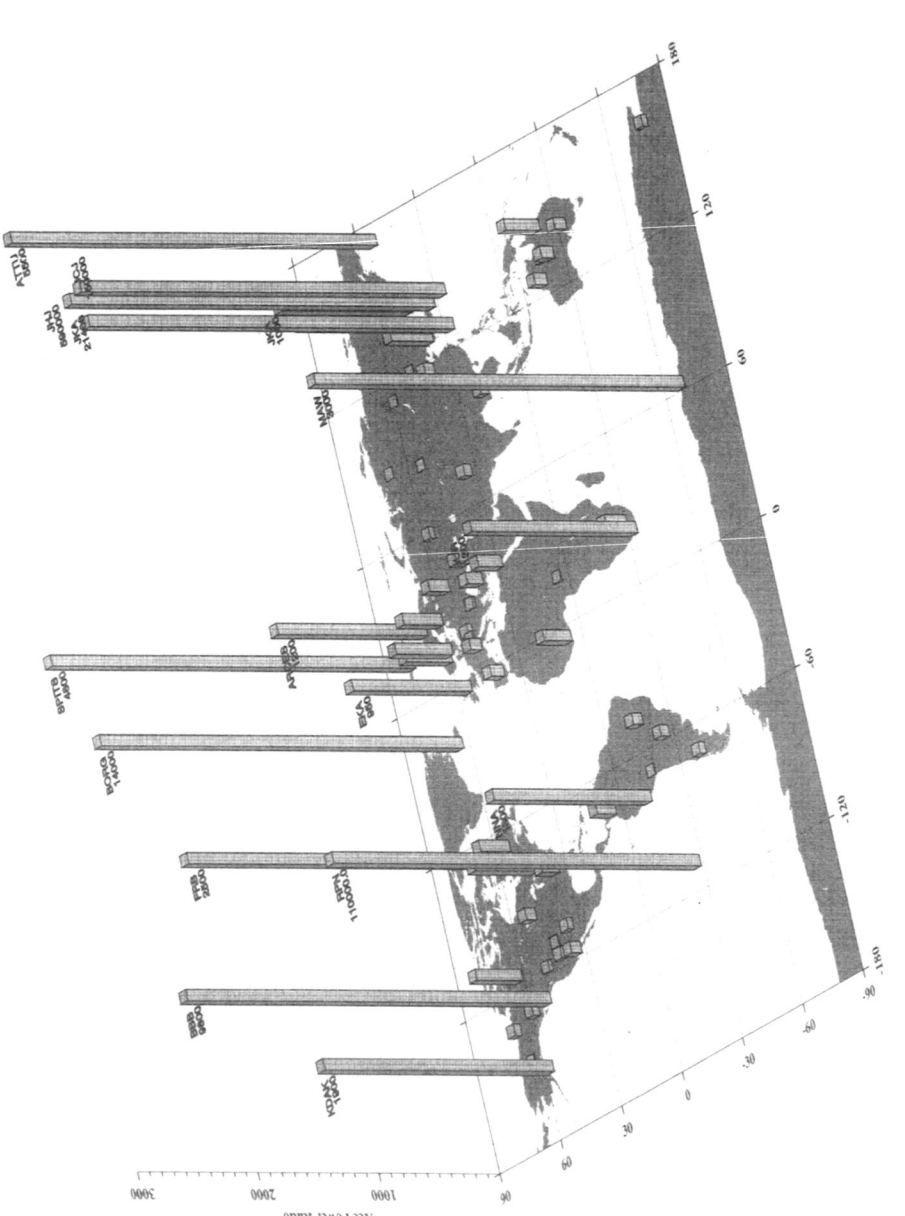

Figure 4

Ratio of average monthly median noise spectra between November 1997 and October 1998 to that of the LNM at 1 Hz.

Estimation of Background Noise

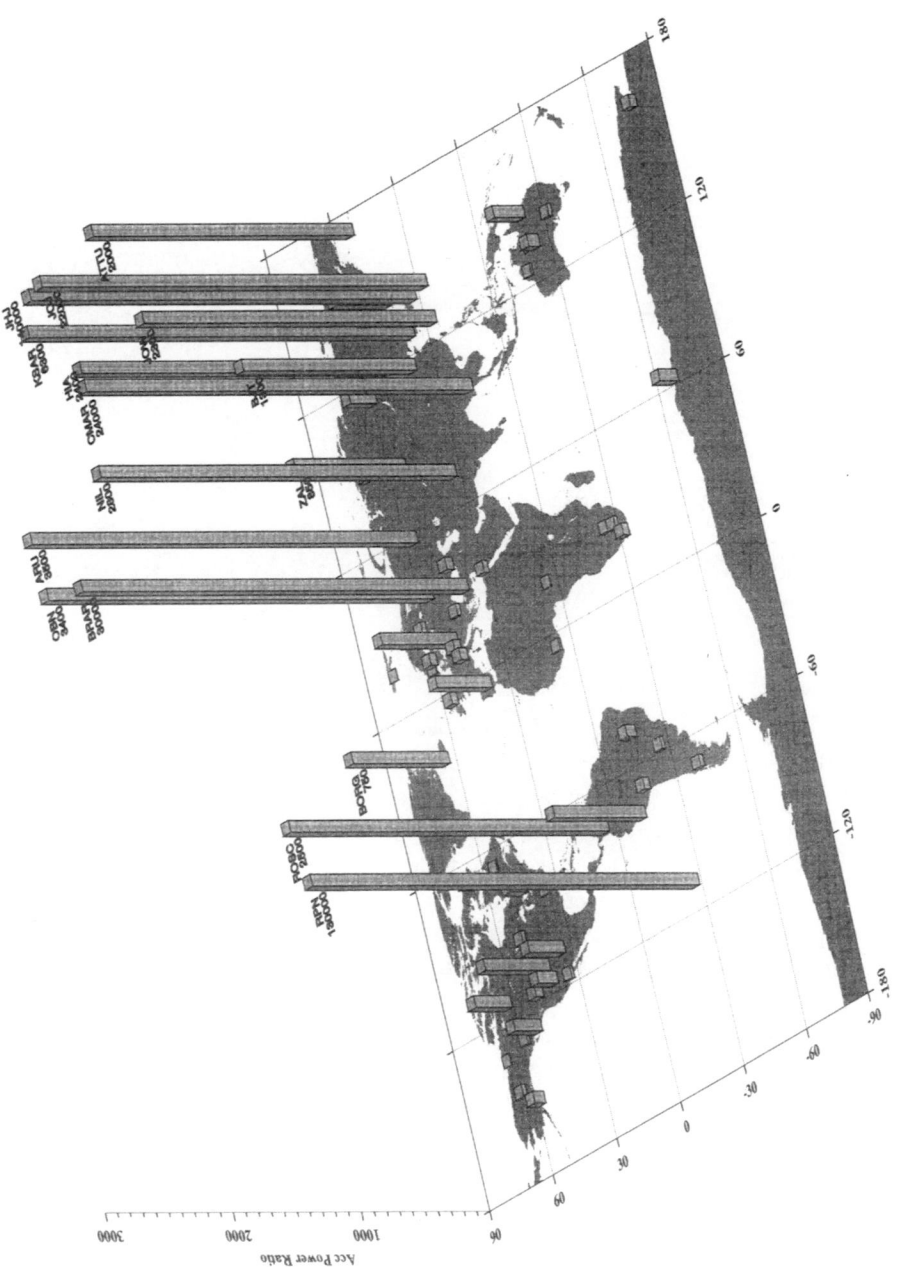

Figure 5

Ratio of average monthly median noise spectra between November 1997 and October 1998 to that of the LNM at 5 Hz.

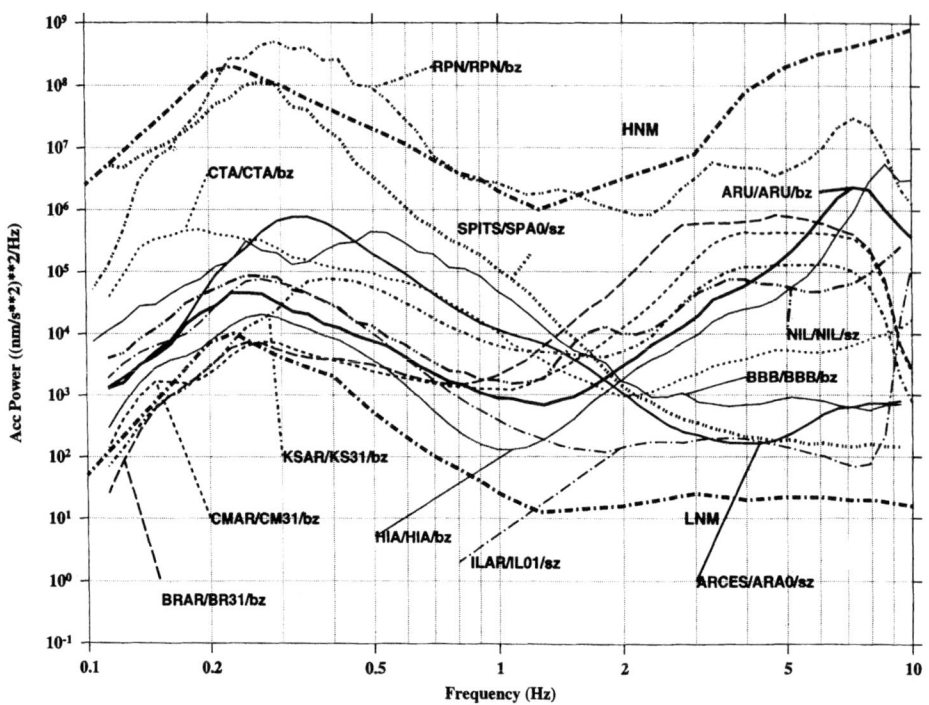

Figure 6
Monthly median noise spectra of selected stations during August 1998 showing relative noise levels.

is from December of 1997 through February of 1998, which coincides with the winter season in the Northern Hemisphere. The second is from June through August of 1998, which coincides with the summer season in the Northern Hemisphere. Averages of the monthly medians during each time period were calculated, and the ratio of these averages for all the stations are shown in Figures 7 and 8 for frequencies of 1 and 5 Hz, respectively.

At frequencies of about one Hertz, background seismic noise is still to some extent influenced by natural sources. Such sources are often most active in winter and as a result noise levels in winter are higher. Figure 7 shows the ratio of the average winter noise level to summer noise level at frequency of about 1 Hz. This figure suggests, as might be expected, that in general stations in coastal regions have a higher winter to summer noise ratio. This change in the noise level is less than an order of magnitude except for station MAW in Antarctica. Note that for this, as for other Southern Hemisphere stations, the seasons are reversed compared to those of the Northern Hemisphere. Noise levels at MAW may be less during the southern winter because the adjacent ocean is frozen; a feature which is also noted for the Northern Hemisphere (see Fig. 15, for station YKA, below).

At frequencies of 5 Hertz, the noise is of more cultural origin and, particularly at high latitudes, higher noise levels are noted in summer. Figure 8 shows the summer

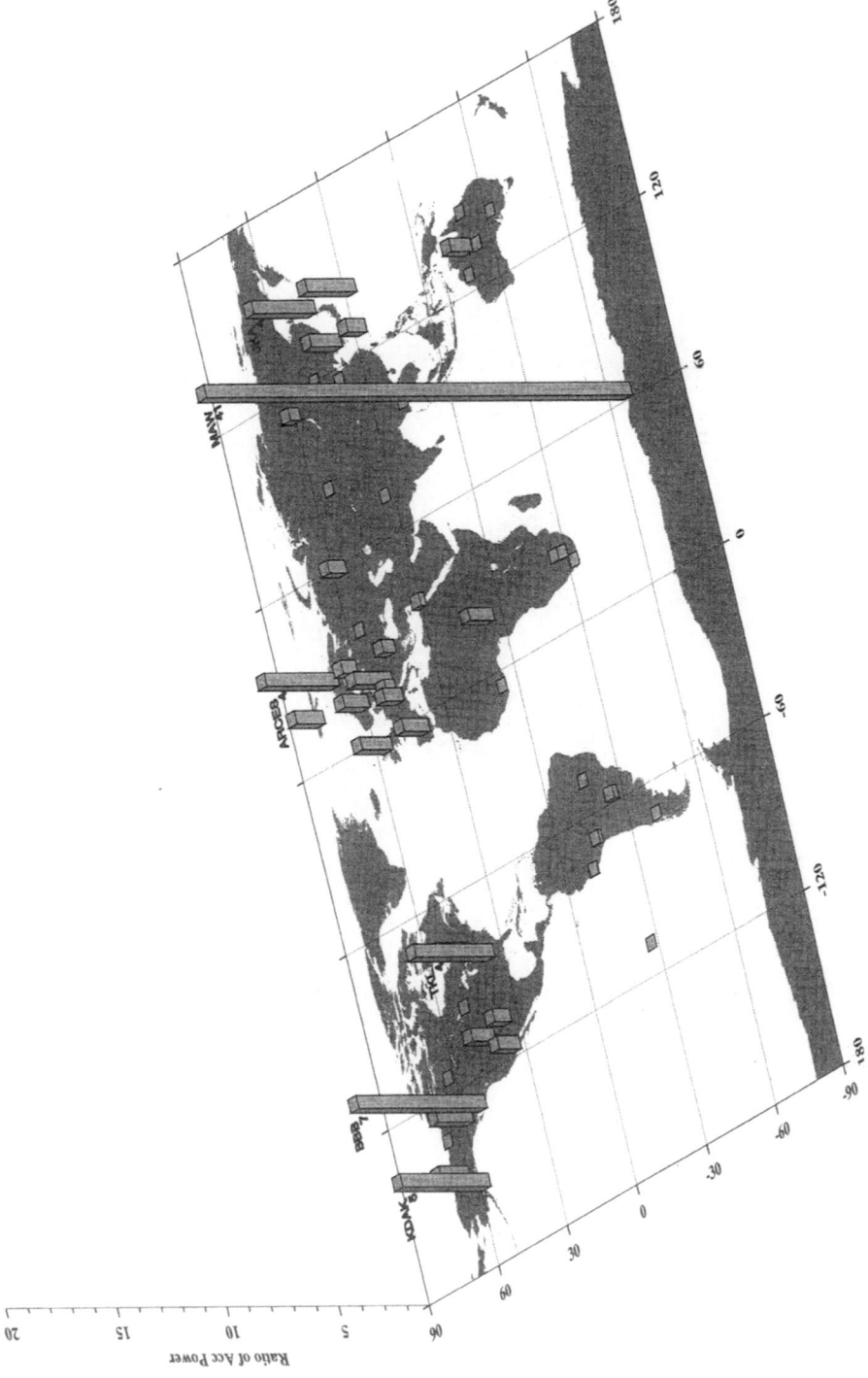

Figure 7

Ratio of average median noise of Northern Hemisphere winter (12/97–2/98) to that of summer (6-8/98) at 1 Hz.

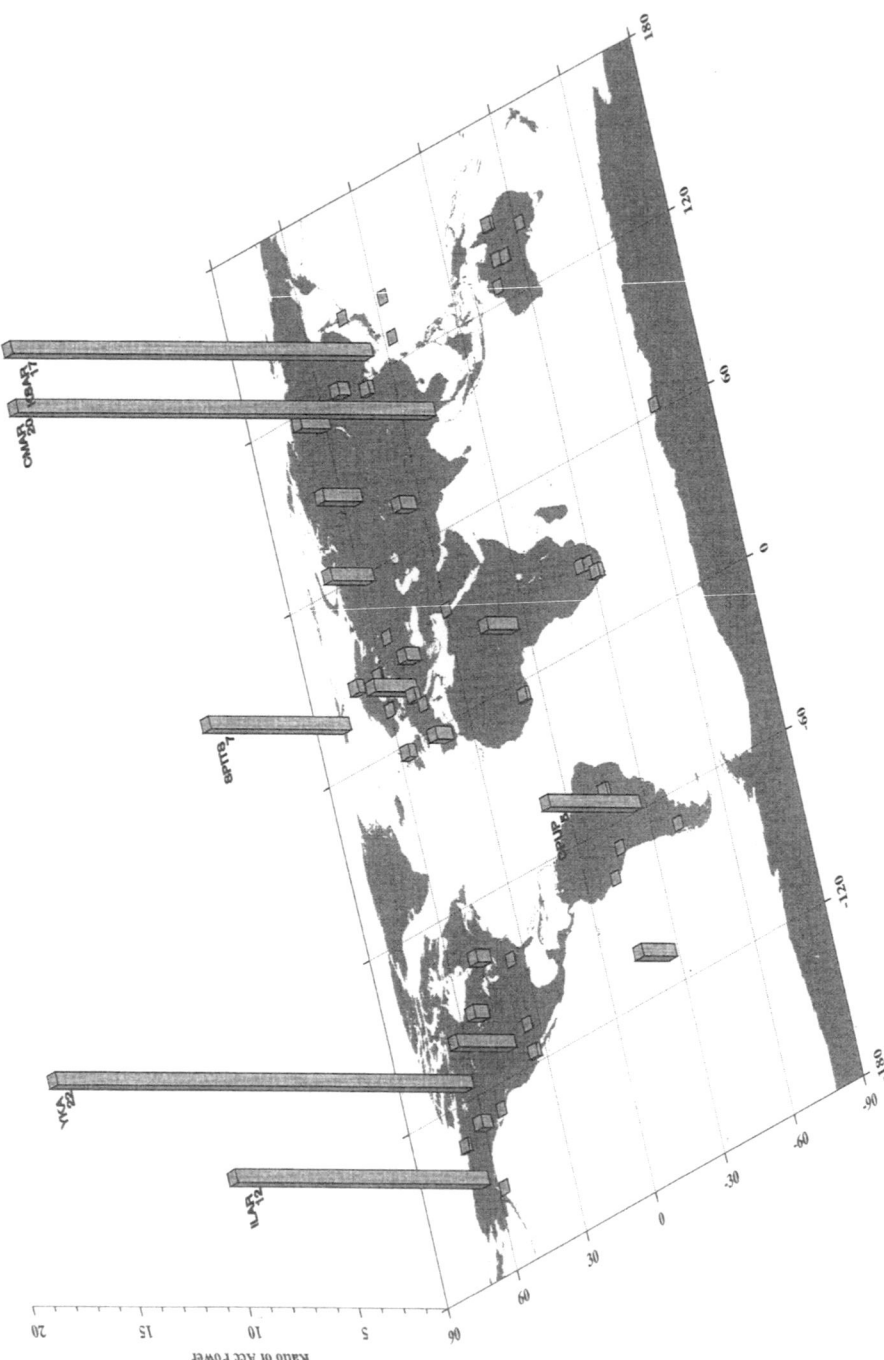

Figure 8

Ratio of average median noise of Northern Hemisphere summer (6-8/98) to that of winter (12/97–2/98) at 5 Hz.

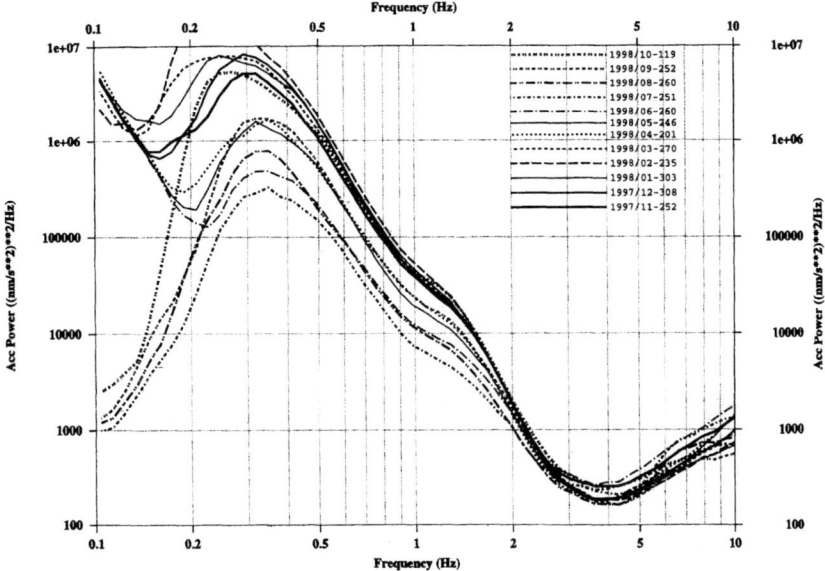

Figure 9
Monthly median noise spectra of ARA0 (Norway) from November of 1997 through October 1998. A decrease in the noise level at low frequencies starts in April and the minimum is reached by July. After September, the station is back at high noise levels. At higher frequencies this change is less significant.

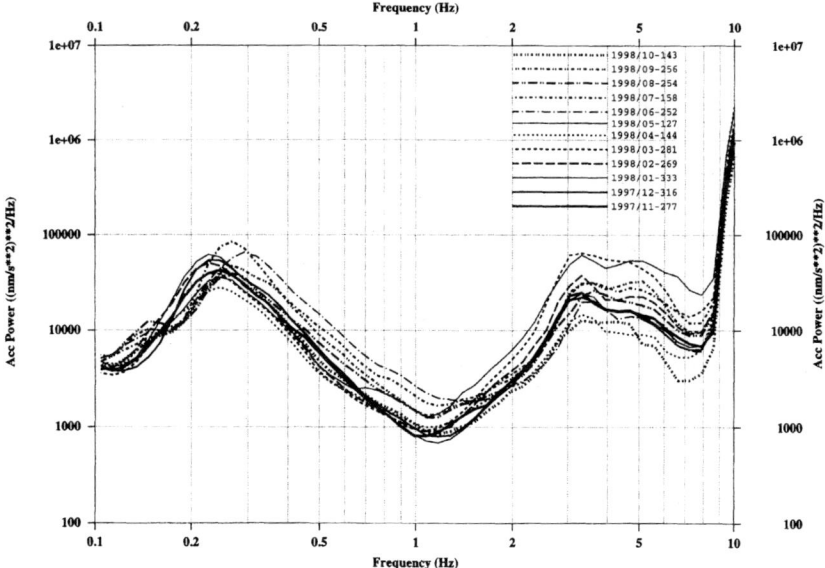

Figure 10
Monthly median noise spectra of CM16 (Thailand) from November of 1997 through October of 1998. At lower frequencies high noise levels are reached in June and by November it is back to normal levels. At higher frequencies, the highest noise levels are observed between March and May.

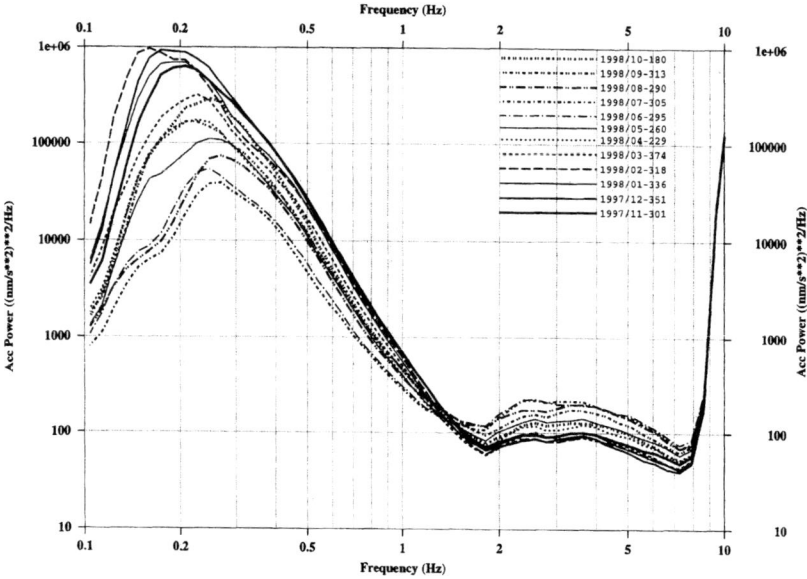

Figure 11
Monthly median noise spectra of IL01 (USA) from November of 1997 through October of 1998. At lower frequencies the minimum noise level is attained by July and the maximum level is reached by December. At higher frequencies the pattern is reversed and high noise levels are observed between June and July.

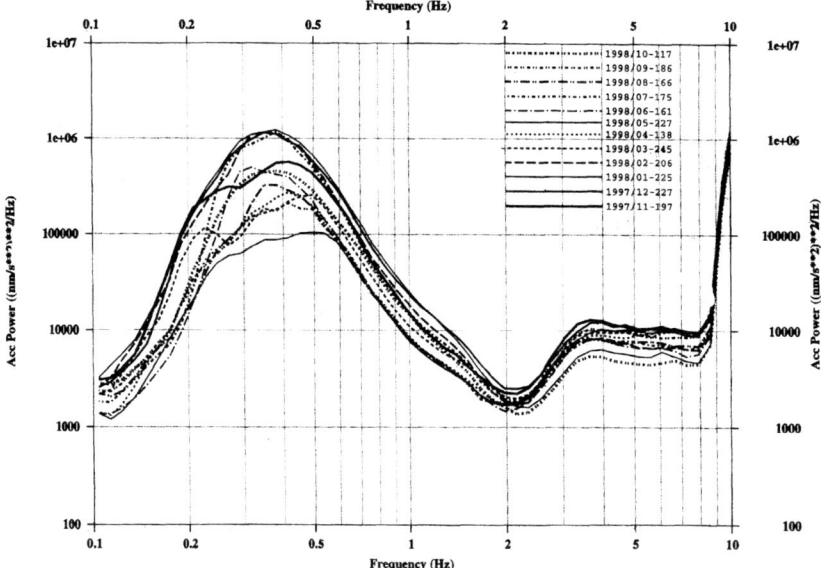

Figure 12
Monthly median noise spectra of KS15 (South Korea) from November of 1997 through October of 1998. At lower frequencies the maximum noise level is observed during January. At higher frequencies no clear pattern is apparent.

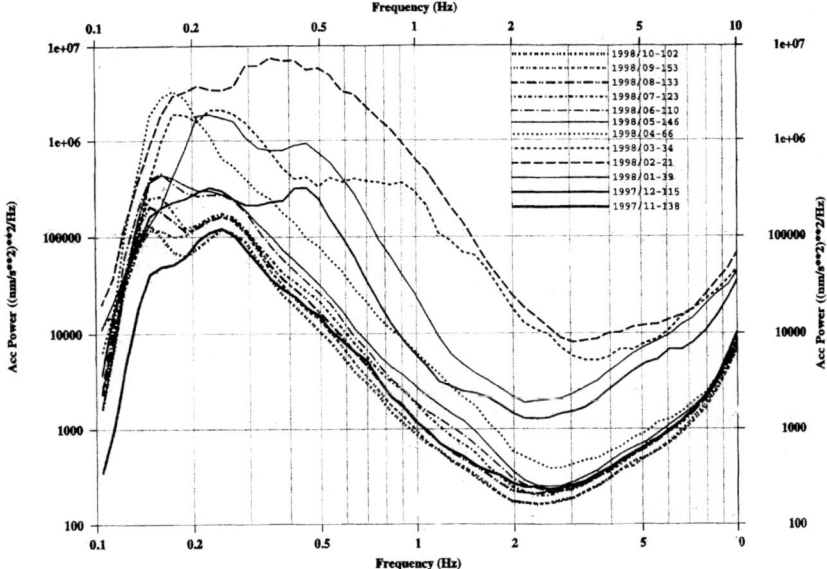

Figure 13

Monthly median noise spectra of MAW (Antarctica) from November of 1997 through October of 1998. At both high and low frequencies higher noise levels are observed between December and March. The change is transitional.

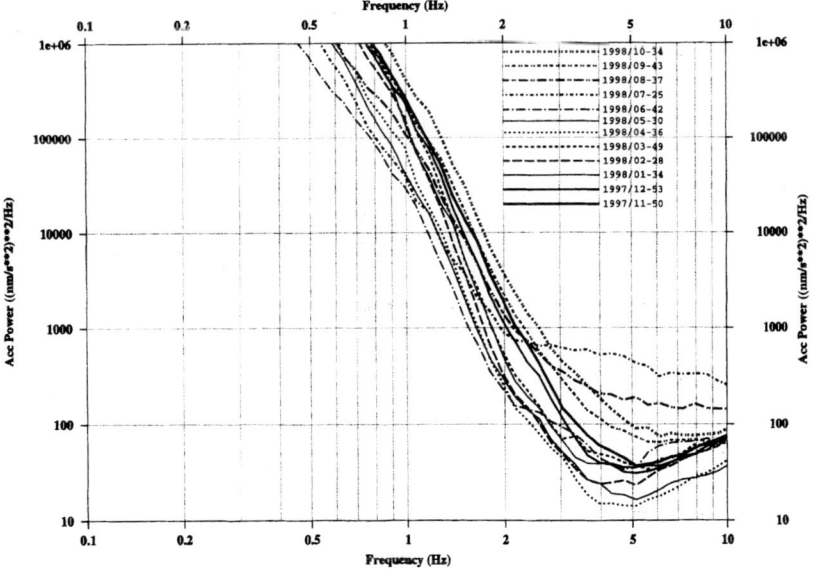

Figure 14

Monthly median noise spectra of SPA0 (Spitsbergen) from November of 1997 through October of 1998. At higher frequencies high noise levels are observed between July and November. The change is transitional.

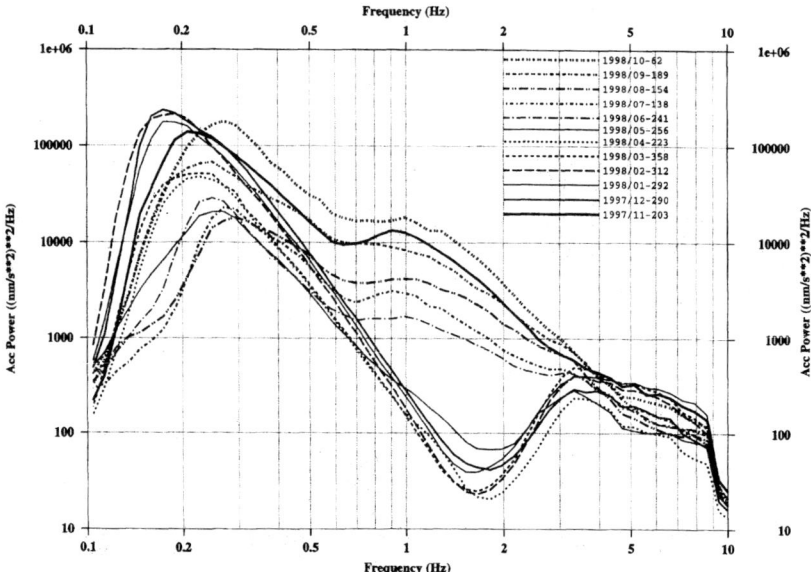

Figure 15

Monthly median noise spectra of YKR8 (Canada) from November of 1997 through October of 1998. At frequencies between 1 and 2 Hz high noise levels are observed between July and November. The low noise period between December and May is due to freezing of the nearby large lake.

to winter ratio at frequencies of about 5 Hertz. Stations like ILAR, YKA, CMAR and KSAR show more than an order of magnitude increase in their noise level during summer.

Median monthly spectra over the 12-month period between November of 1977 and October of 1998 are plotted in Figures 9 through 15 for some representative stations.

Diurnal Variations

Noise sources which have cultural origin often exhibit daily variation. FYEN (1990) has reported extreme diurnal variation of noise levels, exceeding 10 dB, for the NORES array, with peaks at about 6 Hz associated to specific nearby industrial activity. To investigate the presence of such short-term variations in the PIDC noise database, the ratios of median daytime to nighttime noise during June to August 1998 were calculated for frequencies of 1 and 5 Hz. To obtain these values the following procedure has been followed:

• Spectra of vertical component channels of individual stations between June and August 1998 were grouped into day and night. Daytime spectra are those with times between 7 am and 6 pm local time. Local time is calculated for each station by adjusting the GMT time by 4 minutes for each degree of longitude. Due to variations in latitude and local conditions, such a grouping may not completely separate daytime and nighttime.

- Median spectra for each station and each group are calculated.
- At frequencies of about 1 and 5 Hz, ratios of day to night median noise are calculated

Figures 16 and 17 show day-to-night noise ratios for frequencies of 1 and 5 Hz. Since noise at frequency of about one Hz is largely natural in origin, few major variations are evident from Figure 16. On the other hand, for frequencies of about 5 Hz, noise is mostly cultural in origin and differences between daytime and nighttime are quite marked. Figure 17 is the daytime to nighttime ratio map and here stations like BBB, DAVOS (Switzerland), BRAR, ARU, CMAR and JCJ (Japan) show more than an order of magnitude increase in their noise levels from night to day. These results are similar to those obtained by GIVEN (1990) on variations in broadband seismic noise at IRIS/IDA stations in the former Soviet Union. The increase is almost universal and indicates that most of the sites selected for the IMS are less than optimum.

Figures 18 through 21 compare daytime and nighttime noise levels at some representative stations (ASAR (Australia), BRAR, CMAR, ESDC (Spain)) during July 1998. These figures clearly show an increase in daytime noise levels at each station site for frequencies above 1 Hz.

Array Effects

As mentioned earlier, for arrays in addition to collecting noise samples from selected stations and channels, an infinite-velocity beam is formed for each time interval. This is done by taking the average waveforms of all the short-period vertical channels of the array during a time window and calculating the spectrum of the average waveform. A comparison of the beam spectrum with that of the single channels should indicate the ability of the array to suppress noise at various frequencies. MYKKELTVEIT et al. (1990) studied the noise suppression capabilities of the NORES array using computed hourly infinite-velocity beams over a one week interval.

To compare the two, average median spectra of beam and short-period vertical channels during June of 1998 have been calculated and their ratios are displayed in Figures 22 and 23. In these figures a ratio of about 1 indicates no change in the noise level due to beamforming, while a value less than one reflects noise suppression. The results separate into two distinct groups.

Figure 22 shows the spectral ratios for arrays ASAR, CMAR, EKA (Britain), ESDC and ILAR. A common feature of these arrays is their ability to suppress noise with frequencies of more than 1 Hz by almost an order of magnitude. The beamforming has negligible effect on low-frequency noise of about 0.2 Hz, however for frequencies between 0.2 and 2 Hz noise reduction increases with frequency. Figure 23 shows a similar plot for arrays ARCES, BRAR, FINES (Finland), GERES (Germany), HFS (Sweden) and SPITS. Except for BRAR these are all

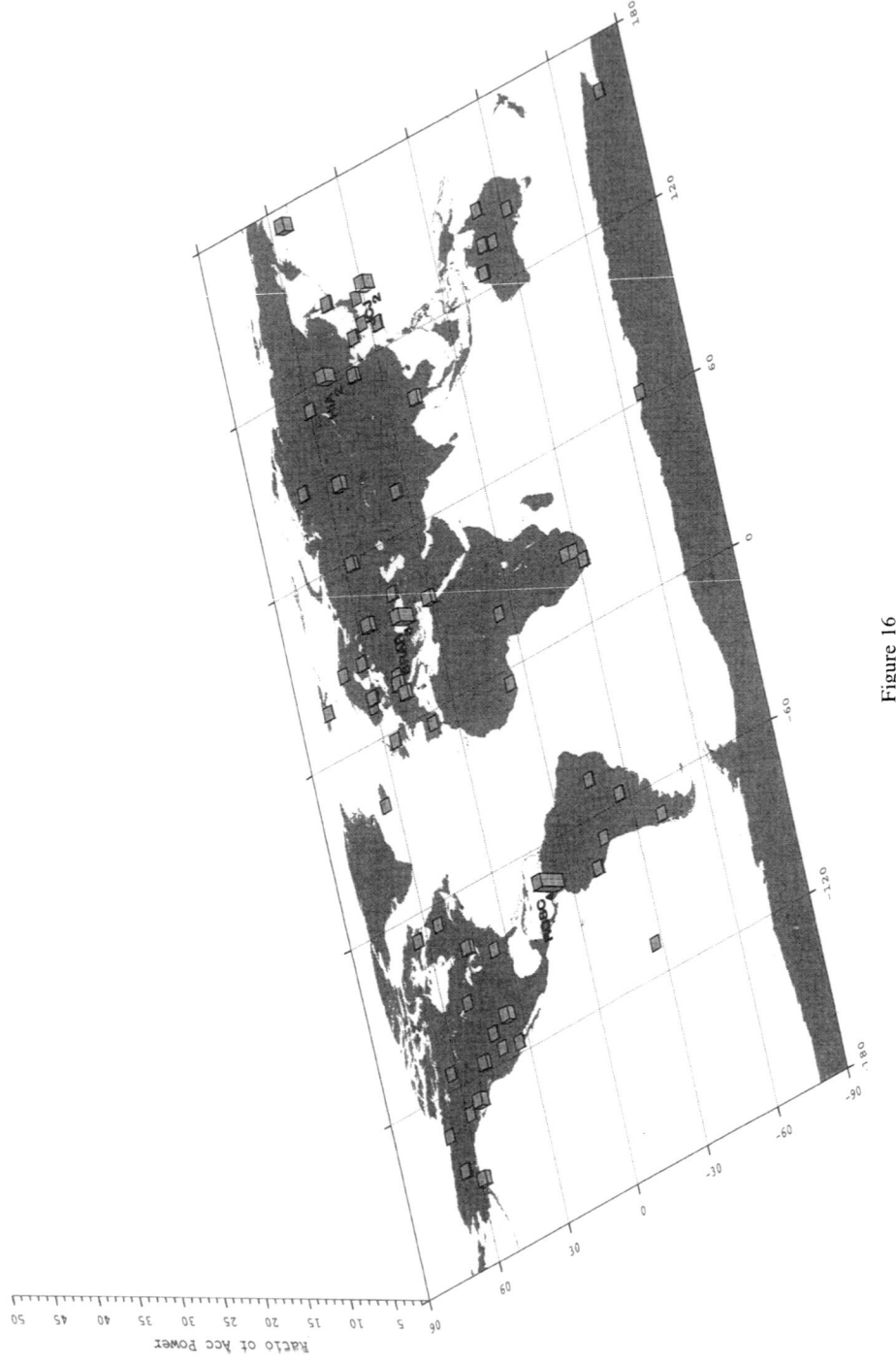

Figure 16

Ratio of median daytime to nighttime noise between June and August 1998 at 1 Hz.

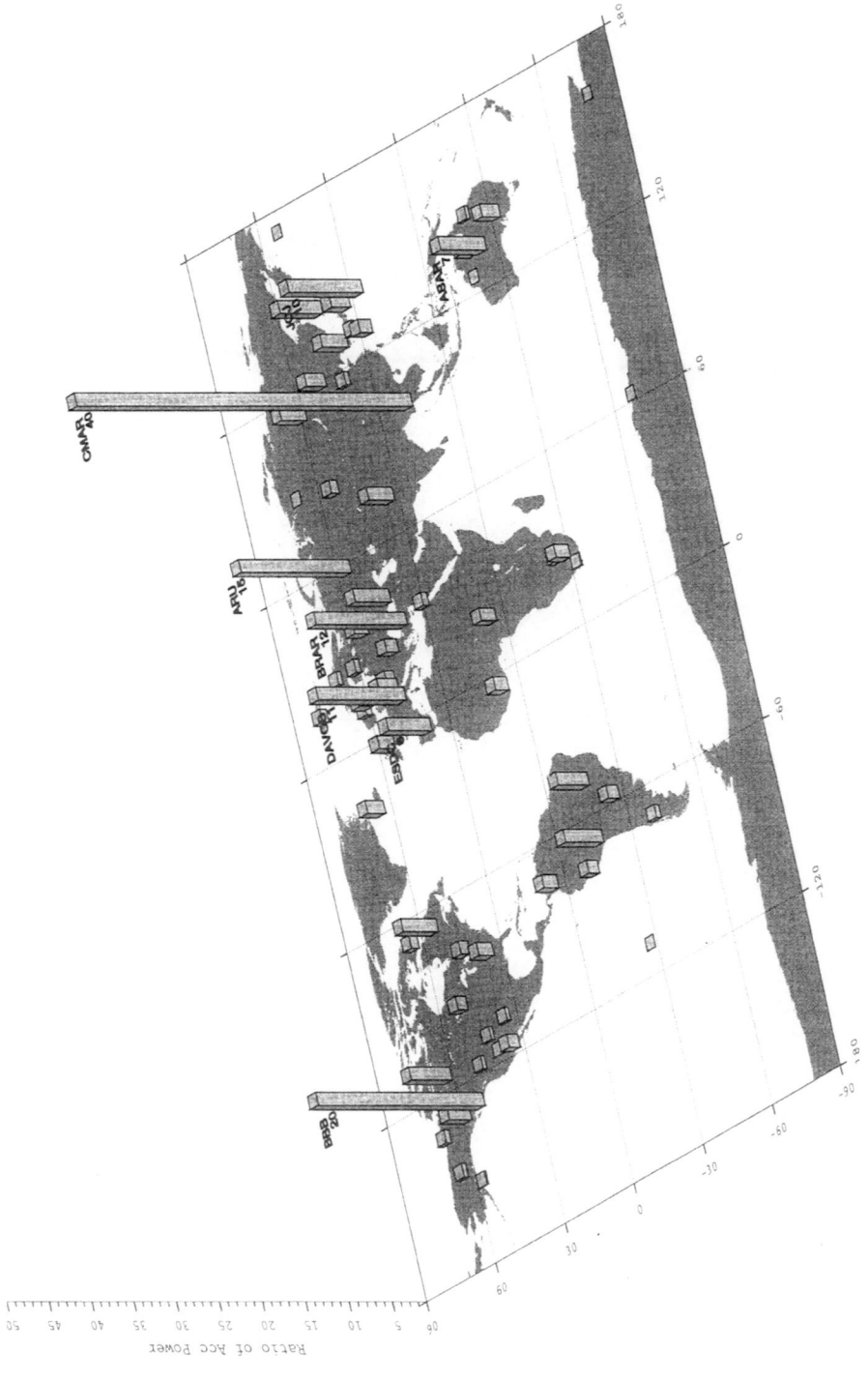

Figure 17

Ratio of median daytime to nighttime noise between June and August 1998 at 5 Hz.

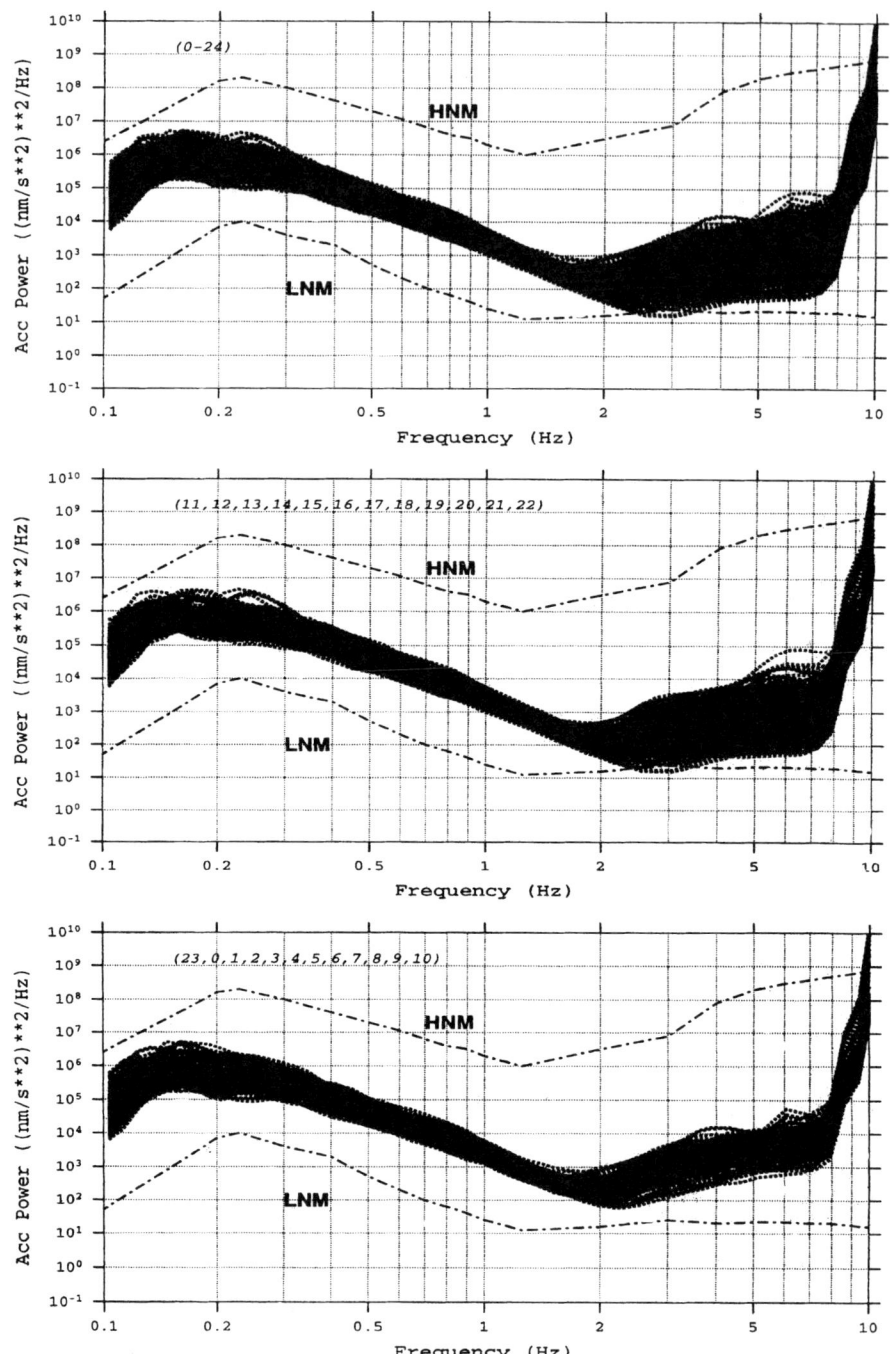

Figure 18
Individual noise spectra for channel sz of element AS12 of array ASAR (Australia) during July 1998, complete collection (top), nighttime (7 pm–6 am, middle) and daytime (7 am–6 pm, bottom).

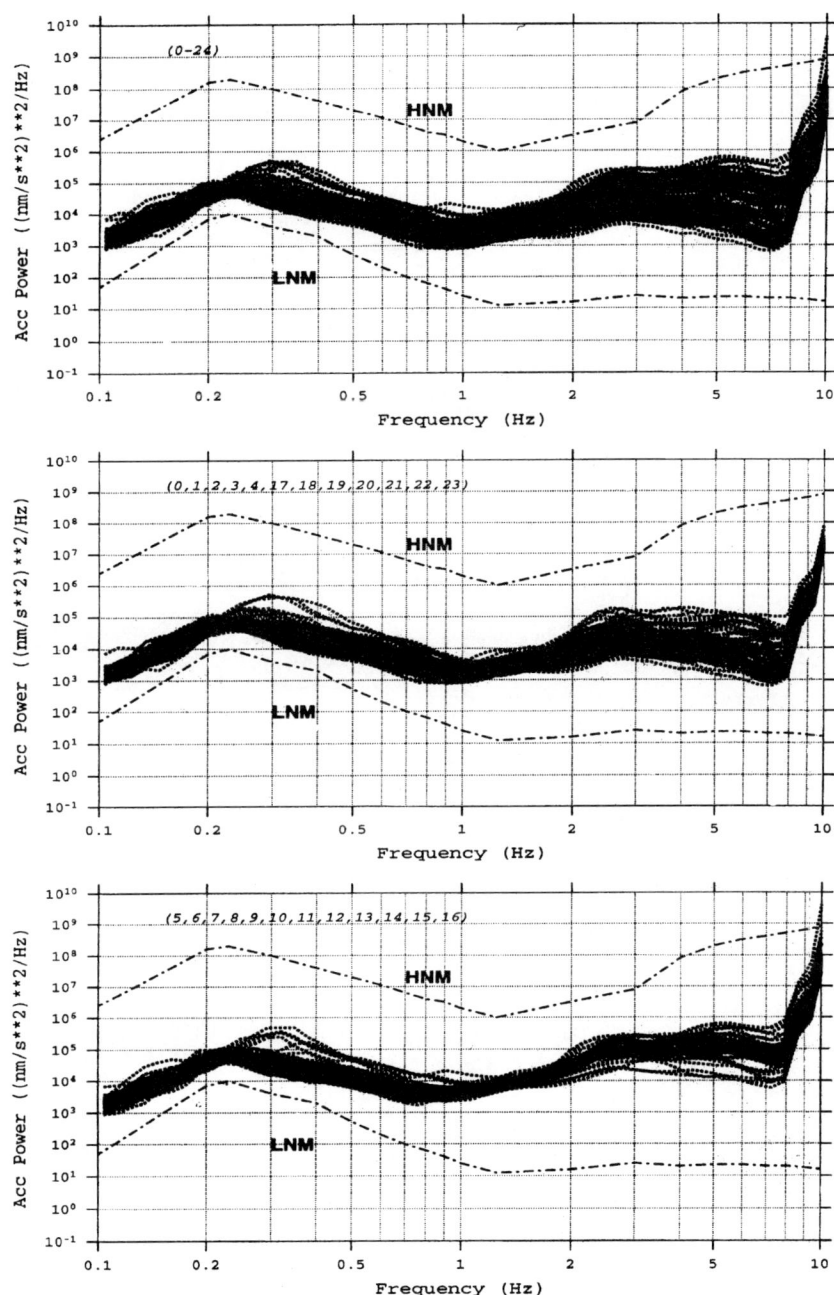

Figure 19
Individual noise spectra for channel sz of element BR01 of array BRAR (Turkey) during July 1998,
complete collection (top), nighttime (7 pm–6 am, middle) and daytime (7 am–6 pm, bottom).

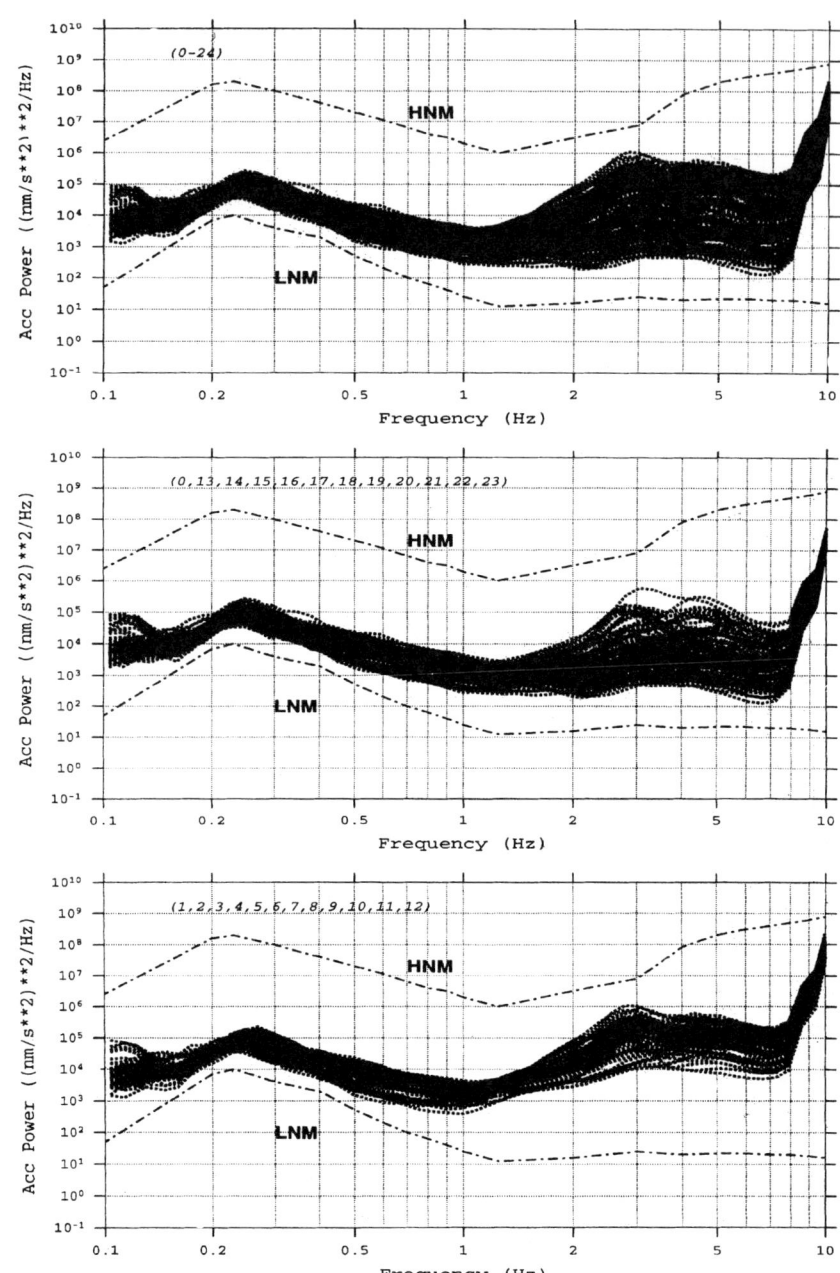

Figure 20

Individual noise spectra for channel sz of element CM16 of array CMAR (Thailand) during July 1998, complete collection (top), nighttime (7 pm–6 am, middle) and daytime (7 am–6 pm, bottom).

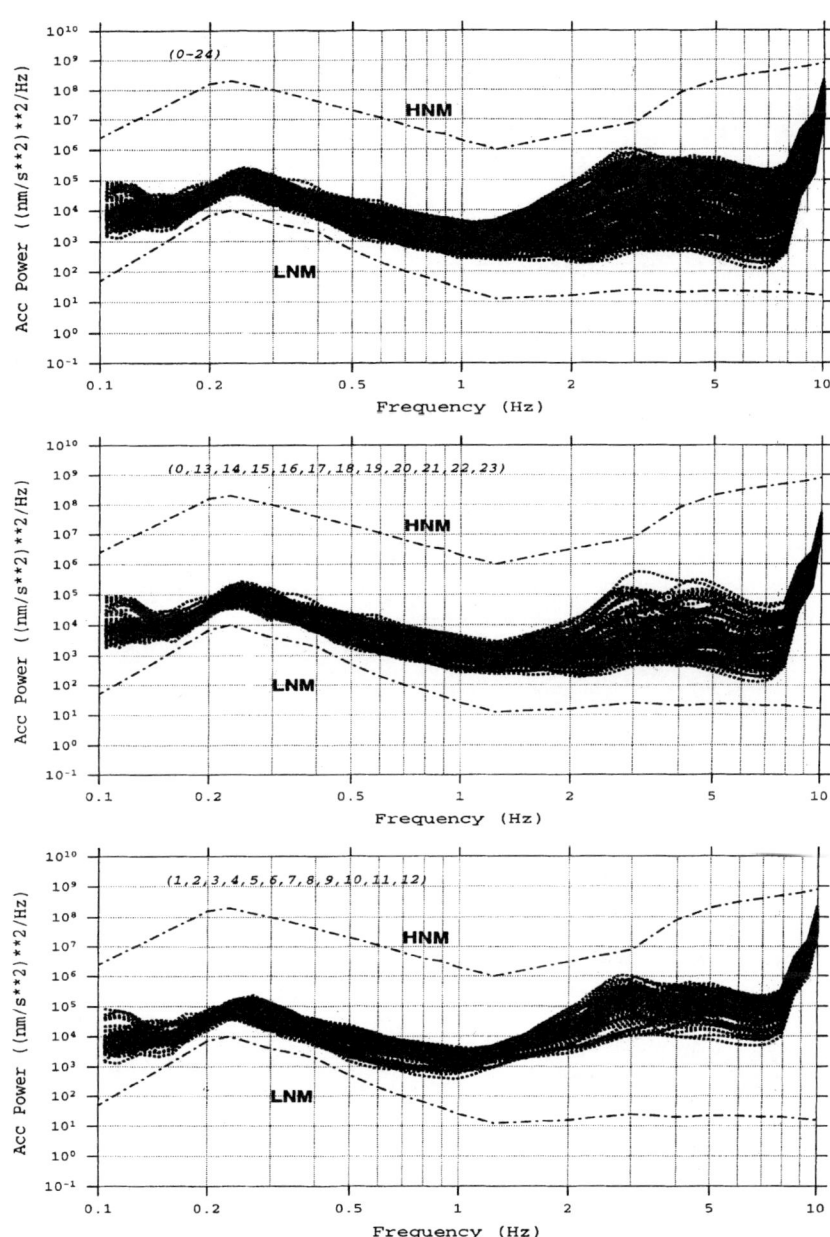

Figure 21
Individual noise spectra for channel sz of element ES01 of array ESDC (Spain) during July 1998, complete
collection (top), nighttime (7 pm–6 am, middle) and daytime (7 am–6 pm, bottom).

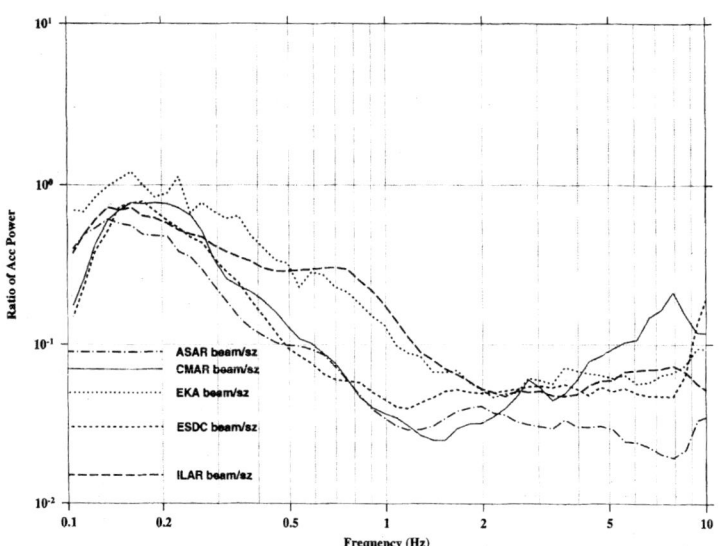

Figure 22
Ratio of infinite-beam to sz channel median noise spectrum during June 1998 for arrays ASAR, CMAR, EKA (UK), ESDC and ILAR arrays.

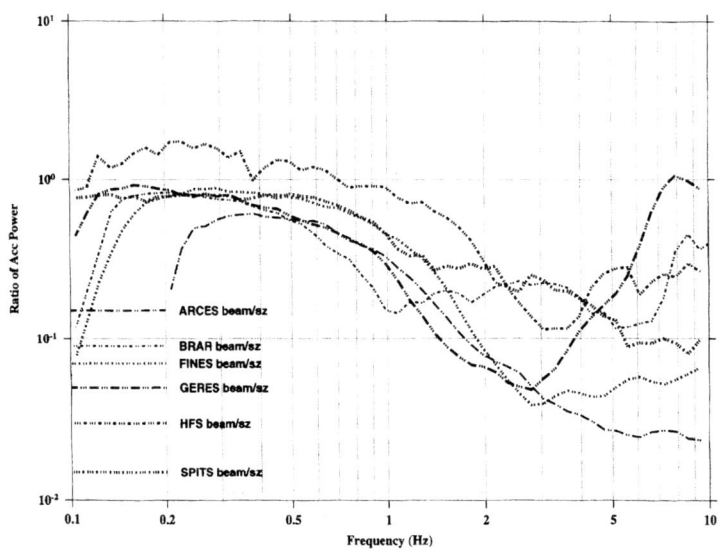

Figure 23
Ratio of infinite-beam to sz channel median noise spectrum during June 1998 for arrays ARCES, BRAR, FINES (Finland), GERES (Germany), HFS (Sweden) and SPITS arrays.

smaller aperture arrays than those of Figure 22, and noise suppression is clearest at higher frequencies.

Conclusions

An automated procedure has been developed to select and calculate spectra of noise samples from data provided by stations at the location given in the CTBT for the IMS seismic network. This procedure has quantified background noise levels as a function of frequency at the sites of 39 primary and 57 auxiliary stations. The results appear to be robust and repeatable. The results of this study form a basis for an improved assessment of the detection capability of the IMS network, and can also be used to improve certain aspects of the automatic processing of IMS data. (We note here that the FDSN station book contains noise spectra from one primary (PPT) and 13 auxiliary station sites (GUMO, KEG, KMI, LZH, MBO, MDT, NOUC, PMSA, PPT, SPA, TLY, SSE, XAN and YSS) from which data have not yet been made available to the PIDC and that thus were not included in the present study.)

A majority of the sites exhibit enhanced levels of noise at higher frequencies, and the observation that in most cases high-frequency noise levels are higher during local daytime than at nighttime demonstrates that the noise is of cultural origin. This effect is particularly marked for most of the stations in Asia. During site surveys, an inventory of local noise sources may indicate how the cultural noise might be reduced by moving the site slightly, and should also prove of value in selecting the optimum location for new sites.

Strong seasonal variations are also apparent at many sites. In general noise levels are higher in the winter than in the summer, however this trend is reversed at the highest latitudes, presumably because adjacent bodies of water that are the predominant noise source are then frozen.

Acknowledgements

The authors would like to thank John Coyne for his considerable assistance with the Geotool program, and Keith McLaughlin for many helpful suggestions.

REFERENCES

COYNE, J. M. and HENSON, I. (1995), *Geotool Sourcebook: User's Manual*, Philips Laboratory Technical Report PL-TR-96-2021.

DARBYSHIRE, J. (1990), *Analysis of Twenty Microseism Storms During the Winter of 1987–1988 and Comparison with Wave Hindcasts*, Phys. Earth Planet. Inter. *63*, 181–195.

Federation of Digital Seismograph Networks Station Book, Incorporated Research Institutions for Seismology (IRIS), 1994 (http://www.iris.edu)

FYEN, J. (1990), *Diurnal and Seasonal Variations in the Microseismic Noise Levels Observed at the NORESS Array*, Phys. Earth Planet. Inter. *63*, 252–268.

GIVEN, H. K. (1990), *Variations in Broadband Seismic Noise at IRIS/IDA Stations in the USSR with Implications for Event Detection*, Bull. Seismol. Soc. Am. *80*, 2072–2088.

GROUP OF SCIENTIFIC EXPERTS (1995), Conference Room Paper 243: GSETT-3 Documentation, Vol. 3: Facilities.

HEDLIN, M. A. H. and ORCUTT, J. A. (1989), *A Comparative Study of Island, Seafloor, and Subseafloor Ambient Noise Levels*, Bull. Seismol. Soc. Am. *79*, 172–179.

LAY, T. and WALLACE, T. C. *Modern Global Seismology* (Academic Press, San Diego 1995).

MYKKELTVEIT, S., FYEN, J., RINGDAL, F., and KVAERNA, T. (1990), *Spatial Characteristics of the NORESS Noise Field and Implications for Array Detection Processing*, Phys. Earth Planet. Inter. *63*, 277–283.

PETERSON, J. (1993), *Observations and Modeling of Seismic Background Noise*, United States Geological Survey Open-File Report 93-322.

(Received August 10, 1999, revised September 15, 1999, accepted November 1, 1999)

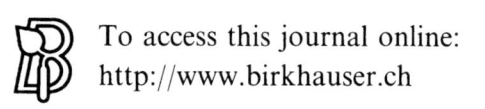

To access this journal online:
http://www.birkhauser.ch

Pure appl. geophys. 159 (2002) 945–967
0033–4553/02/050945–23 $ 1.50 + 0.20/0

❙ Pure and Applied Geophysics

GIS as a Tool for Seismological Data Processing

G. Leonard,[1] Z. Somer,[2] Y. Bartal,[2] Y. Ben Horin,[2]
M. Villagran[2] and M. Joswig[3]

Abstract — A computerized application of an integrated seismological GIS model is presented. An object oriented approach of the GIS topology is introduced and the special functions and features of this system are described. A network topology was selected to simulate the network characteristics of seismological data management and analysis. Each seismological entity is considered as a graphical data object, which is associated to other objects by predefined relationships. The graphical user interface introduced by GIS enables to handle seismological software routines and data in a more intuitive way. Examples of interactive processing of seismic waveforms for detecting, locating and characterizing seismic events using GIS visualization capabilities are presented. The benefits of this system during a passive seismic survey in the framework of the CTBT are highlighted.

Key words: GIS, object oriented, CTBT, NOC, seismology.

Introduction

Within the framework of the Comprehensive Nuclear-Test-Ban Treaty (CTBT), National Data Centers (NDCs) are expected to aid in the evaluation of the true nature of any suspected event. This evaluation will be based on raw and phase data available at the International Data Center (IDC) and also be supported by other informational databases, gathered by the NDCs own technical resources. NDCs can apply their own computer analysis and criteria for distinguishing between CTBT compliance and noncompliance.

A seismic event may lead a signatory state to request further clarification and investigation into the character of a suspected event in the form of an On-Site Inspection (OSI). An inspection team is eligible to enter the signatory state suspected of violating the treaty to clarify any anomaly. The task of the inspection team is to locate the triggering event and verify its character, whether it was a nuclear explosion or a natural event. The search may begin with the rapid deployment of seismic

[1] Israel Atomic Energy Commission, POB 7061, Tel Aviv, Israel.
[2] NDC, Soreq Nuclear Research Center, Yavne, Israel.
[3] Department of Geophysics and Planetary Sciences, Tel Aviv University, Israel.

stations to look for possible aftershocks and to narrow the search area prior to applying other OSI technologies.

The Israeli NDC is the facility responsible for monitoring the treaty in Israel on behalf of the Israeli National Authority. It is responsible for collecting data, analyzing it and developing algorithms for locating and characterizing seismic events in Israel and in the surrounding regions.

The major earthquake sources in Israel are the Gulf of Eilat, the Dead Sea transform fault zone and the Carmel Fara fault (Fig. 1). Most of the current seismic activity from these sources has magnitude (M_L) less than 3 and occurs at focal depths of 5 to 25 km. The seismicity exhibits a pronounced clustering in the Dead Sea fault zone and the Carmel Fara fault (VAN ECK and HOFSTETTER, 1989; HOFSTETTER *et al.*, 1996). Joint focal mechanism analyses carried out on several groups of events along the Dead Sea fault system indicate a north-south strike-slip faulting mechanism (VAN ECK and HOFSTETTER, 1990). Natural seismicity comprises a small fraction of the events recorded by the Israeli Seismic Network (several thousands per year). Most of the events are quarry explosions (Fig. 1). Therefore, the main effort is devoted to the routine discrimination between various event types.

The Israeli NDC includes in its database raw and phase seismic information provided by the IDC and the Geophysical Institute of Israel (which operates the Israeli national seismic network), waveform data from seismic stations designated in the treaty as Cooperating National Facilities (CNF) and bulletin information provided by other sources.

Geographical Information Systems (GIS) techniques offer an environment which simplifies the management of large masses of multi-source and multi-disciplinary information including collection, storage, retrieval, analysis, and interactive visualization. Moreover, GIS provide an attractive environment within which seismologists can evaluate alternative solutions in cases of events in question where further analysis is required. These features will be supportive in the presentation of data and solutions to decisions makers for the assessment of operational strategies during the stage of consultation and clarification. It can also aid the OSI team in the process of evaluating information from various sources, supplied by the Inspected State Party (ISP), such as data from Cooperating National Facilities (CNF), and other auxiliary information.

This manuscript describes the Israeli NDC system from a computer scientist's view of an integrated seismological GIS application. An object oriented approach of the GIS topological model is introduced and the special functions and features of this

▶

Figure 1

Map of Israel and near region showing earthquake and mining activity during the period September–December 1998. Earthquake sources are concentrated in the Gulf of Eilat, along the Dead Sea transform (north and south to the Dead Sea) and on the Carmel – Fara fault (northwest to the Dead Sea).

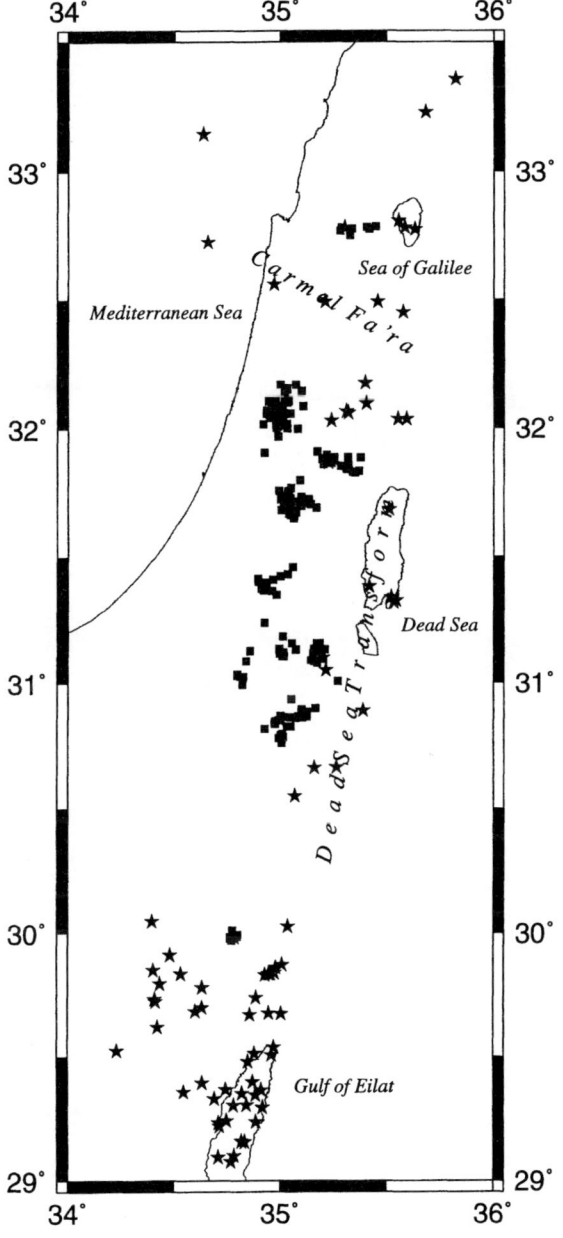

system are described. Interactive visualization and processing of seismic waveforms are presented, including phase picking, hypocenter calculations with associated error ellipses and source identification by pattern recognition techniques.

Principles of Object Oriented GIS in Seismology

By its nature, a GIS application must handle a variety of software codes such as, database management, numerical modeling and optimization, extensive graphic presentation and user-interface techniques. The only practical way to comply with such application complexity is to design and implement an object-oriented methodology. (Extensive overview of this approach can be found in Geographic Facilities Information System Overview, GIS Solution Center IBM Corporation, Document Number GE26-6000-0, January 1990.)

Designing an object-oriented application requires the factoring of objects into classes, defining class interfaces and inheritance hierarchies, and establishing a key relationship among them. An object-oriented algorithm packages both data and procedures that operate on the data. The overall model functionality is accomplished by exploiting Object Oriented Programming Language features like inheritance and interface capabilities. Innovative GIS GUI (Graphical User Interface) enables the user to handle seismological software routines and data in a more intuitive way rather than using exhaustive procedures.

An object oriented GIS is characterized by two major unique features:

1. The integration of interactive graphics with a centralized alphanumeric database system. This means that all data needed for GIS processing can be exchanged between the interactive graphics portion of the application and the database. Thus the graphics are a reflection of the alphanumeric data and *vice versa*.

2. The integration of a basic topological model which determines the relationship between the basic system features. For example, a network topology is suitable for representing seismological objects such as stations and seismic sources in the context of bulletin management.

Seismological GIS application can represent and manage any seismological item of interest (e.g., station, bulletin, etc.). Each seismological object consists of all relevant information and associated relationships to other objects. All objects are visualized in the most intuitive form (are represented graphically and accessible using a simple action of a cursor pointing). The fact that the object information is accessible, enables one to gather all data that are required to process standard seismological routines (signal processing, hypocenter location, etc.), as well as retrieve, store and display the information attached to this object. In order to perform a standard seismological routine, GIS should be fully integrative with the seismological software.

Structure Model Development

A network topology was selected to simulate the environment in which the seismologist operates. The model is a hierarchical network type structure. It is derived and implemented from a general predefined network topology model that consists of four basic feature types (Fig. 2):

Point Feature — An object linked to one Point Connector. Example objects of this class are stations and events.

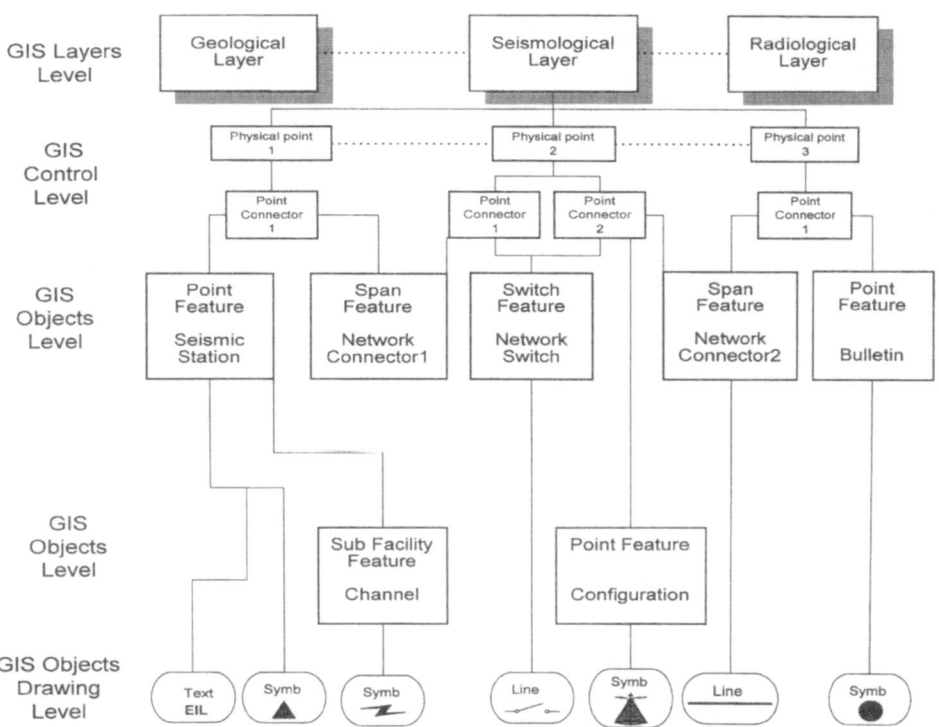

Figure 2

Presents NDCGIS Model as a hierarchical network type structure. At the top are shown the layers that classify the type of objects (e.g., seismological and geological objects). Under each layer a set of unique geographical physical points are presented. Each physical point is a parent of several connectors to which the GIS objects are connected. The way GIS objects are attached to the connectors or to each other defines their topological feature. For example, the configuration object is a "point feature" object because it is linked to "connector 2" under "physical point 2." This configuration is connected to the point feature object "seismic station" via the switch feature object "network switch" and the span feature object "network connector1." It is connected to the point feature object "bulletin" via the span feature object "network connector2." The subfacility object "channel" is a child of the station object. Each described object contains data fields and can be linked to a set of graphical entities that present the object in the most adequate way.

Span Feature — An object linked to two Point Connectors that are attached to different geographical coordinates (Physical Point (PPt)). An example object of this class is a virtual communication line between a data center and a station.

Switch Feature — An object linked to two Point Connectors that are attached to a single geographical coordinate. An example object of this class is a virtual communications switch between an event and a station.

Sub-object Feature — An object linked to all basic object types in a child to parent relationship. An example of this class is a channel that can be regarded as seismic station object child. Deletion of station entity will impose by definition the deletion of all its channels.

Objects derived from these features obtain their special network capabilities by links to virtual Point Connectors and Physical Points. Object visualization on a map is flexible and obtained by the capability to relate various graphic entities for each individual object.

Based on the topological model structure, the GIS system termed NDCGIS is designed to implement several seismological and geological objects. Each object is derived from one of the basic types and thus automatically inherits its topological features. Consequently, specific alphanumeric data and graphical attributes and shapes are defined and assigned to the object. The funtionality of each GIS object is obtained by the definition of its functions. Each object shares common general functions such as: add, delete, edit, move, etc., and in addition unique functions that characterize its role and usage in the context of NDCGIS. This issue will be illustrated later on. The objects of NDCGIS and their application are derived later.

Geographical Information System Environment

NDCGIS was developed as an AutoCADMap based application on a PC/ WIN95. By using AutoCADMap as the main graphical platform we adopted a poweful and standard tool for automated mapping and graphical editing. Moreover, AutoCADMap provides convenient access to the main seismological database by using a standard Open Database Connectivity (ODBC) technique. The major part of the seismological data is stored in an ORACLE database under SUN/Unix operating system. The connection between the NDCGIS PC/WIN95 application and the ORACLE/Unix database is carried out by means of Transmission Control Protocol / Internet Protocol (TCP/IP) standard network communication. Seismological software modules were fused to NDCGIS environment by means of Dynamic Linked Libraries (DLLS) technique.

The seismological modules are:
- MULPLOT — the SEISAN software module for seismological signal processing and phase picking (HAVSKOV, 1999)
- HYPOCENTER — a routine for hypocenter location (LIENERT *et al.*, 1988)

- ELLIPSE — a standard routine for error ellipse calculation according to the International Data Center procedures (IDC, 1999)
- SONODET and COASSEIN — software modules for source identification by pattern recognition, based on a specific master event technique (JOSWIG, 1995)

Figure 3 describes the overall NDCGIS system and associated seismological software modules.

NDCGIS is currently using base maps from ESRI Digital Chart of the World. This source of digital maps covers multiple layers of interest in a vector form longitude latitude rectified coordinate system. It is divided into 5 × 5 degree titles that were produced originally for the U.S Defense Mapping Agency from 1:1000000 scale source maps. This scale covers the world with the highest commercial resolution

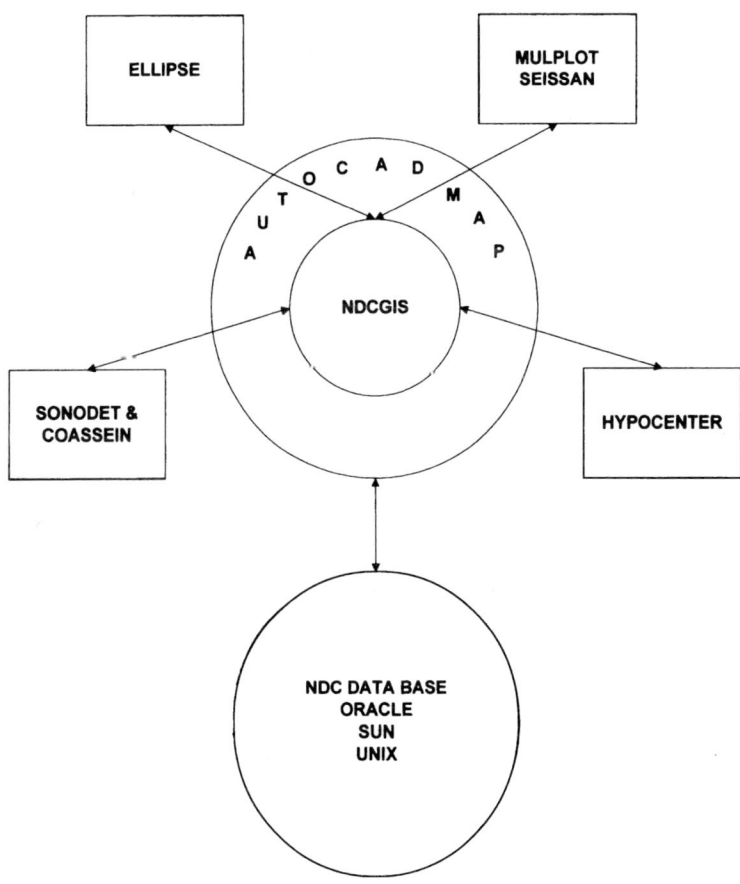

Figure 3
Displays the GIS system which includes the AutoCADMap graphic editor, Oracle Database Manager System and the seismological software modules. The software that implements the topological model of NDCGIS is a sub-module of AutoCADMap.

available, and enables NDCGIS to display and share local and global seismology on the same base maps and coordinate system with sufficient resolution.

Object Implementation

The aim of this section is to describe the objects, highlight their functions and links and demonstrate with examples their respective implementation process.

Station — Belongs to the seismological layer and is derived from a Point Feature.

The data structure follows the definition as defined by the International Data Center (IDC) for their ORACLE seismological database.

Station data include general information (e.g., ID, Name), maintenance information (e.g., ON_DATE, OFF_DATE and importance level) and seismological physical information that plays a role in the procedure of hypocenter location (e.g., longitude, latitude and elevation). A station serves as a parent (host) object for its children channel (z, n, e). Deletion of a station will impose the deletion of all its children channels and their respective data. Station functionality is mostly controlled by basic functions such as: add, edit data, move, delete and display data on a map.

Special station functions include the ability to add to the map a selected group of stations that are imported from external data sources. The station move function assisted the Israeli NDC in the design of an optimal Cooperating National Facilities (CNF) station configuration, within Israel (this issue is elaborated in the section on Application).

Linking a station to a configuration by using the connector object (described later) enables to move it around and still keep the association (logically and graphically) to the same configuration. The same station may be associated with a number of configurations.

Channel — Belongs to the seismological layer and is derived from a Sub-object Feature.

Data fields of a channel include selected general and maintenance information (e.g., component type, instrument name, type and quality) along with other seismic related information like phase type, associated onset time and the waveform file name in which the raw data is stored.

The way phase-related data are stored in a channel object depends upon the function selected. In the case where we decide to construct a configuration consisting of an event and related detecting stations, phase values will find their way to the NDCGIS channel object field from a query result, performed over the ORACLE database. In other cases, we might decide to conduct phase picking for a selected channel. To support this request, the SEISAN signal processing module initiates the current waveform file name stored in the channel. Following the performed picking action, the phase field in the channel object is refreshed with a new value.

The channel objects are moved together with the station when the station move function is applied.

Bulletin — Belongs to the seismological layer and is derived from a Point Feature.

The data structure follows the definition of the International Data Center (IDC). The "Bulletin" object in NDCGIS is a general object, used in the context of bulletin, event or epicenter. Bulletin data fields include standard seismological information such as time, location, error ellipse axes lengths and orientation, M_S and m_b magnitudes etc. The manner in which bulletins are added varies according to the context of the work performed. In the event we would like to display a set of events on a map, we generate a query request for a selected time and location frame. All relevant data are retrieved from the database, and as a result, new bulletin objects are added to the map (Fig. 4). Therefore, bulletin selected data such as error ellipses can be displayed graphically (Fig. 5).

Quarry — Belongs to the geological layer and is derived from a Point Feature.

From a CTBT point of view quarries are considered as important seismic sources. Quarry explosions produce the majority of the local waveform database. Data fields include general information (ID, Name, Type, etc.). Location of quarry sites is kept in the form of air photos.

Master pattern — The Israeli NDC uses a specific master event technique, the Sonogram Detector (SONODET), as an automatic method for screening seismic events in the framework of the CTBT (Fig. 6). This technique is based on the following initial steps:

- predetermination of seismic sources.
- selection of events as master patterns which are imaged on the complete single recorded trace and stored at each seismographic station.

Master pattern is the interface between the seismological and geological layer.

It is derived from a Span feature that links a station to an active zone. The master pattern object represents the master time series (and the related master pattern) originating in the active zone and recorded by the station.

These time series are transformed into sonograms (WÜSTER, 1993; JOSWIG, 1995). In the transformation process the signals are divided into windows of equal length and spectral analyzed. The spectral range is divided into 11 logarithmically spaced bins, and the average spectral density over such a bin, corrected for noise offset, is stored in a sonogram matrix of frequency bin versus time window.

For each seismic source a single pattern is stored in the computer memory as a reference pattern in a 2-D image of signal energy in time and frequency. For each seismic station the selected patterns of all sources of clustered activity (active zones) are stored. The most important master pattern function is the "Sonogram detector." This function evaluates the degree of resemblance between the new imaged signals and those stored as master patterns. The results obtained at individual stations are

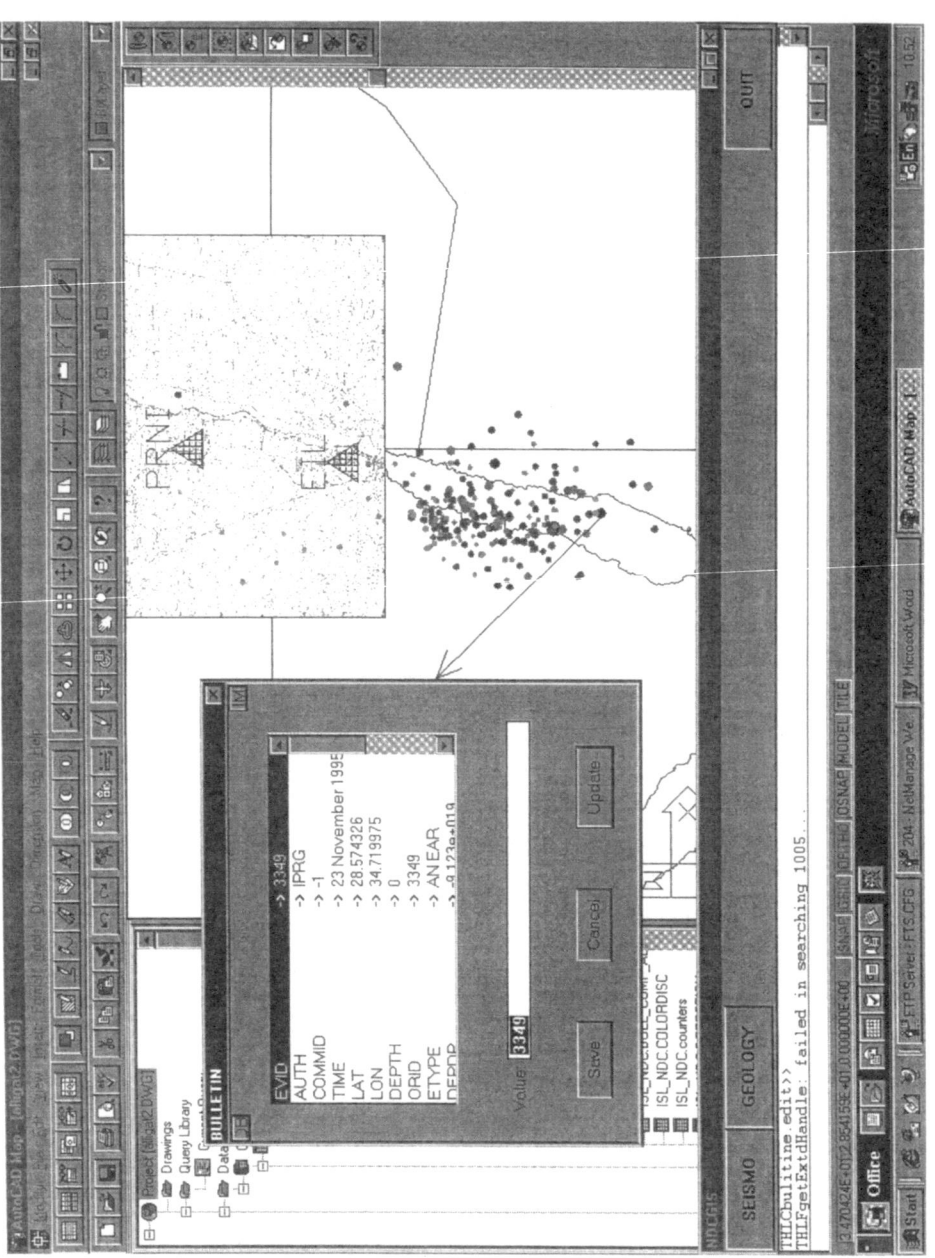

Figure 4

Displayed is a set of aftershocks in a 48-hours time window. The aftershocks followed the 1995, 7.1 M_w earthquake which occurred in the Gulf of Eilat. Clicking on any event opens an NDCGIS standard editor with the event data.

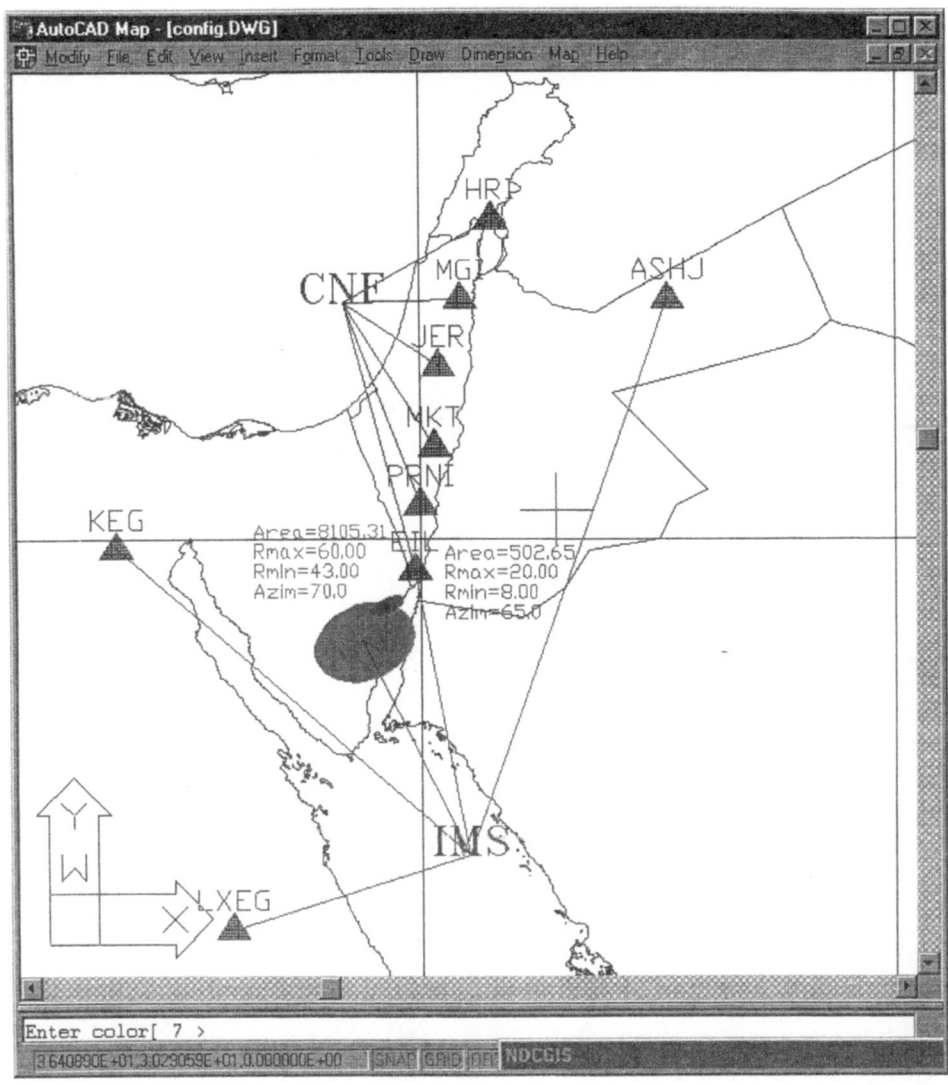

Figure 5
Displays two network configurations and their resultant location (with the associated error ellipse) of the same single event. The larger error ellipse is based on the regional IMS network configuration where the smaller ellipse is a function of the local CNF network configuration.

integrated into the final source identification by a rule-based routine (COASSEIN) (LEONARD *et al.*, 1999). A master pattern object is used in conjunction with the Sonogram detector and it highlights the relationship between an active zone and the recording stations (Fig. 7).

Active Zone — Belongs to the seismological layer and is derived from a Point Feature.

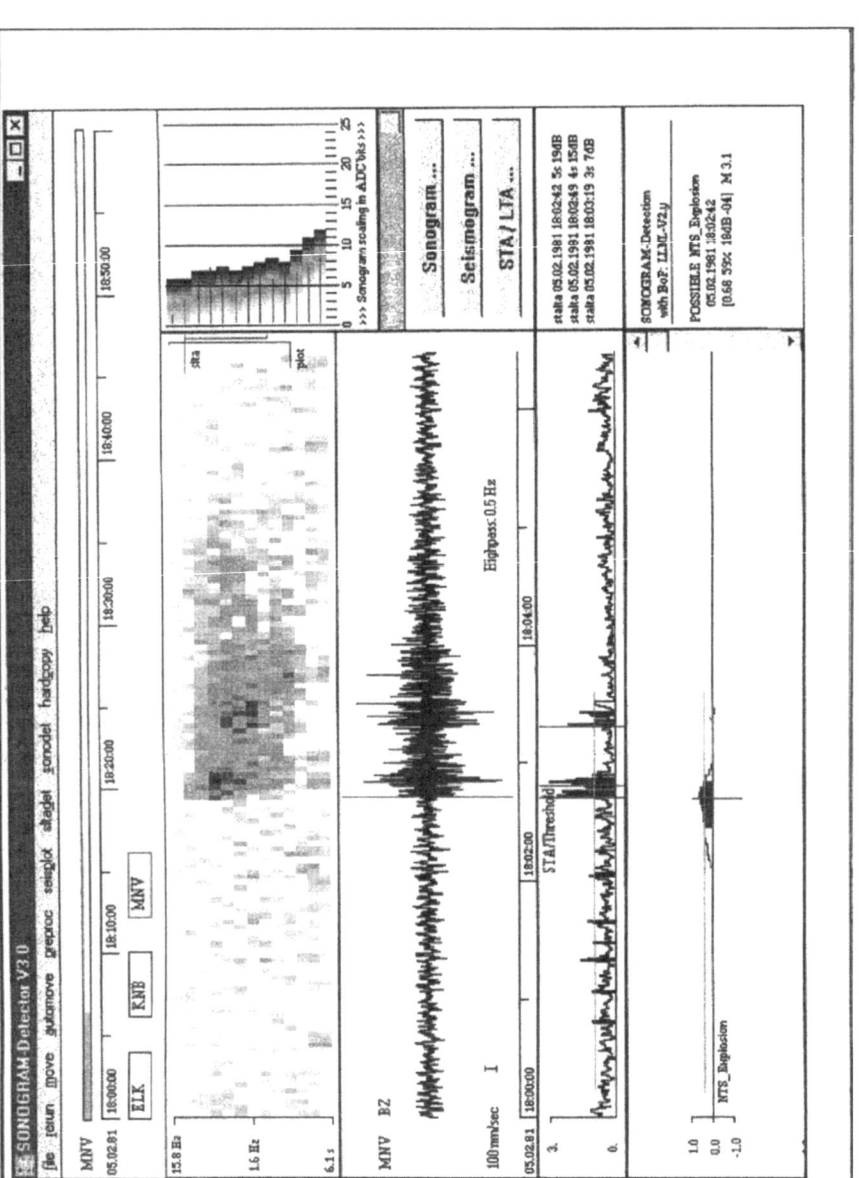

Figure 6

Evaluation of a recorded weak seismic signal at station MNV by the Sonogram detector. The source of the signal is a nuclear test conducted at the Nevada Test Site. The seismogram is complemented by the sonogram (top left window) and by the STA/LTA detector (window below the seismogram). The sonogram displays the power spectral density specifically scaled to enhance the temporary signal energy. The associated noise spectrum is displayed on the top right window; both support the seismologist in detecting weak signals, in determining the optimum filter settings, and in judging the station quality. Also shown is the degree of resemblance of the pattern displayed in the top left window to the most similar predefined pattern(s) (left bottom window). The detection message is displayed on right bottom (POSSIBLE NTS explosion).

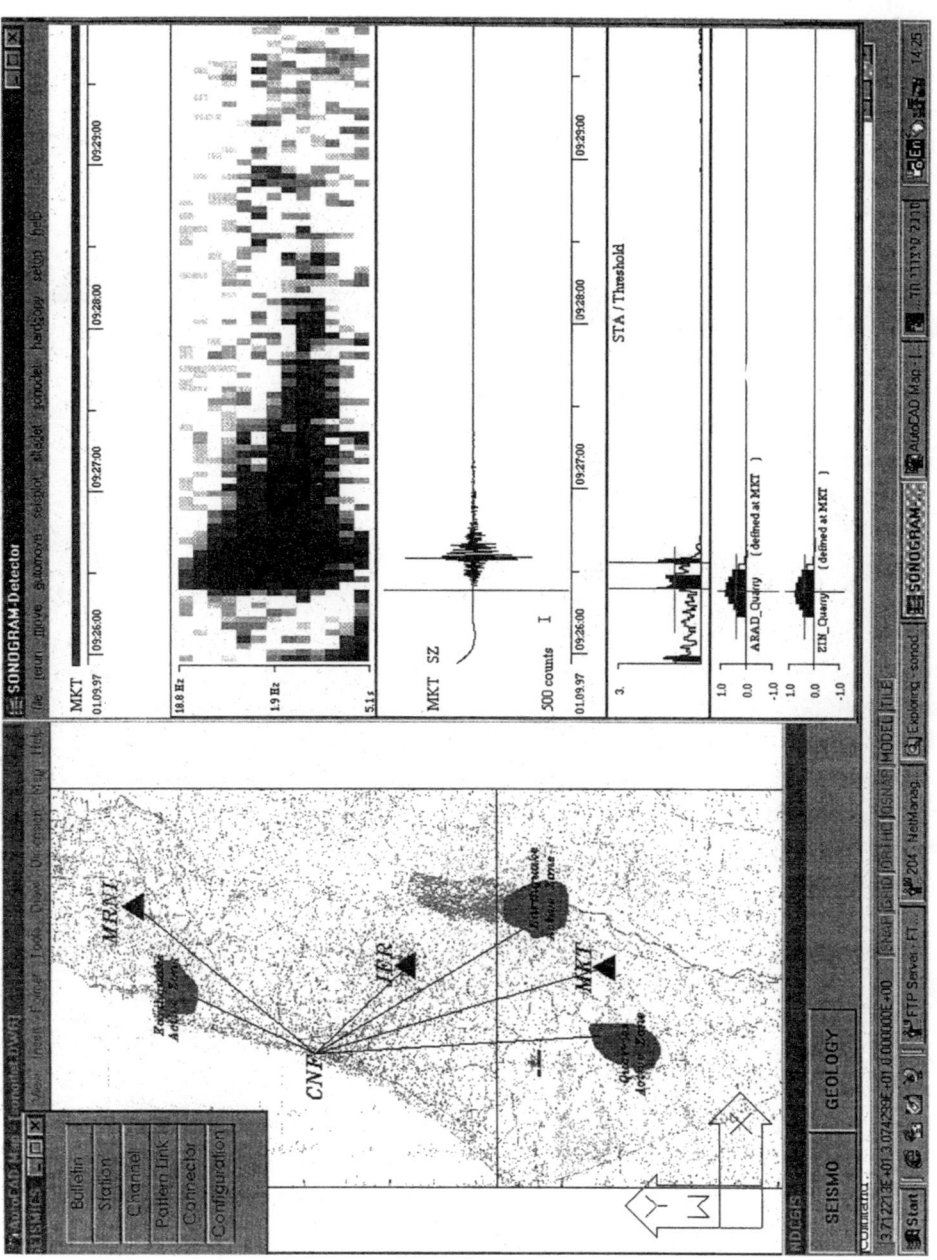

Figure 7

Displays the link between imaged data in the right window to related stations (triangles) and seismic sources (gray patches) in the left window. In this example the image of a recorded signal at station MKT is displayed. The image is correlated to all stored images at this station of known seismic sources (gray polygons termed active zones). The best correlation determines the final identification, Arad Quarry (lower box, right window) located in the active zone close to station MKT.

Active zone object represents a polygon that bounds areas of seismic activity (quarry clusters and fault related seismic activity, Fig. 7). Active zone data fields include information like: ID, Name, Type, Polygon Area and Center, Air photos library and Master Patterns library.

Connector — is derived from a Span Feature.

The connector is a mediator object that marks and displays the existence of relevant links between sets of NDCGIS objects. Connectors are used at present to identify a certain "subject related group" that may include, for example, a specific set of stations that were used to locate and identify a single event. As another example, different sets of stations can be linked to various known seismic sources. These stations are designated to monitor areas of clustered activity (termed "Active Zones") from earthquake and quarry sources for event recognition purposes.

Connector object contains only a few general data fields namely, the From and To connected objects and the connector length. In its common use, a connector object contributes to creating links between one configuration object (described later) and many other NDCGIS objects that relate to that configuration. Because of its important role in assigning special significance to the connected seismological objects, NDCGIS objects cannot be deleted if they are attached to a connector. In order to delete them it is necessary to delete first all the connectors attached to that object.

Configuration — Belongs to the seismological and geological layers and is derived from a Point Feature.

Configuration, the most significant feature, is a "container" object for the "subject related group." Each element of the group is connected to the configuration by a connector object. According to the subject context, a configuration may be linked to local or regional stations, bulletins or active zone objects.

Data fields of a configuration include general information like the number of each object type connected to it, for example, the number of stations, active zones, quarry or fault objects. In addition, included are specific input or output data of configuration functions. This may be in the form of the waveform file that the configuration uses for its phase picking procedure, or the current event date, time and location that resulted from the latest use of the hypocenter location procedure.

The way a configuration is formed in NDCGIS varies according to the context in which the configuration is used. In the simplest way a configuration is added with no connections and as such is regarded as "null configuration." Subsequently, we may create several configurations connected to different sets of stations (networks) by using the connector object. A specific event may be recorded by a few sets of stations. From each configuration a hypocenter location can be calculated.

Figure 5 illustrates two location solutions (and associated error ellipses) of the same seismic event and their corresponding network configuration (local CNF and regional IMS). Each set of seismic stations linked by a set of connectors and their associated location solution is defined as configuration.

Configuration is also added by creating a "subject related group" from the information stored in a phase file. In this case, all stations, channels and bulletin objects of the configuration are automatically added and displayed on the map.

Another way to form a configuration is by the "add by time window selection" function. Once this configuration is constructed, it links, for example, waveform data, which enable phase picking (Fig. 8) and hypocenter location calculation (Fig. 5). This function is elaborated in the section on Application.

Other configuration functions are the Sonodet and Coassein modules described above for station and network identification of a local source.

Application

In this section we will illustrate the method by which NDCGIS can be implemented in a routine procedure of data processing; in the design of seismic station optimal deployment; and the advantage in applying such a tool during an on-site inspection.

Routine Implementation

NDCs can apply their own analysis and criteria for characterizing seismic sources. Such a procedure is summarized through a flow chart in Figure 9.

In this procedure, waveform data are continually retrieved from the main database and an on going STA/LTA program checks for possible events, based on an assigned trigger level from each of the delivering stations. Each triggered event is registered into the database. Then, by applying "add by time window selection" NDCGIS configuration on the map (on the screen) is automatically constructed. It includes all stations and channels that recorded the event at that selected time window. The waveform data file name is stored in the waveform file data field of this configuration.

At this stage the configuration is ready for signal data processing by applying the configuration function "phase picking." This function invokes the signal processing module, enabling it to perform phase picking. Figure 8 shows for example the selected waveforms and phase picks on the background of the related stations. Each phase pick type and time is stored in its corresponding channel data fields "phase ID" and "phase time."

Next, the "hypocenter calculation" function is implemented which generates a new "bulletin" object which reflects the event location and connects it to the configuration. Consequently, the "bulletin" object's functions, "simple ellipse" and "show parameters," display graphically the error ellipse contour and calculated values as shown in Figure 5. Finally, master events are utilized to support the location for event clusters (Fig. 7). If both the image and the location match, the event is characterized and stored in the database.

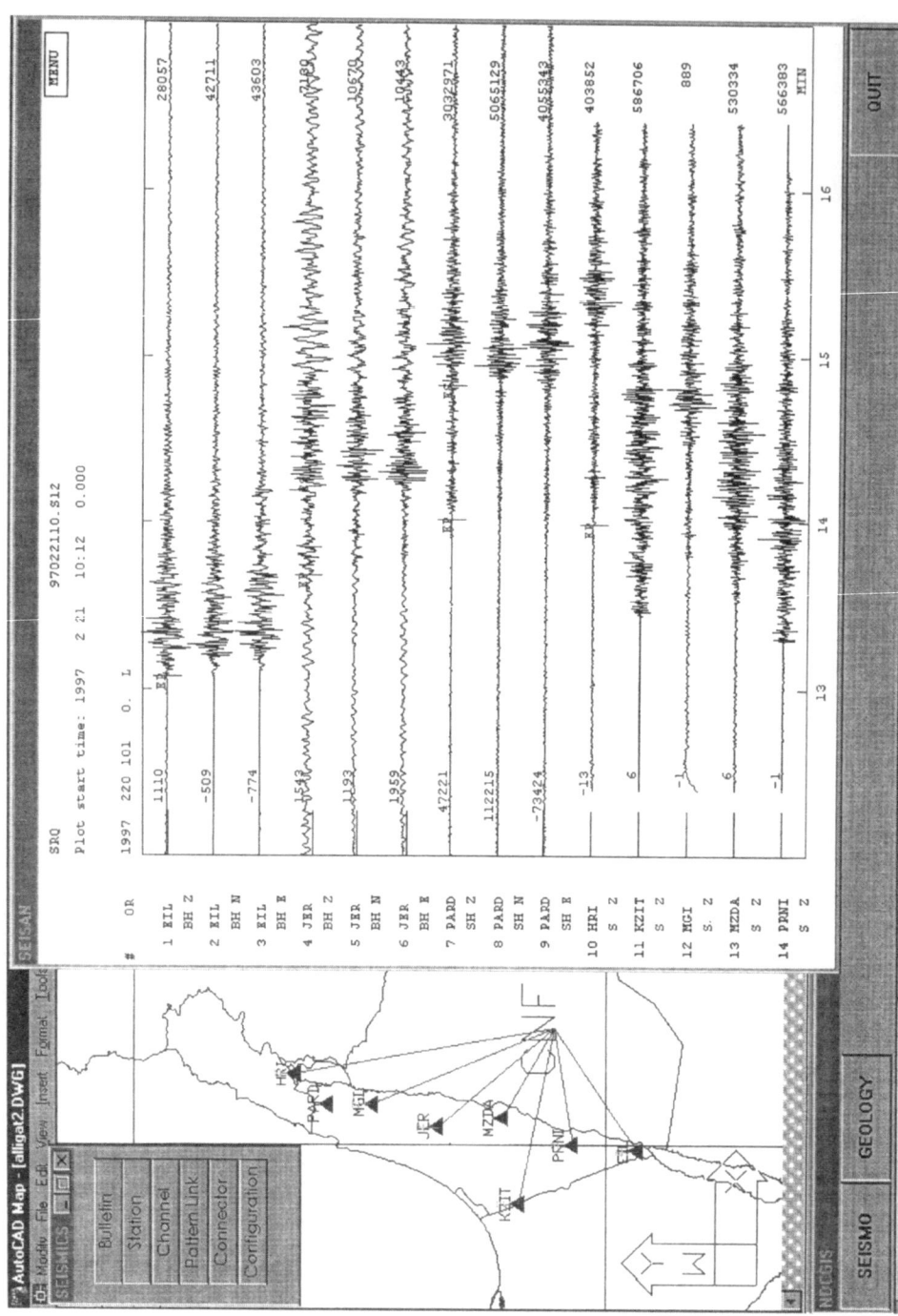

Figure 8

Displays the link between waveform data and related stations. By invoking the "Phase selection" function, picked phases are stored in the channel data fields.

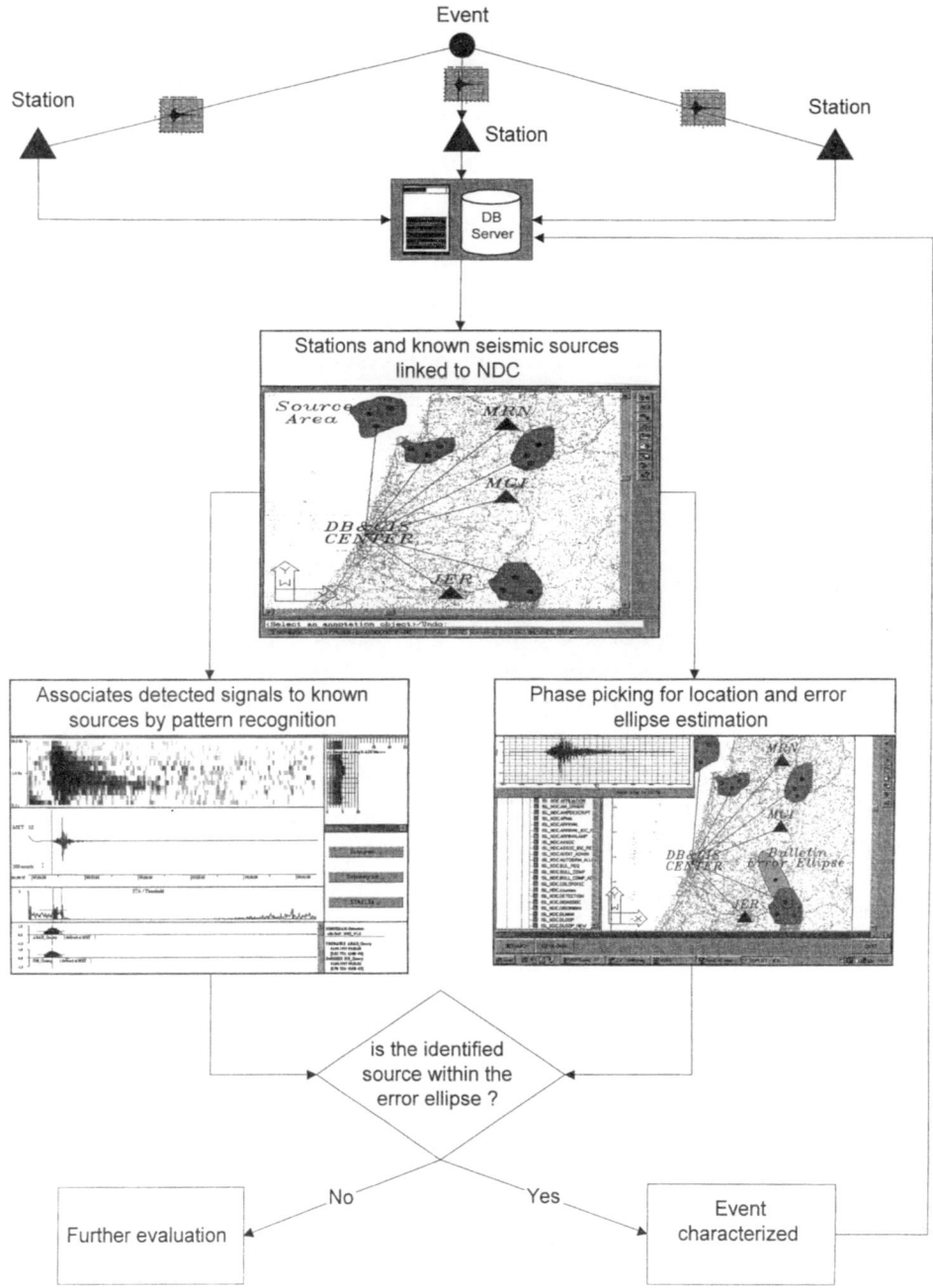

Figure 9
A flowchart demonstrating a routine process for event location and characterization of a local configuration. The process includes procedures for data acquistion, STA/LTA triggering, phase picking, error ellipse analysis and source identification by pattern recognition.

Network Design

The initial task of the inspection team in the case of an OSI is to narrow down the inspection area, beginning with the error region of the event's epicenter, prior to using other OSI technologies. To fulfill this task the inspection team can greatly benefit from auxiliary information provided by certain certified national stations, designated in the treaty as Cooperating National Facilities (CNF). Such a network was optimized using a Genetic Algorithm (BARTAL *et al.*, 2000), for reducing the potential search area in Israel. The goal was to find an optimal network for which the error ellipse area over a specified set of epicenters within Israel would be about 100 square km. The optimization yielded six CNF stations in addition to the two Auxiliary IMS stations. However, the design process of the network should comply with various design constraints such as: state borders, populated areas, access roads, electricity supply, etc.

NDCGIS was used to design the final CNF seismic station optimal deployment. We used the GIS function "move" to locate stations at potential sites on detailed background maps, in proximity to the optimized configuration. Station location adjustments are followed by the display of the calculated error ellipse for each new configuration. From various possible configurations we selected that one which yielded the minimum area of error ellipses at all seismic sources.

Figure 10 demonstrates the change in the error ellipse geometry and dimensions of the same single event occurring in the sea, due to a location change of HRI in the CNF sub-network configuration.

The implied result could have significant implications regarding an accusation, i.e., whether the event has occurred off-shore (smaller ellipse) or whether there is a possibility that it occurred on shore (larger ellipse).

Passive Seismic Survey

Seismic monitoring within the mandated inspection area of a suspect event will be one of the first activities of an OSI inspection team within the framework of the CTBT. The goal of the passive seismic monitoring is to detect aftershock activity that can help determine whether further inspection is justified and, if so, to narrow the search area prior to using other OSI technologies. One of the first major decisions that the OSI team must make is to define the radius of the seismic monitoring area(s). One option is to deploy a focused network that will enable detection of even very small events (down to magnitude −1) in a constrained area. Another option is to deploy a network that covers much of the mandated search area although at the risk of missing small events that occur far from every station. A major issue in this respect is how the rate of station redeployment should be matched to the decay in aftershock activity. The OSI team must weigh, on a continuous basis, the tradeoffs between a discrete reconnaissance and a systematic search. This is mainly imposed by constraints influencing the inspection activity.

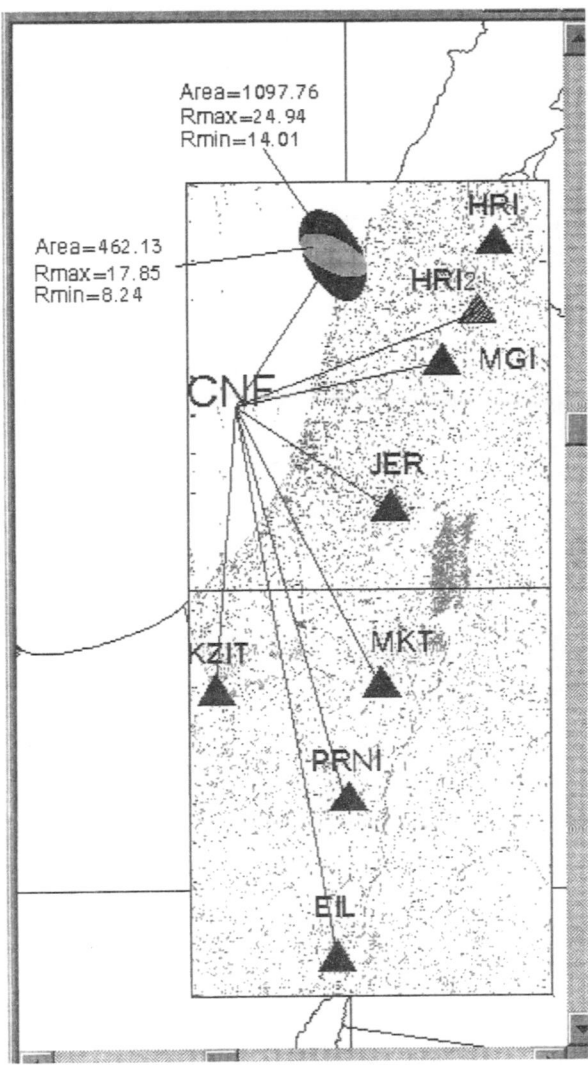

Figure 10
Displays the resultant effect on the error ellipse, due to the adjustment in the location of stations, as part of the design process of the CNF station configuration. The figure presents two error ellipses (black and gray) of the same single event, located in the Mediterranean sea, occurring on the Carmel Fařa fault zone (Fig. 1). Each is associated to a change in the position of station HRI where all other stations (black triangles) are fixed. The black ellipse corresponds to the network configuration with station HRI (black triangle). The gray ellipse corresponds to the network configuration with station HRI2 (hatched triangle).

The incorporation of an integrated GIS system at this stage enables rapid reaction to the accumulating information from the seismic monitoring and other sources. NDCGIS is a sophisticated user interface software tool that can gather and analyze all seismic data using standard seismological routines on one hand, and generate advanced thematic display reports by a mouse click on the other hand. Moreover, seismic data can be integrated with geological and topographical information for an improved overview of the findings. Presentation using GIS is the human intuitive way, and as such will better assist the OSI team in their repeated activity during the inspection period. The repeated activity includes data collection, analysis and evaluation; ranking targets and setting a deployment strategy; deployment of instrumentation according to strategy.

Data Collection

In analyzing the seismic source properties, the OSI team should determine any unusual characteristics in the waveform and whether these correlate from station to station. If the detection parameters have not been properly adjusted, the system may generate many false detections or miss small, particularly low frequency events. Therefore, the seismologist should review continuous data for "missed" events. NDCGIS incorporates the Sonogram detector for scanning continuous data (JOSWIG, 1999).

The Sonogram detector software can assist in on-line tuning of seismic monitoring parameters. The seismogram is complemented by the sonogram (displaying the power spectral density, scaled to enhance the temporary signal energy), the associated noise spectrum and by the STA/LTA detector (Fig. 6). The integration of all displays on top of a GIS desktop will support the seismologist in tuning the detection of weak signals, in determining the optimum filter settings, in judging the station quality, highlight the related station on a map and thus accelerate the response to the acquired field data.

Data Analysis

As introduced previously, the Sonogram detector can evaluate the pictorial information of seismic signals by comparing them to a predefined set of patterns as an automatic method for screening seismic events (Figs. 6 and 7). The degree of resemblance between each new pattern to known master patterns of known sources provides the seismologist with an associated identification message. Each identified source is highlighted on a background map, enabling the seismologist to screen out patterns of frequent noise sources.

Data Evaluation

Auxiliary information provided or gathered by the inspection team will make it possible to examine both areas of natural seismically and unusual activities consistent with a possible treaty violation. Certain facilities may deserve special attention by the inspection team, such as mining districts, areas of recent drilling and areas showing recent ground disturbance (cracking, rock falls, etc.).

At this stage the OSI team may exploit the GIS system to integrate information that includes predefined GIS objects such as mining areas and geological faults located in the inspected area with aftershocks spatial distribution and waveform characteristics. The data retrieval process enables the presentation of detailed documents attached to the objects described above.

This will assist the OSI team in investigating whether the events originate from a natural origin; what their relationships are to traffic and possible cultural sources, and whether these can be verified or isolated in the field.

Redeployment

Following the evaluation stage, specific targets may be selected for further inspection. The GIS system can be used to display detailed background maps presenting the topography, state borders and access roads for designing further redeployment.

As part of seismic data evaluation the OSI team should determine how well the hypocenters are constrained, whether there is a pattern to the locations, how important the specific stations are in the solution, and whether the station locations can be altered to better constrain the solutions. The previous section on "Network Design" explains how NDCGIS can be used in this context.

Discussion

Geographical Information Systems (GIS) serve as a tool to supplement existing ability and knowledge. It is a powerful way to make seismology more relevant when it is linked to policy and decision making. In the framework of the CTBT, GIS encourages a more visual and intuitive approach that recognizes the importance of linking analyzed results to location of events with an emphasis on how results may vary as a function of recording distances.

The goal of this paper is to promote awareness of GIS as an assisting tool in the application to seismological problem solving in general, and specifically within the framework of the CTBT.

An integrated seismological-GIS tool has been developed to support the Israeli National Data Center (NDC) in its activities. The tool provides an attractive environment within which seismologists can manage a large mass of multi source and multi disciplinary information. Seismological objects are defined as part of a general GIS topological network model. Applications are demonstrated focusing on routing processing for locating and identifying seismic events. Using different sources of seismic information, alternative solutions can be presented and visualized simultaneously.

In this study we describe the special functions and features of the integrated seismological-GIS system. Seismological items (e.g., stations, events) are viewed as objects. Each object consists of all relevant information and associated relationships to other objects. All objects are represented graphically and accessible in a simple action of a cursor pointing. In particular we focused in this paper on interactive visualization and the processing of seismic waveforms including phase picking, hypocenter calculations and source identification by pattern recognition techniques.

The most appealing feature of the system is the configuration function. This function makes it possible to analyze, present and visualize simultaneously alternative solutions to decision makers for the assessment of operational strategies in cases where further clarifications are required, regarding events in question, within the framework of the CTBT.

A major issue during an On-Site Inspection is the deployment strategy of seismic stations as a function of the decay rate in aftershock activity. The integrated GIS tool enables the inspection team to rapidly react to the accumulating information of seismic and nonseismic data and thus accelerate the redeployment of instrumentation in a more efficient way.

At present the NDCGIS network model supports seismological activities most adequately. Enhancement of geological entities such as joints and faults is underway to associate between geological structures and seismological information thus facilitating an improved integrated analysis.

Acknowledgments

This study was funded by the Israeli Atomic Energy Commission. We thank Y. Weiler, M. Melamud and S. Lewis for their review of the manuscript.

REFERENCES

BARTAL, Y., SOMER, Z., LEONARD, G., STEINBERG, D. M., and BEN HORIN, Y. (2000), *Optimal Seismic Networks in Israel in the Context of the Comprehensive Test-Ban Treaty*, Bull. Seismol. Soc. Am. *90*(1), 151–165.

HAVSKOV, J. (1999), *The SEISAN Earthquake Analysis Software for the IBM PC and Sun*, Institute of Solid Earth Physics University of Bergen, Version 7.0.

HOFSTETTER, A., VAN ECK, T., and SHAPIRA, A. (1996), *Seismic Activity Along Fault Branches of the Dead Sea Jordan Transform: The Carmel-Tirza Fault System*, Tectonophy. *96*, 317–330.

IDC Processing of Seismic, Hydroacustic and Infasound Data (1999), IDC Documentation, 5.2.1.

JOSWIG, M. (1995), *Automated Classification of Local Earthquake Data in the BUG Small Array*, Geophys. J. Int. *120*, 262–286.

JOSWIG, M. (1998), *Automated Processing of seismograms by SparseNet*, Seism. Res. Lett. *70*, 705–711.

LEONARD, G., VILLAGRAN, M., JOSWIG, M., BARTAL, Y., RABINOWITZ, N., and SAYA, A. (1999), *Seismic Source Classification in Israel by Signal Imaging and Rule-based Coincidence Evaluation*, Bull. Seismol. Soc. Am. *89*, 960–969.

LIENERT, B. R., BERG, E., and FAZER, L. N., (1988), *Hypocenter: An Earthquake Location Method Using Centered Scaled and Adaptively Least Squares*, Bull. Seismol. Soc. Am. *76*, 771–783.

VAN ECK, T. and HOFSTETTER, A. (1989), *Microearthquake Activity in the Dead Sea Region*, Geophys. J. Int. *99*, 605–620.

VAN ECK, T. and HOFSTETTER, A. (1990), *Fault Geometry and Spatial Clustering of Microearthquakes Along the Dead sea Jordan Rift Fault Zone*, Tectonophy. *180*, 15–27.

WÜSTER J. (1993), *Discrimination of Chemical Explosions and Earthquakes in Central Europe – A Case Study*, Bull. Seismol. Soc. Am. *83*, 1184–1212.

(Received June 13, 1999, revised October 1, 1999, accepted November 1, 1999)

 To access this journal online:
http://www.birkhauser.ch

Pure appl. geophys. 159 (2002) 969–987
0033–4553/02/050969–19 $ 1.50 + 0.20/0

Pure and Applied Geophysics

Optimized Seismic Threshold Monitoring – Part 1: Regional Processing

Tormod Kværna,[1] Frode Ringdal,[1] Johannes Schweitzer[1]
and Lyla Taylor[1]

Abstract — Continuous seismic threshold monitoring is a technique that has been developed over the past several years to assess the upper magnitude limit of possible seismic events that might have occurred in a geographical target area. The method provides continuous time monitoring at a given confidence level, and can be applied in a site-specific, regional or global context.

In this paper (Part 1) and a companion paper (Part 2) we address the problem of optimizing the site-specific approach in order to achieve the highest possible automatic monitoring capability of particularly interesting areas. The present paper addresses the application of the method to cases where a regional monitoring network is available. We have in particular analyzed events from the region around the Novaya Zemlya nuclear test site to develop a set of optimized processing parameters for the arrays SPITS, ARCES, FINES, and NORES. From analysis of the calibration events we have derived values for beam-forming steering delays, filter bands, short-term average (STA) lengths, phase travel times (P and S waves), and amplitude-magnitude relationships for each array. By using these parameters for threshold monitoring of the Novaya Zemlya testing area, we obtain a monitoring capability varying between m_b 2.0 and 2.5 during normal noise conditions.

The advantage of using a network, rather than a single station or array, for monitoring purposes becomes particularly evident during intervals with high global seismic activity (aftershock sequences), high seismic noise levels (wind, water waves, ice cracks) or station outages. For the time period November-December 1997, all time intervals with network magnitude thresholds exceeding m_b 2.5 were visually analyzed, and we found that all of these threshold peaks could be explained by teleseismic, regional, or local signals from events outside the Novaya Zemlya testing area. We could therefore conclude within the confidence level provided by the method, that no seismic event of magnitude exceeding 2.5 occurred at the Novaya Zemlya test site during this two-month time interval.

As an example of particular interest in a monitoring context, we apply optimized threshold processing of the SPITS array for a time interval around 16 August 1997 m_b 3.5 event in the Kara Sea. We show that this processing enables us to detect a second, smaller event from the same site (m_b 2.6), occurring about 4 hours later. This second event was not defined automatically by standard processing.

Key words: Seismic event detection, automatic data processing, seismic data analysis.

1. Introduction

Continuous seismic threshold monitoring (TM) is a technique that has been developed at NORSAR over the past decade to monitor a geographical area

---

[1] NORSAR, Post Box 51, N-2027 Kjeller, Norway. E-mail: tormod@norsar.no

continuously in time. Data from a network of arrays and single stations are combined and "steered" toward a specific area to provide an ongoing assessment of the upper magnitude limit of seismic events that might have occurred in that area. The basic principles have been described by RINGDAL and KVÆRNA (1989, 1992), who showed that this method could be useful as a supplement to event detection analysis. Recently, KVÆRNA and RINGDAL (1999) have shown the potential of using the TM method for global network capability estimation.

The main purpose of the TM technique is to highlight instances when a given threshold magnitude is exceeded, thereby helping the analyst to focus on those events truly of interest in a monitoring situation. The analyst can then apply traditional tools in detecting, locating and identifying the source of the disturbance. The capability achieved by the threshold monitoring method is in general dependent upon the size of the target area, and it is convenient to consider three basic approaches:

Site-specific threshold monitoring: A seismic network is focused on a small area, such as a known test site. As discussed in this paper, this narrow focusing enables a high degree of optimization, using site-station specific calibration parameters and sharply focused array beams.

Regional threshold monitoring: Using a dense geographical grid, and applying site-specific monitoring to each grid point, threshold contours for an extended region are computed through interpolation. In contrast to the site-specific approach, it is usually necessary to apply regionally averaged attenuation relations, and the monitoring capability will therefore not be quite as optimized.

Global threshold monitoring: Using a global seismic network, and taking into account that phase propagation time extends to several tens of minutes, global travel-time and attenuation tables are applied, possibly with regional corrections, with a substantially coarser geographical grid than in the regional approach.

The purpose of this paper (Part 1) and a companion paper (Part 2) is to present results from recent developments of the site-specific approach. These developments have been directed towards optimizing the monitoring capabilities for a given target site, by using calibration events to optimize parameters such as bandpass filters, beam-steering parameters, and window lengths for short-term averaging. We thus aim at utilizing the *full resources* of the monitoring network to focus on a specific target site in order to enable monitoring of this target site with as high a capability as the network and available calibration information will allow. The present paper addresses the case of a regional monitoring network, with special application to the Novaya Zemlya nuclear test site.

The area around Novaya Zemlya is characterized by generally low natural seismicity. However, during the last decades a few small events have been observed and several studies have been conducted to locate and characterize these events (e.g., MARSHALL *et al.*, 1989; RINGDAL, 1997; RINGDAL *et al.*, 1997). Some of these events were only detected at the regional arrays operated by NORSAR. We show that threshold monitoring using this network can be used to achieve a high capability for continuous monitoring of this site.

2. Method Description

2.1 Generating the Threshold Trace

Let us assume that a network of seismic stations is available for monitoring a specified target site. For simplicity of presentation, we will assume that these are all array stations, able to provide phase velocity and azimuth information for detected signals. Extension to the single-station case is straightforward. The stations can be located either at regional or teleseismic distances.

Following RINGDAL and KVÆRNA (1989), let us consider a network of seismic stations ($i = 1, 2,...,N$) and a number of seismic phases ($j = 1, 2,...,M$). For a seismic event of magnitude $m_b = m$ an estimate $\hat{m}_{ij}$ of m is given by

$$\hat{m}_{ij} = \log S_{ij} + b_j(\Delta, h) \tag{1}$$

where S_{ij} is a measurement of the signal amplitude of the j-th phase at the i-th station and $b_j(\Delta, h)$ is a distance-depth correction factor for the j-th phase.

In standard formulas for magnitude, the measured amplitude S_{ij} is usually estimated as A/T, i.e., amplitude of ground displacement divided by dominant signal period. In our case, we will use the short-term average (STA) at the expected signal arrival time as a measurement of signal amplitude, so that $\text{STA}_{ij} = S_{ij}$. The value is measured on an array beam or a single channel filtered in an appropriate frequency band.

Traditionally, relation (1) is defined only for the time window corresponding to a detected seismic event. We will now consider the right-hand side of (1) as a continuous function of time. Define the threshold parameter $a_{ij}(t)$ as follows:

$$a_{ij}(t) = \log \text{STA}_{ij}(t) + b_j(\Delta, h) \ . \tag{2}$$

Equation (2) represents a function which can be considered as a continuous representation of the upper magnitude limit for a hypothetical seismic event at a given geographical location (target region). It coincides with the event magnitude estimate if an event occurs at that site. The function is, by definition, tied to a specific station and a specific phase.

The threshold parameter traces are then time-aligned in accordance with the expected travel time of the considered phase, such that the time reference for all threshold parameter traces is the origin time at the target region. For each time sample we obtain a network-based representation of the upper magnitude limit by considering the function:

$$g(m,t) = 1 - \prod_{i,j}\left(1 - \Phi\left(\frac{(m - a_{ij}(t))}{\sigma_{ij}}\right)\right) \tag{3}$$

where m is event magnitude, σ_{ij} is the standard deviation of the assumed magnitude distribution for the i-th station and j-th phase, and Φ denotes the standard $(0, 1)$ normal distribution function. This network-based representation assumes statistical independence among the different station/phase observations, and the function $g(m, t)$ is discussed in more detail by RINGDAL and KVÆRNA (1989).

The function $g(m,t)$ is the probability that a given (hypothetical) seismic event of magnitude m at time t would generate signals that exceed the observed noise values at at least one station of the network. For a given t, the function $g(m,t)$ is a monotonically increasing function of m, with values between 0 and 1. A 90 percent upper limit at time t is defined as the solution to the equation

$$g(m,t) = 0.90 \ . \tag{4}$$

The solution is a function of t, which we will denote $m_{T90}(t)$. We call this the *threshold trace* for the network and target region being considered.

It is important to interpret the 90 percent limit defined above in the proper way. Thus, it should not be considered as a 90 percent network detection threshold, since we have made no allowance for a signal-to-noise ratio which would be required in order to detect an event, given the noise values. Rather, the computed level is tied to the actually observed noise values and to the fact that any hypothetical signal must lie below these values. Our 90 percent limit represents the largest magnitude of a possible hidden event, in the sense that above this limit there is at least 90 percent probability that one or more of the observed noise values would be exceeded by the signals of such an event.

Although not intuitively obvious, it follows from eqs. (3) and (4) that the threshold trace is dominated by the "best" stations in the network (i.e., the stations with the highest sensitivity for events in the target area). Thus, the resulting value of the computed threshold trace remains essentially unchanged if a station with poor detectability is added to the network. In practice, we have found that a cost-effective approach is to select the 5–10 most sensitive stations, preferably with good azimuthal distribution, to monitor a given target area.

An example might serve to illustrate the threshold monitoring principle. Assume that we are monitoring the NZ test site, and thereby aligning the threshold traces of the network stations in accordance with the expected travel times of the considered phases. If an interfering event occurs in Scandinavia, the signals will cause a significant increase on the Scandinavian stations. However due to travel-time differences, the threshold traces will not line up as expected for an event at the target site (NZ), and consequently there will at any time be stations in the network that are little affected by the interfering event. If these stations have normal background conditions, they can be used to show that any hypothetical event at the NZ test site is likely to have been very small.

2.2 Tuning the Threshold Trace

Let us consider threshold monitoring of a specific target area of limited geographical extent. The size of the target area may vary depending upon the application, but typically such an area might be a few tens of kilometers in diameter. A basic assumption is that the target area is defined such that all seismic events within the area show similar wave propagation characteristics.

Parameters such as travel times of the different phases and steering delays for array beam-forming are typically obtained by processing previous events located in the target area. However, our experience has shown that these parameters can be well estimated on the basis of standard earth models. The more critical parameters for threshold monitoring are the filter frequency bands, the STA window lengths and the magnitude calibration, which are discussed below.

Filter bands and STA lengths

The optimum frequency filters and window lengths for signal level estimation (i.e., the length of the short-term average, STA), are derived from analysis of previous events located in the region. Several different filters and STA window lengths are tested by comparing the maximum STA value of the signal to the mean and variance of the background noise. The analysis procedure is as follows:

- Select a number of representative time intervals with background noise, bandpass filter and create STA traces. We assume a lognormal distribution of the STA data, and convert to a normal distribution by taking the logarithm of each sample STA value.
- Determine the mean ($\log(STA)_{mean}$) and standard deviation (σ_{noise}) of the noise observations.
- Determine how many standard deviations the calibration signals are above the noise level by calculating the quantity $nstdev = (\log(STA)_{max} - \log(STA)_{mean})/\sigma_{noise}$, where $\log(STA)_{max}$ is measured at the maximum of the signal. A high nstdev value indicates a low probability of having noise peaks approaching the level of the signals.
- Compare the nstdev measurements for different bandpass filters and STA lengths and use this number as an indicator for selecting the optimum STA length and filter band.

In general, there is a trade-off between the STA window length and the sharpness of the focusing on the target site. A long STA window will make it more likely that off-site events will produce significant peaks on the threshold trace. This trade-off must be seen in conjunction with the desire to maximize the variable nstdev defined above.

Magnitude calibration

The distance-depth correction factors $b_j(\Delta, h)$ in (1) and (2) can either be determined by using globally averaged values (e.g., VEITH and CLAWSON, 1972), by applying a regional model, or by calibration for path effects to the specific target area. The latter method is the most accurate and is preferable, assuming that data from previous calibration events are available. We then obtain the necessary magnitude calibration factors from processing previous events with known magnitudes, using the relation

$$\hat{b}_{ij} = \hat{m}_j - \log(\hat{STA}_{ij}) \quad (i = 1, \ldots, K; \ j = 1, \ldots, L) \tag{5}$$

where $\hat{b}_{i,j}$ is our estimate of the magnitude correction factor for phase i, and event j, $\hat{m}_j$ is the estimate of the magnitude for event j (based on independent network observations), and $\hat{STA}_{ij}$ is our estimate of the signal level at the predicted arrival time of phase i for event j. K is the number of phases considered (there might be several stations and several phases per station), and L is the number of events.

The magnitude correction factor to be used for phase i is then given by

$$b_i = \frac{1}{L} \cdot \sum_{j=1}^{L} \hat{b}_{ij} \ . \tag{6}$$

3. Analysis Results

3.1 Calibrating the Network for Monitoring Novaya Zemlya

We have implemented a site-specific monitoring capability for the Novaya Zemlya test site by deriving optimized processing parameters for the four arrays ARCES, SPITS, FINES and NORES (see Fig. 1). For all arrays except SPITS, the processing parameters could be derived from previous recordings of underground nuclear explosions at the test site, however at SPITS no such recordings are available. A detailed map of the Novaya Zemlya target area and selected seismic events in the region is shown in Figure 2. In the following we describe the calibration data used for the two arrays ARCES and SPITS. A summary of the processing parameters for each of the four arrays is given in Table 1.

Processing parameters for ARCES

In Figure 3 we show the ARCES P and S spectra (measured on the P and S beams) for the 24 October, 1990 event together with an average noise spectrum of the beams. The highest SNR for both phases is found in the passband 3–5 Hz, which also is the filter to be used for TM processing.

At ARCES, both P and S phases (*Pn* and *Sn*) are clearly observable from explosions at the Novaya Zemlya test site. From analyzing the ARCES recordings of

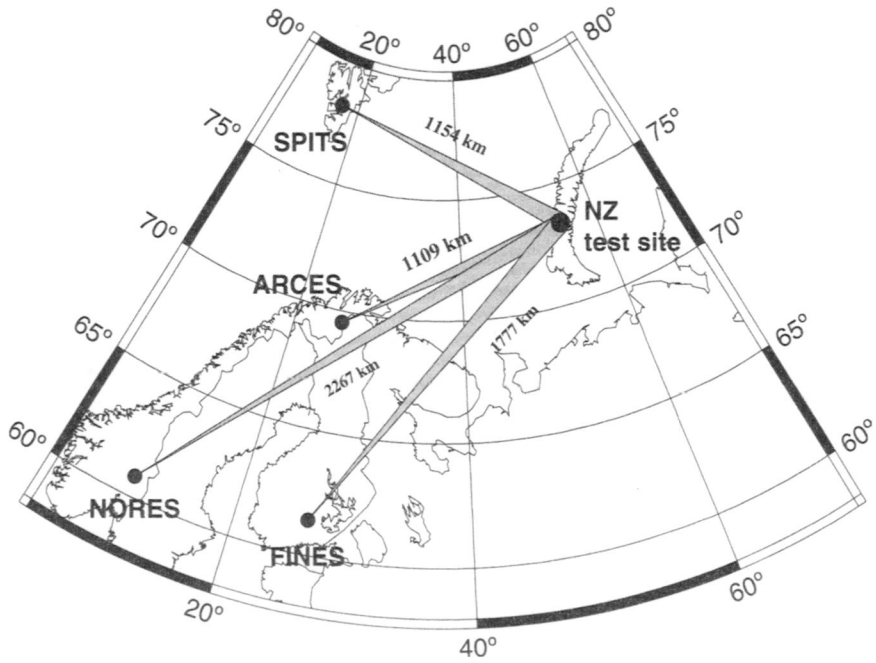

Figure 1
Map of Novaya Zemlya and the locations of the four arrays (SPITS, ARCES, FINES, and NORES) used to monitor the region around the former underground nuclear test site.

the 24 October, 1990 event, we have derived the TM processing parameters given in Table 1. An illustration of the selection of STA processing parameters is shown in Figure 4. The main P-wave energy at ARCES has a relatively long duration, and a 5-second STA length was chosen for TM processing. For the S phase, the main energy lasted about 3-seconds in the 3–5 Hz filter band.

Processing parameters for SPITS

For the SPITS array we have no recordings of known nuclear explosions, and we have therefore had to base the calibration on other events in the Novaya Zemlya region. Key events in this analysis have been the m_b 3.5 event of 13 June 1995, located some 200 km north of the test site, and the m_b 3.5 event of 16 August, 1997, located in the Kara Sea about 140 km southeast of the test site (see Fig. 2).

The SPITS array is located at approximately the same distance from the Novaya Zemlya test site as the ARCES array, and analysis of other events in the Novaya Zemlya region has revealed remarkably similar waveforms both with respect to observable phases (P and S) and frequency content of the signals. As an example we show in Figure 5 the SPITS and ARCES recordings of the m_b 3.5 event of 13 June, 1995, located about 200 km north of the test site. In order to make the waveforms more

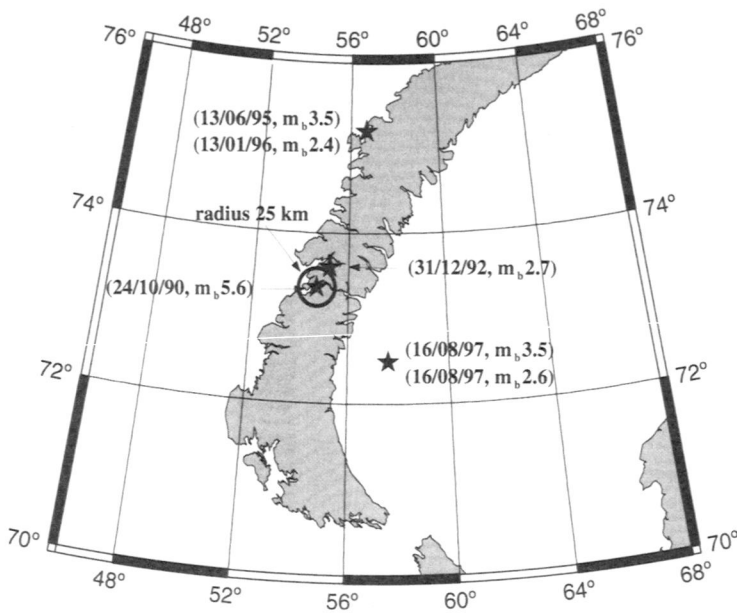

Figure 2

Map of Novaya Zemlya and the monitoring region around the former underground nuclear test site. The radius of the circle is 25 km. Also shown are the locations of selected events in the region.

Table 1

TM processing parameters for the NZ test site

Station	Distance (km)	Phase	Obs. slowness (s/deg)	Obs. back azimuth (deg)	Frequency band (Hz)	STA length (s)	Travel time (s)	Mag. calib.	St. dev. of calib.
ARCES	1108.6	P	11.2	62.2	3.0–5.0	5.0	147.5	2.84	0.3
–	–	S	23.2	64.3	3.0–5.0	3.0	254.2	2.99	0.3
SPITS	1154.2	P	14.8	109.6	3.0–5.0	5.0	152.6	2.95	0.3
–	–	S	23.0	97.6	3.0–5.0	3.0	263.0	3.11	0.4
FINES	1776.9	P	11.6	29.6	2.0–4.0	1.0	224.2	2.78	0.3
NORES	2267.3	P	10.9	33.6	1.5–3.5	1.0	281.4	2.68	0.3

directly comparable, the ARCES recording has been adjusted to the SPITS response function. The distances to SPITS and ARCES are 1065 km and 1210 km, respectively. When comparing the single channel observations SPA0_sz with ARA0_sz (traces nos. 3 and 6 from the top, respectively), we find slightly higher amplitudes at SPITS, although the general seismogram characters with distinct *P* and *S* phases remain the same. We also note that the reduction of the beam amplitudes at ARCES is larger that at SPITS, which is due to the larger aperture of the ARCES array.

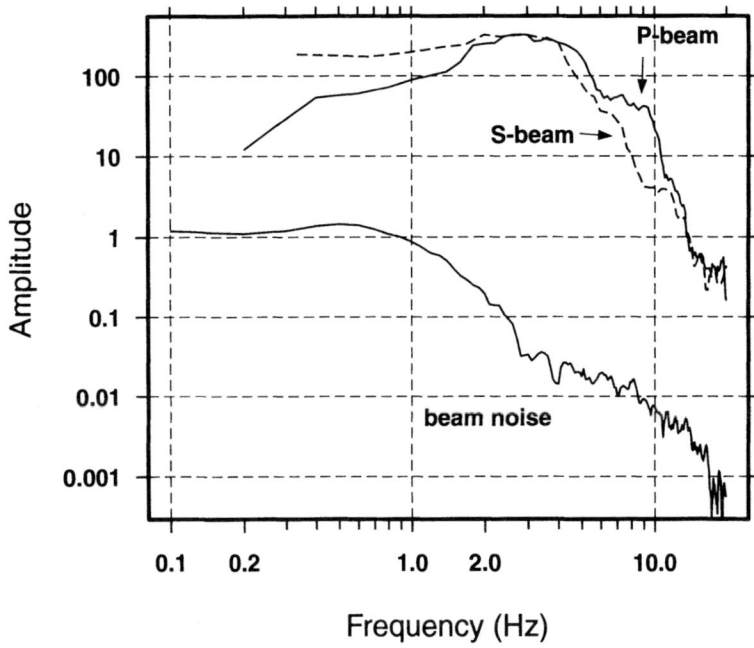

Figure 3
This figure shows beam spectra of the *P* and the *S* phase of the 24 October, 1990 event recorded at ARCES, together with an average beam spectrum of background noise preceding the *P* phase.

For the 13 June 1995 event, shown in Figure 5, the SPITS recording contains clear *P* and *S* phases, whereas the *S* amplitude slightly exceeds the *P* amplitude. For the other reference event of 16 August, 1997, shown in Figure 6, the *S* amplitude is significantly smaller than the *P* amplitude, and the *P* amplitude is about half the amplitude of the 13 June, 1995 event.

The propagation path from the Novaya Zemlya test site to the SPITS array is quite similar to the propagation path from the 13 June, 1995 event. Both sites are located on the western side of Novaya Zemlya and the propagation paths to SPITS do not include crossings of any major geological structures. The 13 June, 1995 event is, however, located about 90 km closer to SPITS than the Novaya Zemlya test site. The propagation path from the 16 August, 1997 event crosses the structure of the Novaya Zemlya island, and the distance to SPITS is about 135 km longer than for events at the test site.

In order to be conservative with respect to the estimation of the SPITS magnitude thresholds at the Novaya Zemlya test site, we have added $0.2m_b$ units to the mean calibration factors (i.e., decreased the detectability) derived from the two reference events. Additionally, we have increased the standard deviation associated with the *S* amplitude-magnitude relation from 0.3 to $0.4m_b$ units.

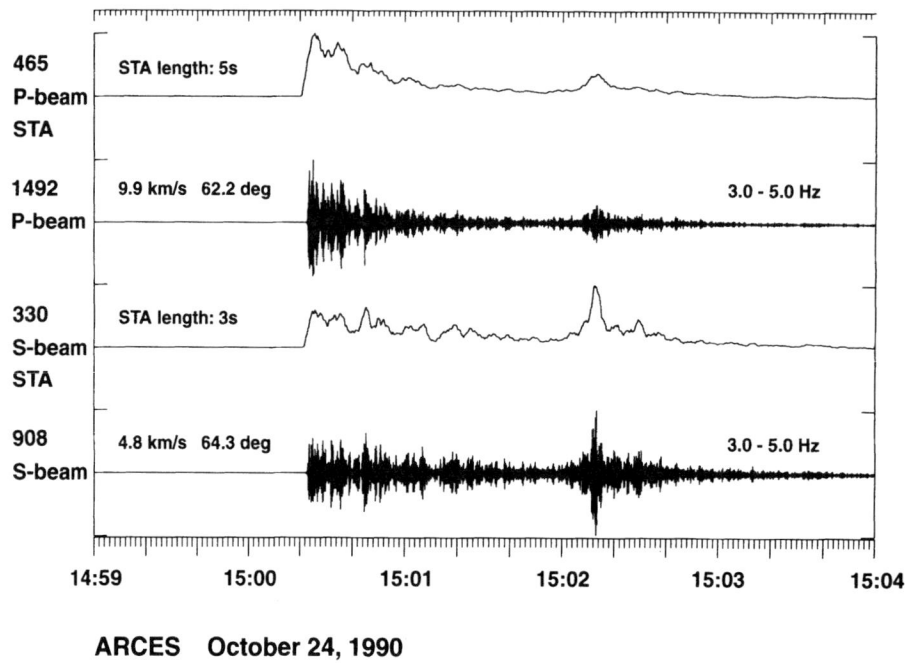

ARCES   October 24, 1990

Figure 4

Illustration of the STA calibration procedure for ARCES. Traces no. 2 and 4 from the top of the figure are ARCES *P* and *S* beams from the 24 October, 1990 nuclear explosion at Novaya Zemlya, filtered in the optimum frequency band 3.0–5.0 Hz. Traces no. 1 and no. 3 are short-term averages (STAs) of the filtered beams.

3.2 *Monitoring Examples for Novaya Zemlya*

In order to investigate the utility of the TM method in an operational environment, we have implemented continuous calculation of the threshold level for the NZ test site using the four arrays shown in Figure 1. Plots have been generated for each day processed, beginning 1 November, 1997. Figure 7 shows as an example results for 9 February, 1998. The plot shows the magnitude thresholds for *P* phases at each of the four arrays, with the combined network threshold monitoring trace on top. We have also included *S* phases for the two closest arrays (SPITS and ARCES) in the threshold monitoring calculations for Novaya Zemlya, however these traces are not shown on the plot. The network trace is a composite trace taking into account the individual traces for *P* and *S* on SPITS and ARCES, *P* for FINES and *P* for NORES.

We note that the individual arrays have a number of peaks corresponding to both regional and teleseismic events. For example, a sequence of peaks during the middle of the day (marked on the figure) is especially pronounced on FINES, ARCES, and NORES. These peaks are caused by a sequence of presumed underwater explosions

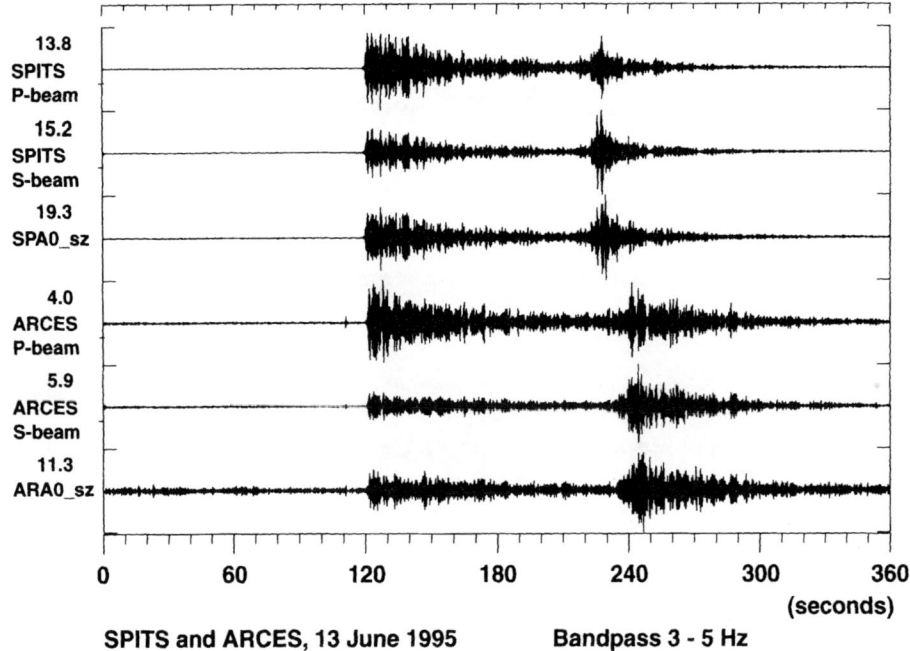

13.8
SPITS
P-beam

15.2
SPITS
S-beam

19.3
SPA0_sz

4.0
ARCES
P-beam

5.9
ARCES
S-beam

11.3
ARA0_sz

0   60   120   180   240   300   360
(seconds)

SPITS and ARCES, 13 June 1995   Bandpass 3 - 5 Hz

Figure 5

SPITS and ARCES recordings of the 13 June, 1995 event (m_b 3.5) located about 200 km north of the Novaya Zemlya test site. To make the data directly comparable, the ARCES recording has been converted to the SPITS response. The maximum amplitudes given together with the trace labels are the original digital counts multiplied by the calibration constant at 1 Hz. The distances to SPITS and ARCES are 1065 km and 1290 km, respectively.

in the Baltic Sea. We note that this sequence of peaks is effectively suppressed on the combined network threshold trace, since the phase arrival times do not correspond to the predicted time pattern for the target area. In fact, the network threshold trace has only three significant peaks, which can all be associated with seismic events detected and located by conventional processing. The peaks numbered 1 and 3 result from earthquakes at teleseismic distances (Sea of Okhotsk and Hindu Kush, respectively). Peak number 2 corresponds to a presumed underwater explosion in the Barents Sea, near the northern coast of the Kola Peninsula. Otherwise, the threshold trace is well below m_b 2.5, thus showing that the monitoring capability is below this level for essentially the entire time period. Within the uncertainties inherent in the statistical formulation, and taking a reservation for the short time instance surrounding the interfering event, we can therefore conclude that no seismic event of m_b 2.5 or larger occurred at the test site for this day.

Figure 8 shows a second example, which covers 16 December, 1997. Two important features are illustrated in this figure. First, the key array SPITS happened to be out of operation, resulting in a general deterioration of the combined network

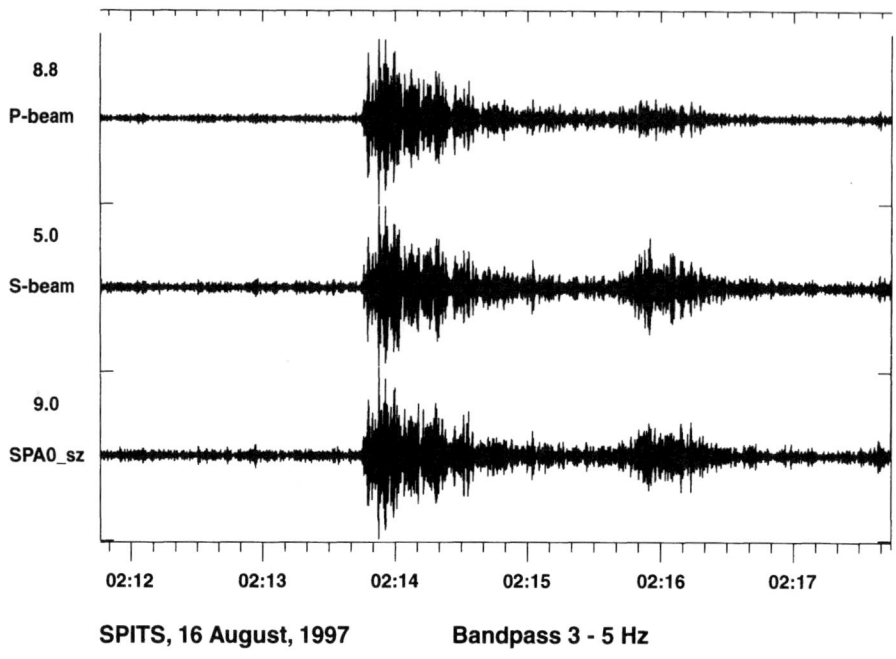

Figure 6

SPITS recording of the 16 August, 1997 event (m_b 3.5) located in the Kara Sea, southeast of the Novaya Zemlya test site. The maximum amplitudes given together with the trace labels are the original digital counts multiplied by the calibration constant at 1 Hz.

capability. Second, there was an unusually large increase in the background noise level at the other key array, ARCES. This increase was caused by a very strong storm system moving through northern Norway at that time, producing increased microseismic noise at ARCES over the entire frequency spectrum. In spite of the coincidence of these two unfavorable factors, we note that the network threshold trace still, in general, remains below magnitude 2.5. There are about 10 peaks slightly exceeding 2.5 on this day, but they can all be "explained" as resulting from interfering events.

During a two-month period (November and December, 1997), we analyzed the results in detail, and found 90 peaks on the network threshold trace that exceeded m_b 2.5. Of these, 73 were caused by teleseismic earthquakes, and in particular by a large aftershock sequence near Kamchatka. The remaining 17 peaks were correlated with small earthquakes close to SPITS and some local events in Fennoscandia (mostly mining explosions).

During these two months, the continuous TM method was able to provide results that enabled monitoring of the NZ test site down to m_b 2.0 for most of the time period. All peaks exceeding m_b 2.5 were correlated to events outside the target region,

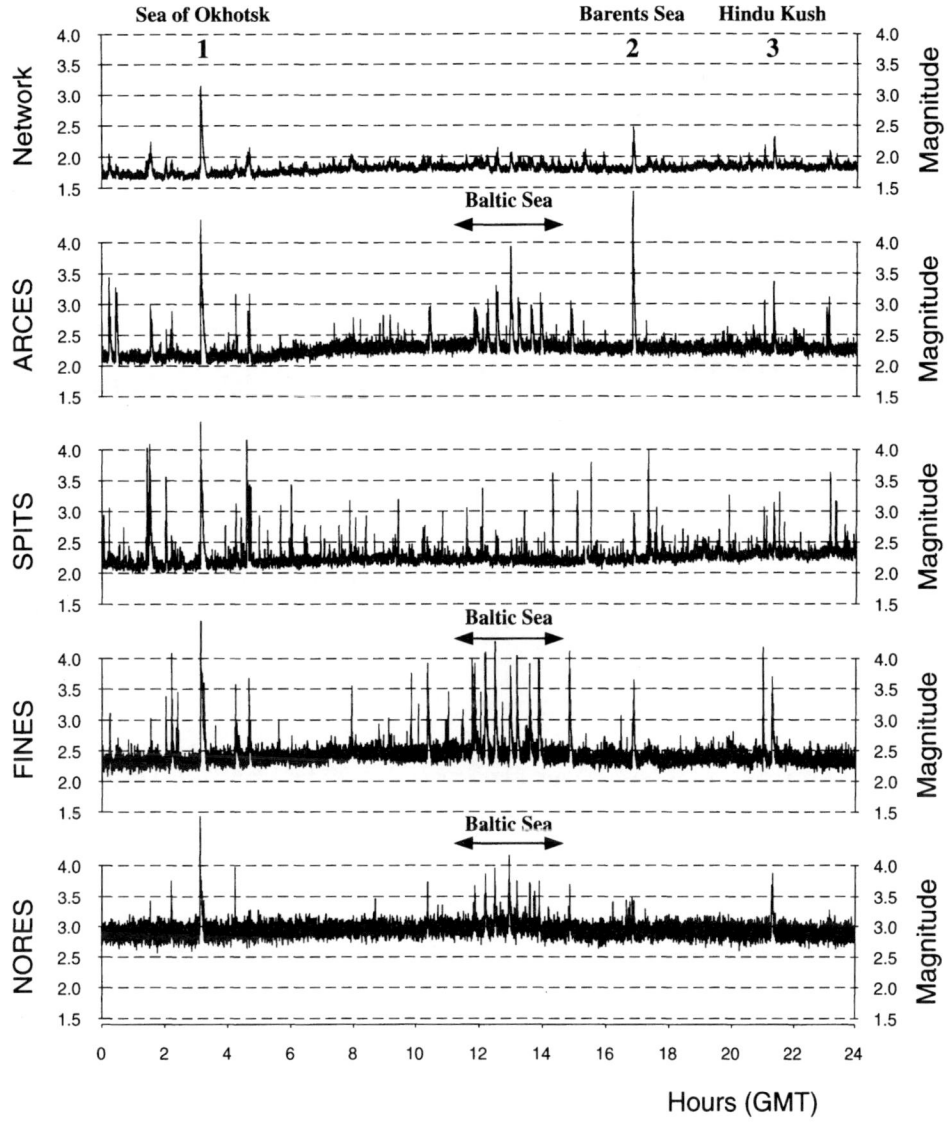

Figure 7
Example of site-specific threshold display of the Novaya Zemlya test site for 9 February, 1998. See text for detailed explanation.

so we can therefore conclude at the confidence level inherent in the method that no seismic event of magnitude exceeding 2.5 occurred at the NZ test site during this time period. Although we have not continued to analyze subsequent months in the same

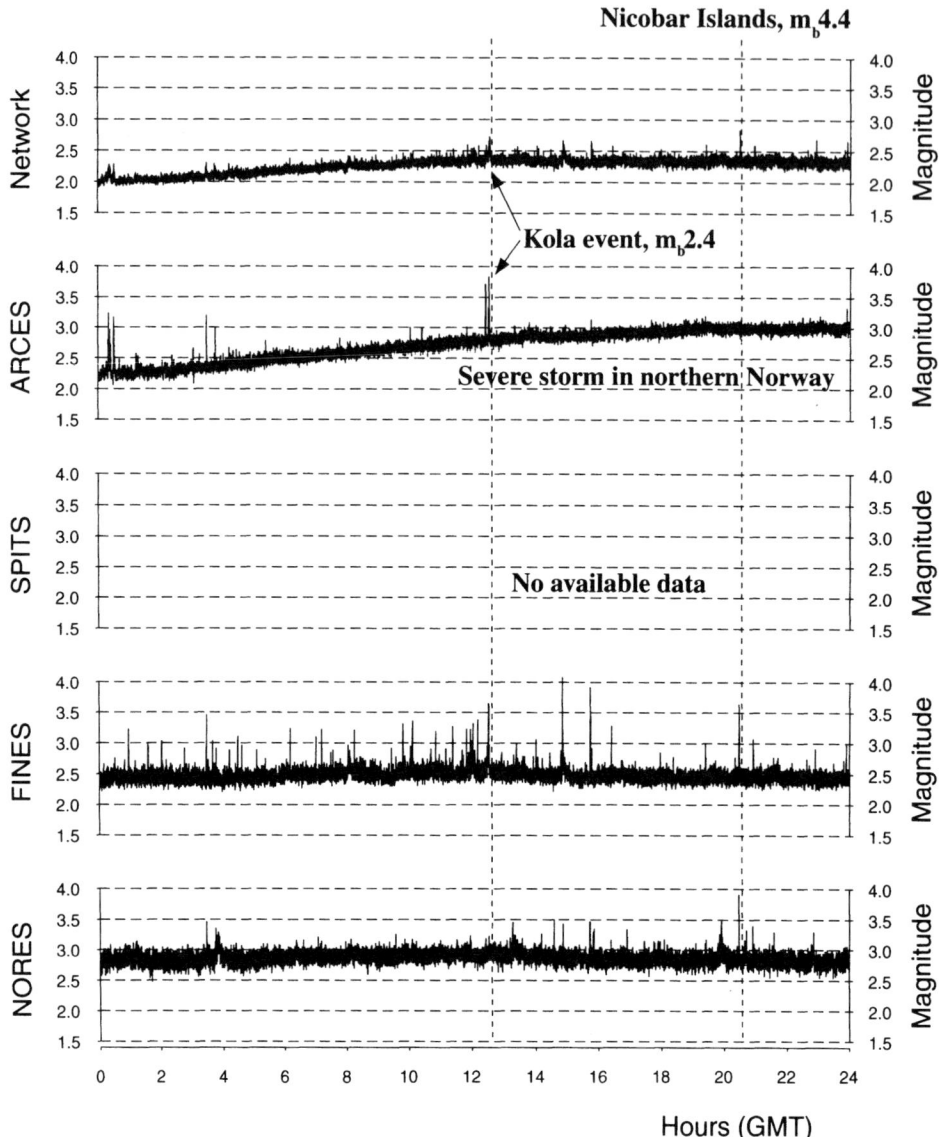

16 December 1997

Figure 8
Example of site-specific threshold display of the Novaya Zemlya test site for 16 December, 1997. See text
for detailed explanation.

detail, we have been able to confirm, within the uncertainty previously discussed, that
no seismic event significantly above m_b 2.5 occurred near the test site during all of
1998.

The site-specific threshold monitoring technique can be successfully applied even when using only a single station. This is of particular interest in cases where one station has a much higher capability than other network stations. As an example, we show in Figure 9 the results of optimized threshold monitoring, using the SPITS array, of the region around the location of the Kara Sea event on 16 August, 1997 (HARTSE, 1998; RICHARDS and KIM, 1997). Our main reason for using SPITS only is that the other key array, ARCES, was out of operation at the time. Furthermore, we have calibrated the SPITS array to the main event (m_b 3.5) at 02.11 GMT, which was detected and located by the prototype International Data Center in Arlington, Virginia.

The traces in the figure show the magnitude thresholds for the first 12 hours of 16 August, 1997. Four peaks stand out on the trace, and the causes of these peaks are the following:

(1) The main Kara Sea event (the strongest peak)
(2) A series of small disturbances very close to the SPITS array

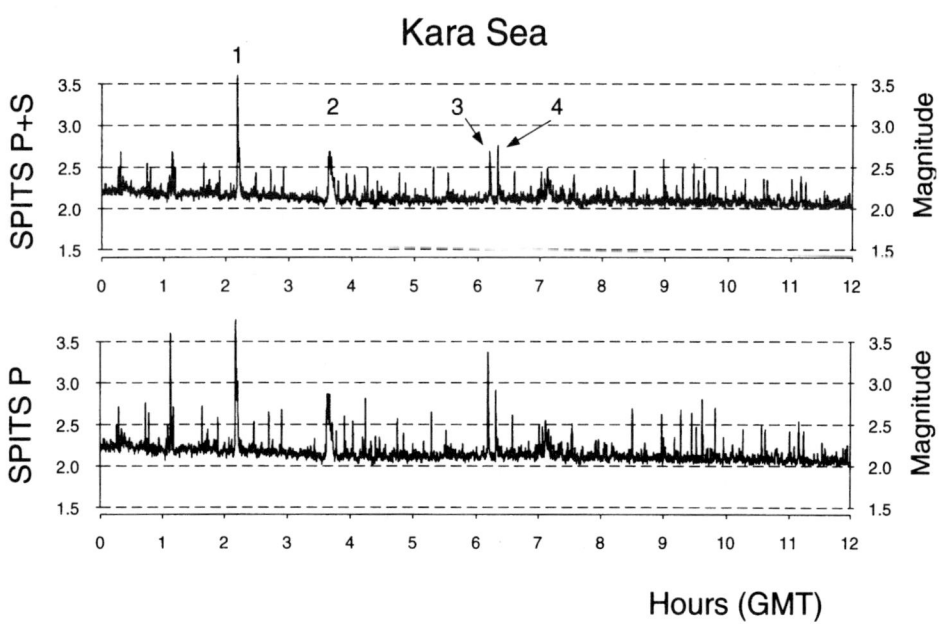

Figure 9

An example of optimized threshold monitoring of the region around the location of the Kara Sea event, using data from SPITS alone for the first 12 hours of 16 August, 1997, calculated when using parameters derived from the event on 02.11 GMT. Four peaks stand out on the trace, and the causes of these peaks are the following: 1) The main Kara Sea event. 2) A series of small disturbances very close to the SPITS array. 3) A local/regional event at a distance of approximately 250 km from SPITS. 4) The Kara Sea aftershock.

(3) A local/regional event at a distance of about 250 km from SPITS
(4) A second, smaller event from the same location as the main Kara Sea event.

Thus, we are able in this case, by carefully analyzing the peaks on the threshold trace, to detect and locate the second, smaller Kara Sea event (m_b 2.6), which was not automatically associated and located by either the Prototype IDC (BACHE *et al.*, 1990) or by the regional detection processing routinely carried out at NORSAR (MYKKELTVEIT *et al.*, 1990). The reason that this event was not in the NORSAR list of located events is that only the SPITS *P* onset was automatically detected, while the *S* phase did not exceed the detection threshold. Thus the event did not satisfy the event definition criterion used at NORSAR, which requires at least two matching phases (1*P* and 1*S*) from the same array or detections at two different arrays.

Figure 10 shows waveforms of the two Kara Sea events, as recorded by SPITS. With the appropriate filters and beamsteering, it is quite clear that these two events are in very close proximity. This conclusion was further confirmed after analyzing data from the Amderma station in northern Russia (RINGDAL *et al.*, 1997). Both the *P* and *S* phases at SPITS can be identified visually, and with the aid of additional *f-k* analysis the coinciding locations of these two events can be

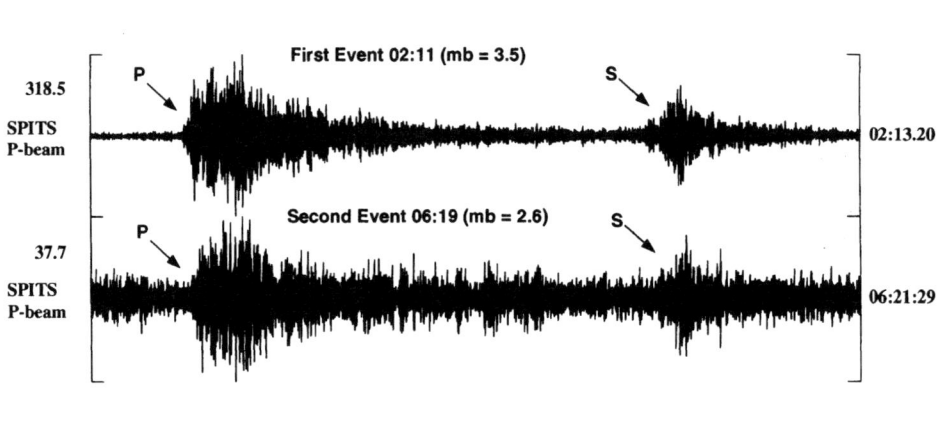

Figure 10

SPITS array beams filtered in the band 4–16 Hz for the two seismic events in the Kara Sea on 16 August, 1997. In order to enhance the *S* phase, both traces have been steered using a *S* type velocity (4.5 km/s) together with the appropriate azimuth.

further confirmed. We might note that such additional *f-k* analysis is a useful supplement in the threshold monitoring procedure, and such analysis has indeed confirmed that the additional peaks in Figure 9 are not consistent with a Kara Sea location.

4. Discussion

As stated by KVÆRNA and RINGDAL (1999), it is important to be aware that the main purpose of the threshold monitoring method is to call attention to any time instance when a given threshold is exceeded. This will enable analysts to focus their efforts on those events that are truly of interest in a monitoring status. The analyst will then apply other, traditional analysis tools in detecting, locating and characterizing the source of the disturbance. Thus, the threshold monitoring method is a supplement to, and not a replacement of, traditional methods.

In principle, site-specific threshold monitoring, given enough calibration data and computer resources, could be expanded to be applied on a global level. However, in practice, there will be a need to apply both the site-specific and the regional or global approaches in day-to-day monitoring. Nevertheless, the site-specific method could be further optimized, e.g., by considering different filter bands in parallel and applying specially generated digital filters to search for signals conforming to predetermined characteristics. We are currently investigating the feasibility and benefits of this type of optimization.

In studies of network detection capability, the question of "false alarms" is an important consideration, and in general there is a trade-off between detectability and false alarm rate. In the threshold monitoring application, no "detection threshold" is set, and the question of "false alarms" therefore has a different meaning. The "peaks" seen on the threshold plots could be defined as false alarms if they correspond to seismic events outside the target area. However, it is important to remember that these peaks correspond to time intervals of high "noise" values due to signals from interfering events, and therefore are an integral part of the overall background threshold level. The possibility that on-site events could be "hidden" during such times of interference must be considered, and it is a question of judgement whether realistic "evasion scenarios" could be provided in practice in this connection.

The excellent capability of the site-specific threshold monitoring technique as demonstrated in this paper is due, to a large part, to our emphasis on high-frequency passbands in the regionally based site-specific monitoring. High-frequency arrays as used in our example from Novaya Zemlya advantageously suppress the noise (or signal coda) from interfering events, and retain signal coherency even at high frequencies. This adds to the capability of detecting small events in the background of interfering events.

   The optimized site-specific threshold monitoring technique is especially suited to monitor earthquake activity at low magnitudes for sites of special interest, and could also be useful for monitoring earthquake aftershock sequences. This paper has focused on the application of the method to a regional seismic network. As will be discussed in a companion paper (Part 2), the method is equally applicable to a teleseismic network, or a combination of the two. We will continue these studies in order to expand the scope and characterize the long-term capabilities of the threshold monitoring method. At the same time, we are working on streamlining and optimizing the technique, in order to further improve performance.

   To obtain a fully automatic monitoring procedure, we have begun investigating the possibility of utilizing detector information to label the threshold peaks. Current results indicate that the azimuth and slowness estimates of the detected phases at the individual arrays can be effectively used for such labelling. It is, however, important to quantify the quality of these azimuth and slowness estimates, to take into account the possibility of incorrect estimates. These efforts will be thoroughly documented in a separate paper.

Acknowledgment

   This research has been sponsored by the Defense Threat Reduction Agency, U.S. Department of Defense, under contract no. DSWA01-97-C-0128. This is NORSAR contribution No. 670.

REFERENCES

BACHE, T. B., BRATT, S. R., WANG, J., FUNG, R. M., KOBRYN, C., and GIVEN, J. W. (1990), *The Intelligent Monitoring System*, Bull. Seismol. Soc. Am. *80*, Part B, 1833–1851.
HARTSE, H. E. (1998), *The 16 August 1997 Novaya Zemlya Seismic Event as Viewed from GSN Stations KEV and KBS*, Seism. Res. Lett. *69*(3), 206–215.
KVÆRNA, T. and RINGDAL, F. (1999), *Seismic Threshold Monitoring for Continuous Assessment of Global Detection Capability*, Bull. Seism. Soc. Am. *89*(4), 946–959.
MARSHALL, P. D., STEWART, R. C., and LILWALL, R. C. (1989), *The Seismic Disturbance of 1986 August 1 near Novaya Zemlya: A Source of Concern?*, Geophys. J. Int. *98*, 565–573.
MYKKELTVEIT, S., RINGDAL, F., KVÆRNA, T., and ALEWINE, R.W. (1990), *Application of Regional Arrays in Seismic Verification Research*, Bull. Seism. Soc. Am. *80*, Part B, 1777–1800.
RICHARDS, P. G. and KIM, W.-Y. (1997), *Testing the Nuclear-Test-Ban Treaty*, Nature *389*, 781–782.
RINGDAL, F. (1997), *Study of Low-magnitude Events near the Novaya Zemlya Nuclear Test Site*, Bull. Seismol. Soc. Am. *87*, 1563–1575.
RINGDAL, F. and KVÆRNA, T. (1989), *A Multichannel Processing Approach to Real-time Network Detection, Phase Association and Threshold Monitoring*, Bull. Seismol. Soc. Am. *79*, 1927–1940.
RINGDAL, F. and KVÆRNA,T. (1992), *Continuous Seismic Threshold Monitoring*, Geophys. J. Int. *111*, 505–514.
RINGDAL, F., KVÆRNA, T., KREMENETSKAYA, E. O., and ASMING, V. E. (1997), *The Seismic Event near Novya Zemlya on 16 August 1997*, Semiannual Tech., Summ., 1 April – 30 September 1997, NORSAR Sci. Rep. 1-97/98, 110–119, NORSAR, Kjeller, Norway.

VEITH, K. F. and CLAWSON, G. E. (1972), *Magnitude from Short-period P-wave Data*, Bull. Seismol. Soc. Am. *62*, 1927–1940.

(Received June 11, 1999, revised October 2, 1999, accepted November 1, 1999)

 To access this journal online:
http://www.birkhauser.ch

Pure appl. geophys. 159 (2002) 989–1004
0033–4553/02/050989–16 $ 1.50 + 0.20/0

❘Pure and Applied Geophysics

Optimized Seismic Threshold Monitoring – Part 2: Teleseismic Processing

TORMOD KVÆRNA,[1] FRODE RINGDAL,[1] JOHANNES SCHWEITZER[1]
and LYLA TAYLOR[1]

Abstract — This second paper (Part 2) pertaining to optimized site-specific threshold monitoring addresses the application of the method to regions covered by a teleseismic or a combined regional-teleseismic network. In the first paper (Part 1) we developed the method for the general case, and demonstrated its application to an area well-covered by a regional network (the Novaya Zemlya nuclear test site). In the present paper, we apply the method to the Indian and Pakistani nuclear test sites, and show results during the periods of nuclear testing by these two countries in May 1998. Since the coverage by regional stations in these areas is poor, an optimized approach requires the use of selected, high-quality stations at teleseismic distances.

To optimize the threshold monitoring of these test sites, we use as calibration events either one of the nuclear explosions or a nearby earthquake. From analysis of the calibration events we derive values for array beamforming steering delays, filter bands, short-term averages (STA) lengths, phase travel times (*P* waves), and amplitude-magnitude relationships for each station. By applying these parameters, we obtain a monitoring capability of both test sites ranging from m_b 2.8–3.0 using teleseismic stations only. When including the nearby Nilore station to monitor the Indian tests, we show that the threshold can be reduced by about 0.4 magnitude units. In particular, we demonstrate that the Indian tests on 13 May, 1998, which were not detected by any known seismic station, must have corresponded to a magnitude (m_b) of less than 2.4.

We also discuss the effect of a nearby aftershock sequence on the monitoring capability for the Pakistani test sites. Such an aftershock sequence occurred in fact on the day of the last Pakistani test (30 May, 1998), following a large (m_b 5.5) earthquake in Afghanistan located about 1100 km from the test site. We show that the threshold monitoring technique has sufficient resolution to suppress the signals from these interfering aftershocks without significantly affecting the true peak of the nuclear explosion on the threshold trace.

Key words: Seismic event detection, automatic data processing, seismic data analysis.

1. Introduction

In the first of a sequence of two companion papers on optimized site-specific seismic threshold monitoring (hereafter referred to as Part 1), we gave a general description of the continuous seismic threshold monitoring technique that has been

[1] NORSAR, Post Box 51, N-2027 Kjeller, Norway. E-mail: tormod@norsar.no

developed at NORSAR over the past several years to monitor a geographical area continuously in time. Data from a network of arrays and single stations are combined and steered toward a specific area to provide an ongoing assessment of the upper magnitude limit of seismic events that might have occurred in that area. The main purpose of the threshold monitoring (TM) technique is to highlight instances when a given threshold magnitude is exceeded, thereby helping the analyst to focus on those events truly of interest in a monitoring situation. The analyst can then apply traditional tools in detecting, locating and identifying the source of the disturbance.

While the case of optimized seismic threshold monitoring using a regional network was discussed in Part 1, the purpose of the present paper (Part 2) is to present results from applying the method to a teleseismic network or a combined regional-teleseismic network. In particular, we apply the method to selected time intervals during India's and Pakistan's nuclear tests in May 1998 (BARKER *et al.*, 1998; SCHWEITZER *et al.*, 1998; WALLACE, 1998) using data from the GSETT-3 network (RINGDAL, 1994) transmitting data to the Prototype International Data Center in Arlington, VA for joint processing. We carefully select a subset of the stations in the GSETT-3 global network which have the highest detection capabilities for these regions, and calibrate these stations with regard to the basic processing parameters. We show that the network can be used to achieve a high capability for continuous monitoring of the test sites, and we include a brief discussion of the relationship between the traditional detection capability estimates and the threshold monitoring levels.

2. Method Description

2.1 Selecting the Network

The basic principles for the site-specific threshold monitoring method were given in Part 1, and are equally applicable to the regional case discussed in that paper and to the teleseismic or combined approach discussed here. There are, however, differences in selecting the network to be used. In either case, it is naturally important to select a network composed of stations with high detection capability for the site to be monitored, and with a good azimuthal distribution around the site. In the teleseismic case, there is, nevertheless, a need to limit the number of stations to be included in the threshold computations. This is for three main reasons:

- The main contribution to lowering the threshold trace will come from the stations with the highest sensitivity to the target area. There is no need to include stations that will contribute essentially nothing.
- The optimization of the processing parameters is a very time-consuming task, and even small mistakes can cause erroneous results when combining data from several stations.

• The projected International Monitoring System will comprise 170 primary and auxiliary stations worldwide. Making probability calculations based upon such a large number of stations could easily lead to calculations at the tail of the magnitude distribution, where the basic normality assumptions may not be valid.

For these reasons, we recommend limiting the number of stations to around 10, and we will stay with approximately this number of stations in the examples shown in this paper.

2.2 Tuning the Threshold Trace

As in the regional case discussed in Part 1, we consider a specific target area of limited geographical extent. The size of the target area may vary, although typically such an area might be a few tens of kilometers in diameter. A basic assumption is that the target area is defined such that all seismic events within the area exhibit similar wave propagation characteristics. When using a teleseismic network, a typical target area may usually be somewhat larger than in the regional case, because small shifts in travel-time patterns or azimuths may correspond to a relatively larger shift of the geographical aiming point.

As in the regional case, parameters such as travel times of the different phases, steering delays for array beamforming, filter frequency bands, the STA window length and the magnitude calibration values are obtained on the basis of processing results for a set of calibration events. The procedure for developing such optimized parameters is similar to the regional case, and will not be repeated here.

3. Analysis Results

In May 1998, India and Pakistan conducted several underground nuclear tests. Figure 1 shows the station network used in this study, the locations of the Pakistani and Indian nuclear tests, and the location of the Afghanistani earthquake (main shock) occurring about 30 minutes ahead of the second Pakistani test. The corresponding event information from the Reviewed Event Bulletin (REB) of the Prototype International Data Center is given in Table 1.

We have used available data from these nuclear explosions to test the performance of the threshold monitoring technique for selected time intervals. Both the Indian and Pakistani explosions provide interesting case studies, for a number of reasons:

• The Indian explosions on 11 May, 1998 occurred at the same site as the nuclear explosion on 18 May, 1974. This provides for a very detailed and instructive comparison between the recorded waveforms for the 1974 and 1998 events.

• The second set of Indian nuclear explosions, on 13 May, 1998, were not detected seismically by any station available to us. The threshold monitoring technique can therefore be used to provide an estimate of the upper magnitude limit of these explosions.

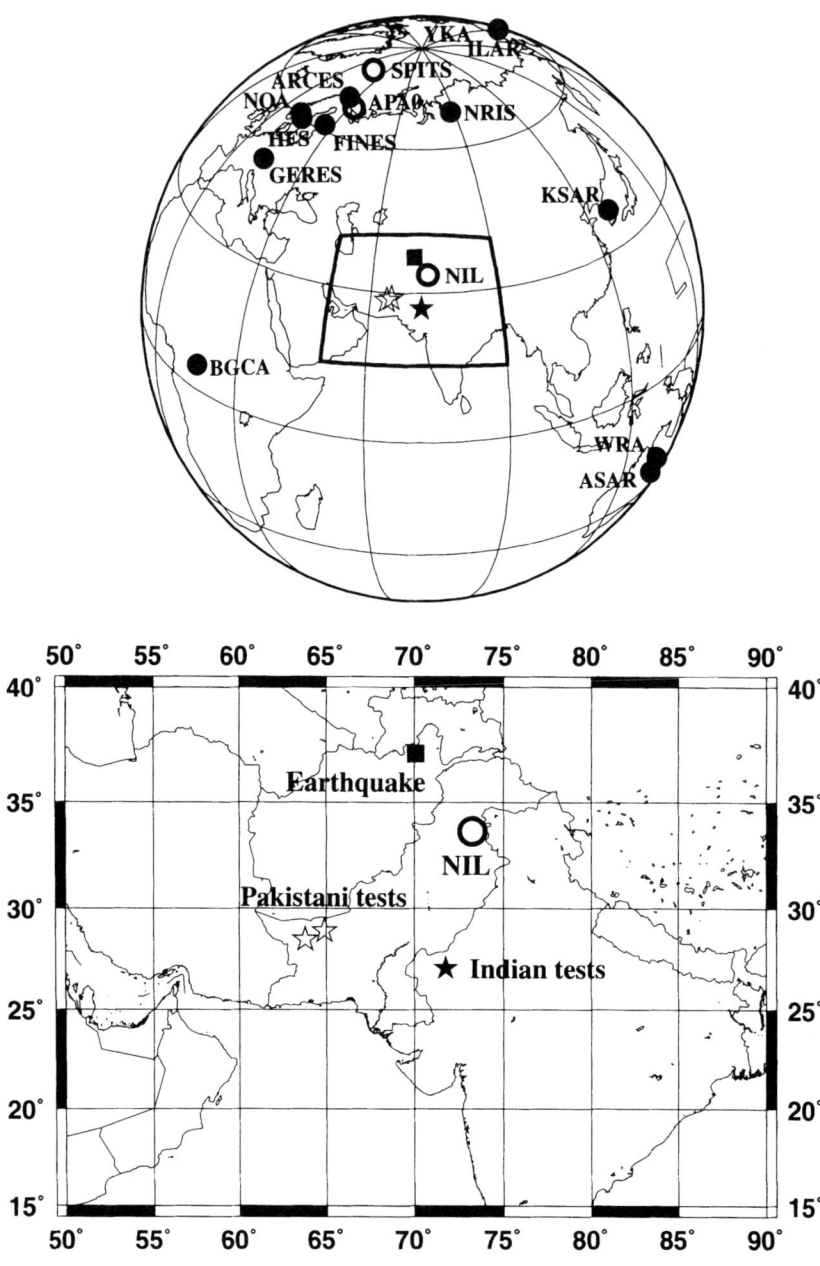

Figure 1

The upper map shows the locations of the stations used in this study. The filled circles show GSETT-3 seismic stations used for monitoring of the Indian nuclear test site. These stations all provide continuous data to the Prototype International Data Center (PIDC) in Arlington, Virginia. The stations NIL, HFS, SPITS and APAO (open circles) do not provide continuous data to the PIDC. The area within the rectangle is expanded in the lower map, where the filled star indicates the location of the Indian explosions, and the open stars indicate the locations of the Pakistani explosions. The filled square shows the location of the main shock of the earthquake sequence in northern Afghanistan.

Table 1

Event information from the Reviewed Event Bulletin (REB)

Origin time	Lat.	Lon.	Depth	m_b	Nsta	Flinn-Engdahl Region
1998/05/11 10:13:44.2	27.0716	71.7612	0.0	5.0	50	India–Pakistan Border Reg.
1998/05/28 10:16:17.6	28.9032	64.8933	0.0	4.9	60	Southwestern Pakistan
1998/05/30 06:22:25.7	37.1570	70.0682	0.0	5.5	33	Afghanistan–Tadjikistan Bord. Reg.
1998/05/30 06:54:57.1	28.4948	63.7814	0.0	4.6	51	Southwestern Pakistan

- The two sets of Pakistani explosions, on May 28, 1998 (Explosion P1) and May 30, 1998 (Explosion P2) were located about one degree apart, and it is interesting to investigate how the TM method performs for Explosion P2 when the tuning parameters from Explosion P1 are used.
- The origin time of Explosion P2 was about 38 minutes after the origin time of an m_b 5.5 earthquake in Afghanistan, followed by numerous large aftershocks, and this gives us an opportunity to study the performance of the TM method during a strong interfering earthquake sequence. The earthquakes were located 9–10 degrees away from the explosion sites.

Very limited data at local and regional distances were available for these explosions. In fact, for the Indian explosions, only one such station (NIL in Pakistan) was in operation, whereas for the Pakistani explosions, only teleseismic data were available. The available station geometry should provide us with important information regarding the characteristics of site-specific monitoring at teleseismic distances.

3.1 Monitoring India's Nuclear Test Site

During the Indian nuclear tests on 11 and 13 May, 1998, the GSETT-3 monitoring network operated numerous sensitive stations at teleseismic distances. However, only one station was available within the local or regional distance range. This station, a GSETT-3 auxiliary station in Nilore, Pakistan (NIL), is located 6.7 degrees away from the Indian test site and provided the P phase with the highest SNR (937.4) of any station in the global network for this event. The NIL vertical-component recording of the Indian test is shown in Figure 2. The highest SNR relative to the background noise level was found between 1 and 2 Hz for both the P and the Lg phase, and this filter was used prior to the calculation of STA traces. A 1.5 second STA length was used for P, and for the longer duration Lg phase an STA length of 8 seconds was used.

The remaining 12 stations used for monitoring were all located at teleseismic distances, and only P phases were considered for calculation of the magnitude thresholds. A typical example of teleseismic recording is the observations at the large-aperture NORSAR array (NOA). The recordings of the 11 May, 1998 test and the test conducted at the same site on 18 May, 1974 are illustrated in Figure 3, which

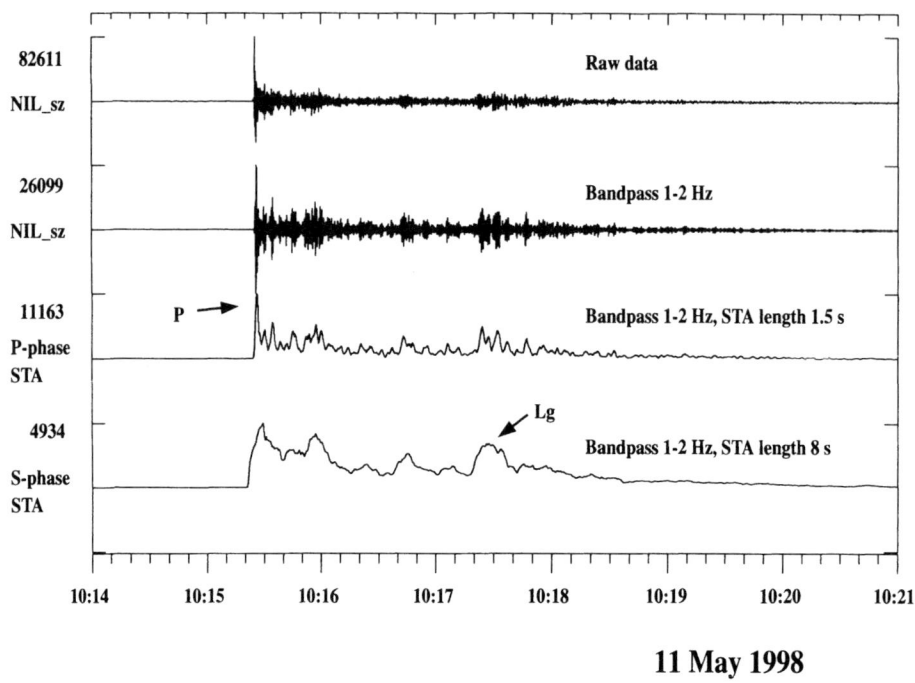

11 May 1998

Figure 2

Panel showing NIL recording of the Indian nuclear test of 11 May, 1998. The upper trace shows the raw data of the vertical component, and the second trace shows the same data filtered in the band 1–2 Hz. The two lower traces show the STA traces used for representing the amplitudes of the *P* and *Lg* phases. Notice that different STA lengths were used for *P* and *Lg*.

shows a pairwise comparison of NOA *P*- wave recordings for both events. The data were band pass filtered between 1 and 3 Hz; the traces were aligned visually and sorted by the NOA sites, with the upper bold trace for each site corresponding to the 1998 event and the lower trace corresponding to the 1974 event.

The 1974 event was a single explosion, whereas the 1998 event comprised three separate explosions, apparently detonated simultaneously. It is interesting to note the remarkable similarity of the two event recordings at each NOA site. In contrast, the variability of the waveform shape across the NOA array is rather large, and we attribute this variation to near receiver scattering/focusing effects. Another interesting observation from the NOA recordings is that the amplitudes of the two events are not very different, with the 1998 event reaching true amplitudes 1.5–2 times larger than the 1974 event.

The list of stations and the TM processing parameters derived from the recordings of the 11 May, 1998 explosion are given in Table 2. Figure 4 shows the results from site-specific threshold monitoring of a five-hour time interval around the 11 May, 1998 Indian nuclear test, using the processing parameters derived from the nuclear test itself. The top trace shows the combined network thresholds, and the following

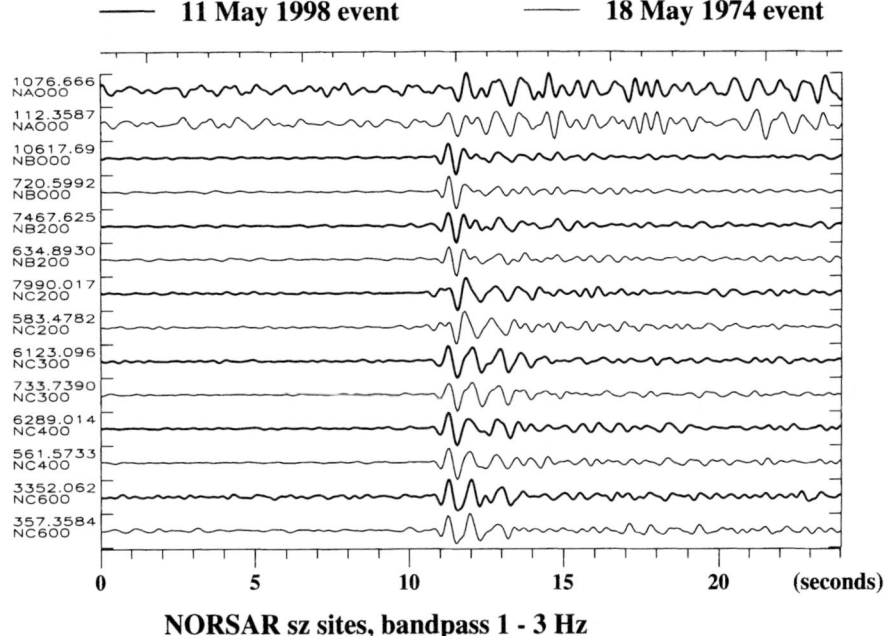

NORSAR sz sites, bandpass 1 - 3 Hz

Figure 3

Observations of the 18 May, 1974 and of the 11 May, 1998 explosions at the Indian nuclear test site. Shown are pairwise seismograms at single sites of NORSAR. The upper bold trace consistently shows the 11 May, 1998 and the lower trace the 18 May, 1974 explosion. All data were 1–3 Hz bandpass filtered, the traces were normalized with the given maximum amplitudes (in counts), and the traces were aligned visually to a common onset time.

eight traces show the thresholds derived from each of eight station selected (*P* phase only).

The time tolerances were set to accommodate a target area with a radius of 25 km around the explosion site. Several distinct peaks are seen on the threshold traces for the individual arrays, however for the network trace the only significant peak corresponds to the nuclear test. The 90% magnitude thresholds during noise conditions vary around m_b 2.7–2.8. We would also like to emphasize that the peak on the network threshold trace caused by the nuclear test has a value that is slightly lower than the actual event magnitude. In cases when an event occurs in the target region, the threshold calculations should be replaced by the maximum likelihood estimate of the event magnitude (RINGDAL and KVÆRNA, 1992; KVÆRNA and RINGDAL, 1999).

According to the Indian authorities, two explosions of 0.5 and 0.3 kt took place on 13 May, 1998, with origin time 06:51 GMT, however, no signals were detected by the GSETT-3 stations, and we have calculated the magnitude threshold (90% upper magnitude limit) of the reported event, using the processing parameters derived from the Indian test of 11 May, 1998. Our estimated upper magnitude limit of the reported

Table 2

TM processing parameters derived from recordings of the 11 May 1998 Indian nuclear test

Station	Distance (deg)	Phase	SNR in REB	Theo. ray parameter (s/deg)	Obs. slowness (s/deg)	Obs. back azimuth (deg)	Frequency band (Hz)	STA length (s)	Travel time (s)	Mag. calib.	St. dev of calib.
NIL	6.68	P	937.4	13.73	–	–	1.0–2.0	1.0	102.0	1.55	0.15
–	–	Lg	(3.8)	33.04	–	–	1.0–2.0	8.0	223.5	2.18	0.15
NRIS	43.05	P	191.1	8.10	–	–	2.0–4.0	1.0	481.6	3.16	0.15
FINES	45.87	P	80.3	7.90	7.34	120.37	2.0–4.0	1.0	504.5	3.48	0.15
KSAR	47.96	P	51.4	7.75	7.66	269.29	1.5–3.0	1.0	521.3	3.93	0.15
GERES	49.39	P	43.3	7.65	6.95	95.05	1.0–2.0	1.0	532.3	4.08	0.15
ARCES	50.16	P	182.6	7.59	7.53	125.88	2.0–4.0	1.0	538.3	3.29	0.15
NOA	52.49	P	48.0	7.41	7.52	101.84	1.2–3.2	1.0	554.3	3.63	0.15
BGCA	55.19	P	174.0	7.23	–	–	1.5–3.5	1.0	575.6	3.59	0.15
WRA	76.54	P	314.1	5.66	5.35	318.10	1.0–3.0	1.0	713.8	3.59	0.15
ASAR	78.39	P	199.3	5.53	5.67	307.30	1.0–3.0	1.0	724.1	3.52	0.15
ILAR	83.65	P	157.0	5.12	3.93	323.11	1.5–3.5	1.0	750.3	4.19	0.15
YKA	90.60	P	238.0	4.65	5.02	349.59	1.5–3.0	1.0	785.6	3.89	0.15

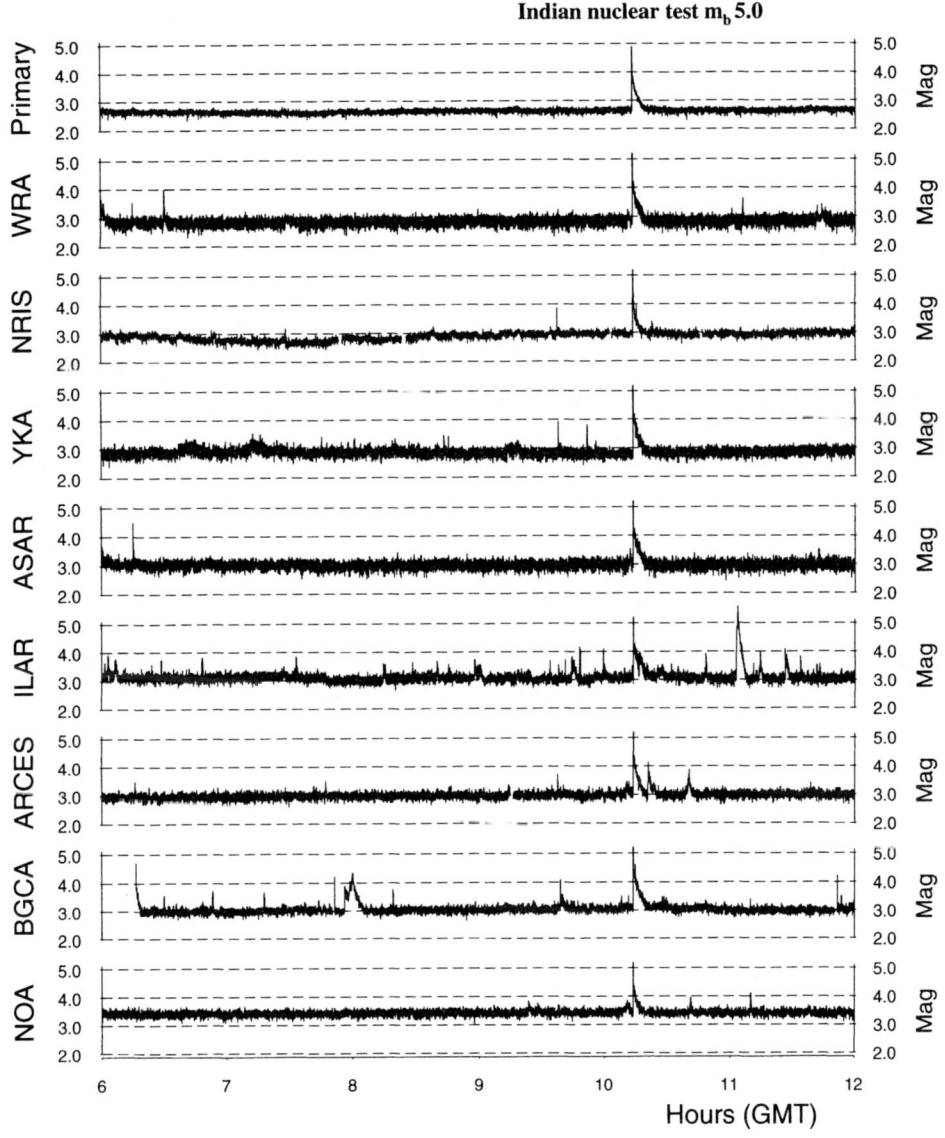

11 May 1998

Figure 4
Site-specific threshold monitoring of a 6-hour time interval around the Indian nuclear test, using the processing parameters given in Table 2. The plot shows the individual *P* phases (STA traces) for 8 selected stations, with the combined network threshold trace on top (Primary). The time tolerances were set to accommodate a target area with a radius of 25 km around the explosion site. The only significant peak on the network threshold trace corresponds to the nuclear test.

explosions is m_b 2.4, which is consistent with the value (m_b 2.5) obtained by SCHWEITZER *et al.* (1998), calculated from NIL data alone. Using the same NIL data, WALLACE (1998) estimated a slightly smaller value of m_b 2.2.

Figure 5 illustrates results from analysis of a four-hour time interval around the announced nuclear test, and it is instructive to compare the two traces on the figure. The upper trace corresponds to the GSETT-3 90% network detection capability, requiring at least 3 P detections (KVÆRNA and RINGDAL, 1999). The lower trace is the TM result for the same 11 stations (i.e., the 90% upper limit of any event that could have occurred at the site). For the purpose of this comparison, we included the NIL station in the detection capability estimation during the two hours of available data. It is clear that the inclusion of one excellent station does not significantly improve the three-station network detection capability, which is in effect governed not by the best station, but by the third-best station in the network. In contrast, the TM approach takes full advantage of NIL capabilities, which cause the monitoring threshold to be lowered by about 0.4 magnitude units when data from this station are available.

3.2 Monitoring Pakistan's Nuclear Test Sites

According to the official Pakistani reports, their first nuclear test consisted of one large explosion and four detonations of small tactical weapons. The seismic

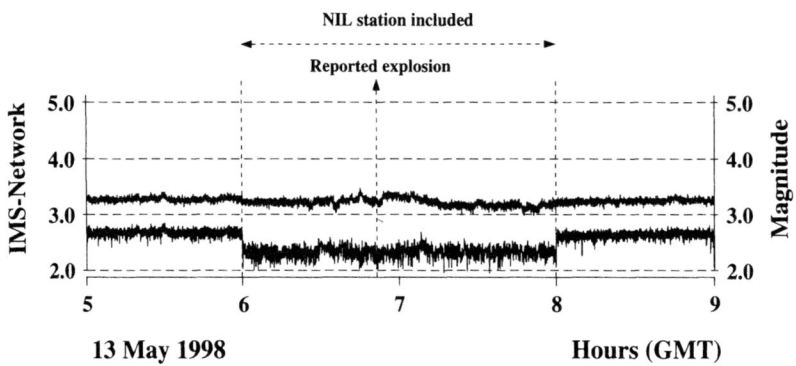

Upper trace - Three-station detection threshold of IMS network

Lower trace - Threshold monitoring limit of IMS network (largest possible event)

Figure 5

The plot shows magnitude thresholds for a four-hour time interval around the announced Indian nuclear test of 13 May, 1998. The upper trace corresponds to the GSETT-3 90% network detection capability (requiring at least 3 P detections), whereas the lower trace is the TM result (i.e., the 90% upper limit of any event that could have occurred at this site). For the time interval 06:00 to 08:00 data from the Pakistani station NIL are included in both calculations. The largest TM peak around 07:10 is caused by the P phase from an m_b 4.5 event located in Java, Indonesia.

observations do not reveal any separate signals, therefore we assume in the following discussion that Explosion P1 is a single event with a network m_b of 4.9. Except for the Australian station ASAR, we only used stations with continuous data available at NORSAR in this threshold monitoring study. The parameters derived from analysis of this event are listed in Table 3.

Figure 6 shows the results from site-specific threshold monitoring of a 7-hour time interval around the first Pakistani nuclear test, using the processing parameters derived from the nuclear test itself. The plot shows the individual P phases (STA traces) for each of the 8 arrays, with the combined network threshold trace on top. The time tolerances were set to accommodate a target area with a radius of 25 km around the explosion site. Several distinct peaks are seen on the threshold traces for the individual arrays, although for the network trace the only significant peak corresponds to the nuclear test. During noise conditions on this day, the 90% magnitude threshold varies around m_b 3.0.

The best performance of the threshold monitoring method is obtained by using processing parameters derived from recordings of previous events located in the target region. In our case we will use the processing parameters derived from Explosion P1 to obtain a nearly optimum monitoring of the target region surrounding the explosion site. An interesting question that arises in this context is how distant from the location of Explosion P1 can the processing parameters effectively be used?

In Figure 7 we display the monitoring results of a two-hour time interval around Explosion P2. In order to investigate how sensitive the network magnitude thresholds are to the definition of the target area, we have calculated several network threshold traces for this two-hour time interval using different parameter settings:

- In the upper panel, the processing parameters derived from Explosion P1 are used and the time tolerances correspond to a region with 25 km radius around the

Table 3

TM processing parameters derived from recordings of the 28 May, 1998 Pakistani nuclear test

Station	Distance (deg)	Phase	Theo. ray parameter (s/deg)	Obs. slowness (s/deg)	Obs. back azimuth (deg)	Frequency band (Hz)	STA length (s)	Travel time (s)	Mag. calib.	St. dev of calib.
FINES	41.413	P	8.21	7.91	127.7	1.5–3.0	3.0	469.6	3.71	0.3
APAO	43.167	P	8.09	7.73	140.79	2.0–4.0	2.0	483.4	4.20	0.3
GERES	43.015	P	8.06	6.84	93.20	1.0–2.0	4.0	487.7	4.22	0.3
HFS	46.269	P	7.87	5.54	146.35	1.5–3.0	4.0	508.4	3.66	0.3
ARCES	46.492	P	7.85	8.30	130.59	2.0–4.0	3.0	507.0	3.82	0.3
NORES	47.472	P	7.79	7.75	99.08	2.5–5.0	2.5	516.5	3.54	0.3
SPITS	53.825	P	7.33	8.89	129.44	3.5–7.0	4.0	568.6	5.18	0.3
ASAR	84.490	P	5.05	3.13	312.13	1.5–3.0	2.0	755.4	3.71	0.3

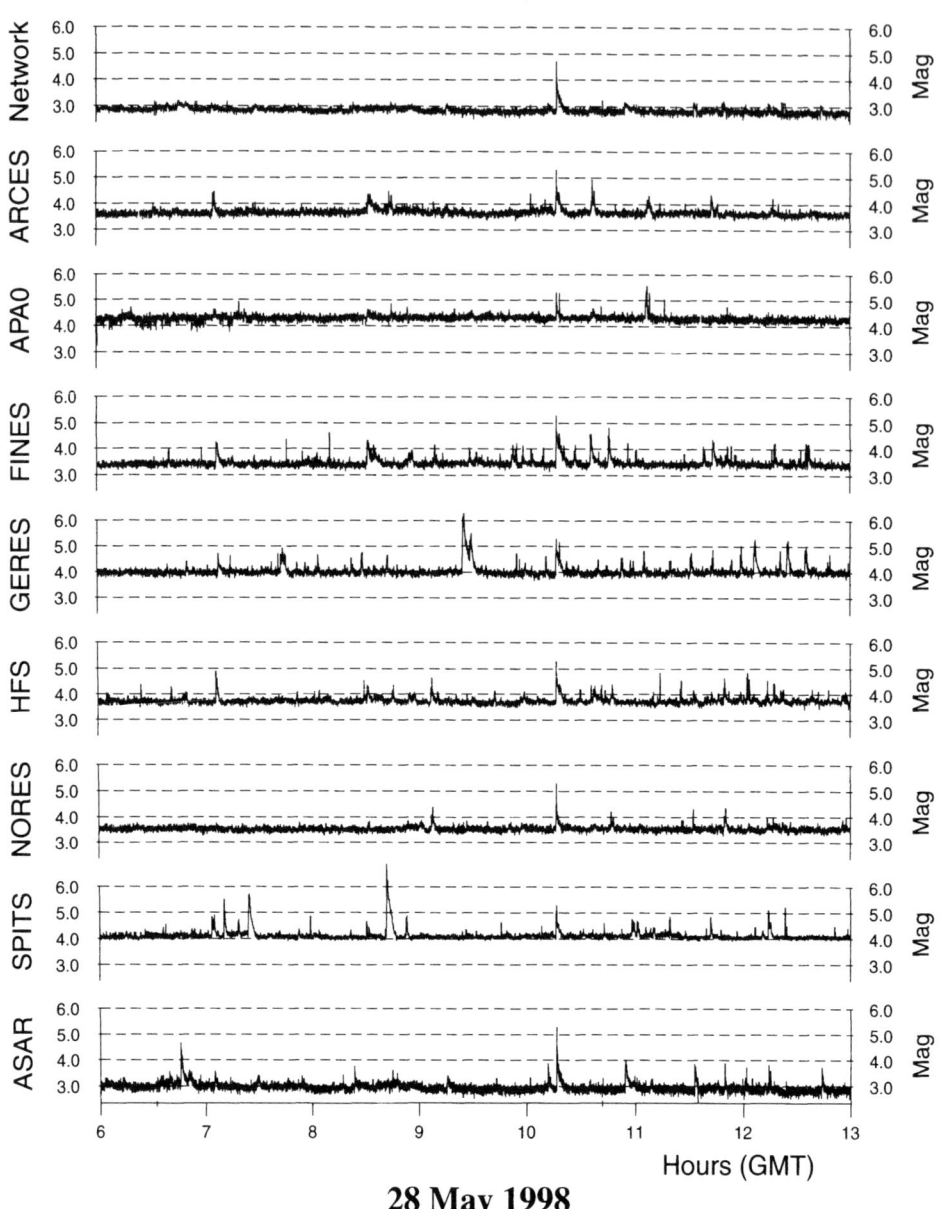

28 May 1998

Figure 6

Site-specific threshold monitoring of a 7-hour time interval around the first Pakistani nuclear test, using the processing parameters given in Table 3. The plot shows the individual *P* phases (STA traces) for each of the 8 arrays, with the combined network threshold trace on top. The time tolerances were set to accommodate a target area with a radius of 25 km around the explosion site. Notice that for the network trace the only significant peak corresponds to the nuclear test.

location of Explosion P1. The peak of Explosion P2, located about 100 km from the center of the target region, is clearly visible.

- In the second panel we have again used the processing parameters of Explosion P1, however the time tolerances are now increased to include a region with 100-km radius. To accommodate an expected larger amplitude variation within such a large region, the assumed standard deviation of the magnitude estimates are increased from 0.3 to 0.4. The threshold peaks are marginally higher and broadened relative to the peaks shown in the upper panel.
- In the third panel we have used processing parameters derived from Explosion P2 itself. The threshold peak for Explosion P2 is now slightly sharper than in the upper two panels, and the peak value is increased by about 0.3 m_b units.

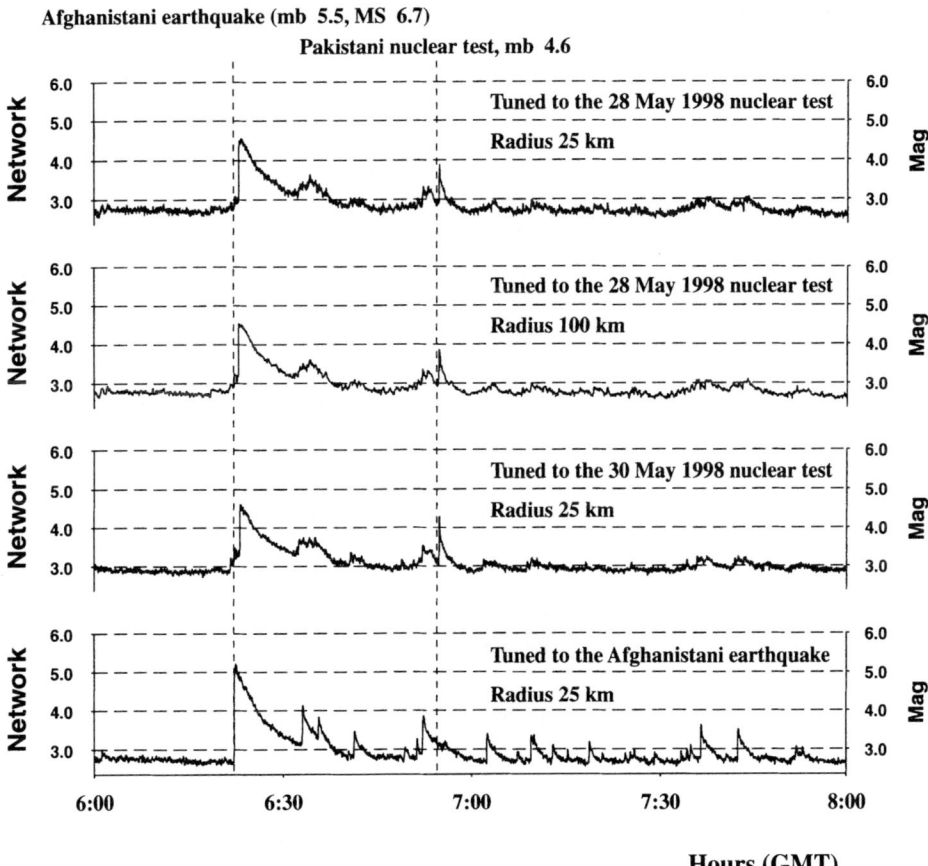

30 May 1998

Figure 7
The panels illustrate the difference in 90% network magnitude thresholds for a two-hour time interval around the second Pakistani nuclear test when using different processing parameters. See text for details.

- In the lower panel we have used the processing parameters derived from the main shock of the Afghanistani earthquake sequence, located approximately 1100 km away from Explosion P2. A target region with a radius of 25 km is assumed. Through detailed analysis we found 16 distinct peaks that have sharp signatures consistent with aftershocks at this location. In this case, the peak caused by Explosion P2 is smeared out and is clearly lower than several of the aftershock peaks.

The results shown in Figure 7 indicate that for the given teleseismic station geometry, a shift of 100 km relative to the target location has a relatively small influence on the definition on the threshold peaks. The same applies to changing the radius of the target area by increasing the time tolerances. This implies that the processing parameters derived from Explosion 1 can be effectively used to monitor a radius of at least 100 km from the location of Explosion 1. Nevertheless as seen from the lower panel in Figure 7, efficient monitoring can not be conducted when the location difference between the target site and the calibration site is of the order of 1000 km.

4. Discussion

The results presented in this paper indicate that the site-specific threshold monitoring method can be effectively used at teleseismic distances. From observations of the 11 May, 1998 Indian nuclear test we have derived optimum processing parameters for the 11 GSETT-3 stations assumed to have the best detection capability for the Indian test site. Our results can be summarized as follows:

- The TM magnitude threshold of the GSETT-3 primary network for the Indian test site is around m_b 2.8 during normal noise conditions. The stations of this network are located at teleseismic distances from the test site.
- During background noise conditions, regional data from the Nilore (NIL) station alone provides TM magnitude threshold of about m_b 2.4 for the Indian test site (SCHWEITZER *et al.*, 1998). Supplementing NIL data with data from the other teleseismic GSETT-3 stations does not lower the TM magnitude thresholds during normal noise conditions, nonetheless they are important if interfering events occur.
- During background noise conditions the GSETT-3 three-station detection capability varies around m_b 3.3, both with and without the use of NIL data. This illustrates that supplementing a network with one additional good station does not necessarily significantly improve the three-station detection capability of the network.

Using data from the arrays located in northern Europe supplemented with data from the Australian array ASAR, the Pakistani test area can be monitored down to a magnitude of 3.0 during background noise conditions. We have also verified that the monitoring performance is only marginally reduced for a test site located as far as 100 km from the location of the tuning event.

The benefit from using a network for monitoring becomes particularly evident during an earthquake sequence located as close as 10 degrees from the target region. The signals from the aftershocks are suppressed by up to 0.5 m_b units, making the peak of the nuclear test within the aftershock sequence stand out clearly on the network threshold trace.

An interesting application not throughly discussed in this paper is to provide monitoring of an aftershock sequence. In fact, we analyzed a five-hour interval of the Afghanistani earthquake sequence, using the sharpness of the network threshold peaks as criteria for declaring an aftershock. When compared to the aftershocks reported in the REB, we concluded that all events were found by our method. For the five-hour time interval, we additionally identified more that 10 possible aftershocks that were not reported in the REB. The confirmation of these events will, however, require further data analysis.

Another application of the TM approach would be to determine consistent magnitudes for the aftershocks. After introducing the region-specific magnitude calibrations from the main events, and subsequently using the maximum likelihood method for calculating the network magnitudes, we expect to achieve decidedly more precise estimates than those calculated by traditional averaging.

We conclude that the optimized site-specific threshold monitoring technique is well suited to monitor seismic activity at low magnitudes for sites of special interest, and could also be useful for monitoring earthquake aftershock sequences. This paper has focused on the application of the method to a teleseismic seismic network. As discussed in Part 1, the method is equally applicable to a regional network. We will continue these studies in order to expand the scope and characterize the long-term capabilities of the threshold monitoring method. At the same time, we are working on streamlining and optimizing the technique, in order to further improve its performance.

Acknowledgment

This research has been sponsored by the Defense Threat Reduction Agency, U.S. Department of Defense, under contract no. DSWA01-97-C-0128. This is NORSAR Contribution No. 671.

REFERENCES

BARKER, B., CLARK, M., DAVIS, P., FISK, M., HEDLIN, M., ISRAELSSON, H., KHALTURIN, V., KIM, W.-Y., MCLAUGHLIN, K., MEADE, C., MURPHY, J., NORTH, R., ORCUTT, J., POWELL, C., RICHARDS, P. G., STEAD, R., STEVENS, J., VERNON, F., and WALLACE, T. (1998), *Monitoring Nuclear Tests*, Science *281*, 1967–1968.

KVÆRNA, T. and RINGDAL, F. (1999), *Seismic Threshold Monitoring for Continuous Assessment of Global Detection Capability*, Bull. Seismol. Soc. Am. *89*(4), 946–959.

RINGDAL, F. (1994), *GSETT-3: A Test of an Experimental International Seismic Monitoring System*, Annali Geofis. *37*, 241–245.

RINGDAL, F. and KVÆRNA, T. (1992), *Continuous Seismic Threshold Monitoring*, Geophys. J. Int. *111*, 505–514.

SCHWEITZER, J., RINGDAL, F., and FYEN, J. (1998), *The Indian Nuclear Explosions of 11 and 13 May, 1998*, Semiannual Tech. Summ., 1 October 1997–31 March 1998, NORSAR Sci. Rep. 2-97/98, 121–130, NORSAR, Kjeller, Norway.

WALLACE, T. C. (1998), *The May 1998 Indian and Pakistan Nuclear Tests*, Seism. Res. Lett. *69*(5), 386–393.

(Received June 11, 1999, revised October 2, 1999, accepted November 1, 1999)

To access this journal online:
http://www.birkhauser.ch

Pure appl. geophys. 159 (2002) 1005–1020
0033–4553/02/051005–16 $ 1.50 + 0.20/0

| Pure and Applied Geophysics

A Scheme for Initial Beam Deployment for the International Monitoring System Arrays

Jin Wang[1]

Abstract — The International Monitoring System (IMS) includes a diverse set of seismic arrays with different configurations. These configurations have apertures ranging from less than 1 to more than 25 km and minimum interelement spacings varying from 0.1 to 3.6 km. This paper presents a scheme for initial beam deployment for this variety of seismic arrays. Beamforming is equivalent to a spatiotemporal bandpass filter of which passband is defined by the minimum and maximum wavenumbers, which are functions of the geometry configuration of the array. Deployment of steered-beams for signal detection is based on the wavenumber resolution of the array, slowness and frequency distributions of seismic phases, and coherence properties of seismic signals and noises among sensors. Within the wavenumber passband, all possible slowness values are determined by the resolution for each frequency band, and those that are outside the range of seismological interest are excluded. The appropriate azimuthal distribution for each selected slowness is determined from the azimuthal resolution. Using this approach, detection beams for each array are rationally deployed in the slowness-azimuth and frequency domain.

Key words: Seismic array, beamforming, detection, nuclear monitoring.

1. Introduction

Seismic arrays play an important role in the verification of the Comprehensive Nuclear-Test-Ban Treaty (CTBT), which was signed in 1996. Seismic array signal processing has been researched extensively in past years. Detailed reviews of the development of seismic arrays were given by RINGDAL and HUSEBYE (1982) and MYKKELTVEIT *et al.* (1990). The International Monitoring System (IMS) for the CTBT includes many seismic arrays. These arrays have different geometry configurations, apertures ranging from 1 to 25 km, and minimum interelement spacing varying from 0.1 to 3.6 km. The primary objectives of seismic arrays are to detect, locate low-magnitude events, and improve the capability of signal detection. A fundamental assumption in array signal processing is that the signals across the array are coherent whereas noises between any two sensors are incoherent. Under this assumption, the delay-and-sum beamforming algorithm

[1] Center for Monitoring Research, Science Applications International Corporation, 1300 N. 17th Street, Suite 1450, Arlington, VA 22209, U.S.A. E-mail: wangang2000@yahoo.com

can be employed to improve the signal-to-noise ratio for signal detection. In reality, however, the coherency of signal and noise across an array departs from the simple model and is a function of sensor spacing, frequency, and local geology.

Improving detection capability at seismic arrays has long been a research topic. For example, ANGLIN (1971), WEICHERT and HENGER (1976), and BACHE et al. (1986) discussed the detection capabilities, noise characteristics, and optimum beam deployment at the teleseismic array YKA, which has an aperture of about 20 km. DER et al. (1988) presented a general frequency domain processing scheme at small arrays for a given signal and noise power spectral matrix. KVAERNA (1989) and RINGDAL (1990) discussed the detection capabilities, noise characteristics, and optimum beam deployment at the regional array NORESS, the aperture of which is 3 km. MYKKELTVEIT et al. (1995) proposed a new detection parameter control file to improve detection performance at the small Spitsbergen array SPITS, which has an aperture of 1 km. WANG et al. (1996) presented preliminary work on how to deploy detection beams for arrays with different geometries. For the array NORSAR (NOA), a very large aperture array, a specially designed procedure that applies corrections to each component as a function of slowness and azimuth has been used (FYEN, 1997). The purpose of this paper is to summarize the strategy that is used at the Prototype International Data Centre (PIDC) for initial beam deployments for a variety of arrays.

2. Seismic Array Signal Processing

An array consists of a group of closely spaced passive receivers or sensors. The geometry configuration of an array constrains its detection capability; the finite aperture limits the capability to distinguish propagating waves; the discrete, non-continuous sensor spacing introduces spatial aliasing. The data received at an array are equivalent to the output of a spatiotemporal bandpass filter (JOHNSON and DUDGEON, 1993). The passband of this spatiotemporal filter is defined by the minimum and maximum resolvable wavenumbers which are constrained by the geometry configuration of an array.

For a uniformly spaced linear array, the resolution and the Nyquist frequency in the wavenumber domain are determined by the length of the linear array and the sensor spacing. The one-dimensional Sampling Theorem can be extended to multidimensional signal processing for regular arrays; the individual sensors that are on a regular grid. For an irregular array, however, the resolution and aliasing characteristics are less obvious, but can be analyzed by the so-called array pattern. For a planar array with a given geometry, its array pattern, **B**, is defined by the following equation:

$$\mathbf{B}(k_x, k_y) = \sum_i \sum_i e^{j(k_x(x_i - x_j) + k_y(y_i - y_j))} \ , \tag{1}$$

where (x_i, y_i) and (x_j, y_j) are coordinates of sensors relating to a reference sensor, and k_x and k_y are x and y components of the wavenumber. Array pattern is equivalent to the wavenumber response of an array as a spatiotemporal filter to a monochromatic plane wave. The pattern constrains the performance of the array in terms of resolution and aliasing in the two-dimensional wavenumber domain. The main lobe of the array pattern represents the resolution for the array. The larger the aperture of the array, the narrower the main lobe, which indicates better resolution. The grating lobes, with amplitude at higher wavenumbers equal or close to that of the main lobe, indicate aliasing in the wavenumber space. More discussion regarding array signal processing can be found in WHALEN (1970), HAYKIN (1991), and JOHNSON and DUDGEON (1993).

As an example, Figure 1 shows the geometry configuration of array YKA in Canada, one of the large aperture arrays of the IMS. Each circle in Figure 1 represents a seismic sensor. YKA consists of two linear arrays, which are perpendicular to each other. The aperture of YKA is about 20 km, and the sensor spacing is 2.5 km. Figure 2 shows the related array pattern of this evenly spaced array. The normalized contours start from 0.3 at the outside ring and increase in steps of 0.1 toward the center. The main lobe is in the center, and the eight grating lobes, repetitions of the main lobe, are at higher wavenumbers.

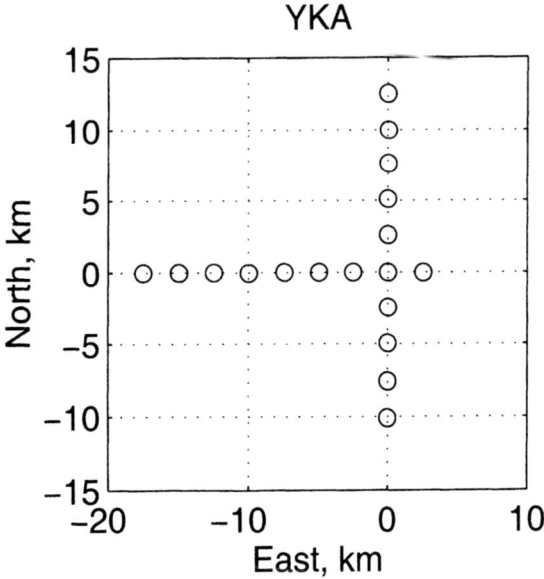

Figure 1
Geometry of the array YKA. The open circles represent sensor locations relative to the reference station (0,0).

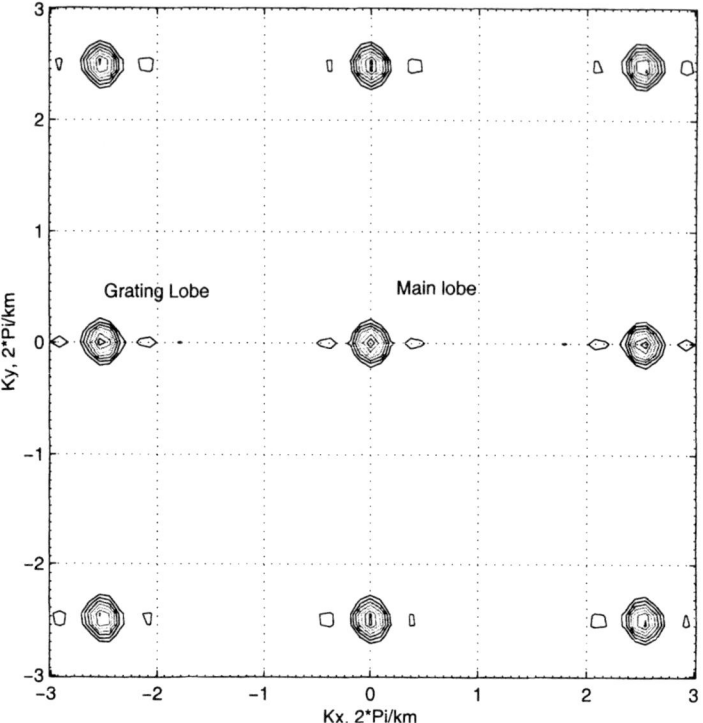

Figure 2

Contour map of the array pattern for YKA. The main lobe is located in the center. There are eight grating lobes in the figure. The contours are normalized values starting from 0.3 at the outside with 0.1 intervals toward the center.

Figure 3 shows the geometry of another IMS array, ARCES, which is located in Norway. It consists of four concentric rings. The aperture of ARCES is about 3 km, and the sensor spacing is not uniform. For this kind of irregular array, the array pattern may not have grating lobes in the range of wavenumbers of seismic signals of interest. If the maximum slowness of seismic body waves is 0.35 sec/km (2.86 km/s in velocity), and the highest frequency is 20 Hz, the wavenumber will be about 44 radian/km $(2*\pi*0.35*20)$. Figure 4 shows the array pattern of ARCES. In this case, no grating lobe exists within the range of wavenumbers of interest.

The width of the main lobe can be defined and then measured in several ways (JOHNSON and DUDGEON, 1993). The 3-dB drop of the main lobe, R_3dB, is used in this study as a measure of the wavenumber resolution of the array:

$$\Delta k = R\_3dB \ .$$

R_3dB is estimated from the radius of the contour with 0.7 of the maximum of the array pattern. If the contour is not a circle, the R_3dB is estimated from the average

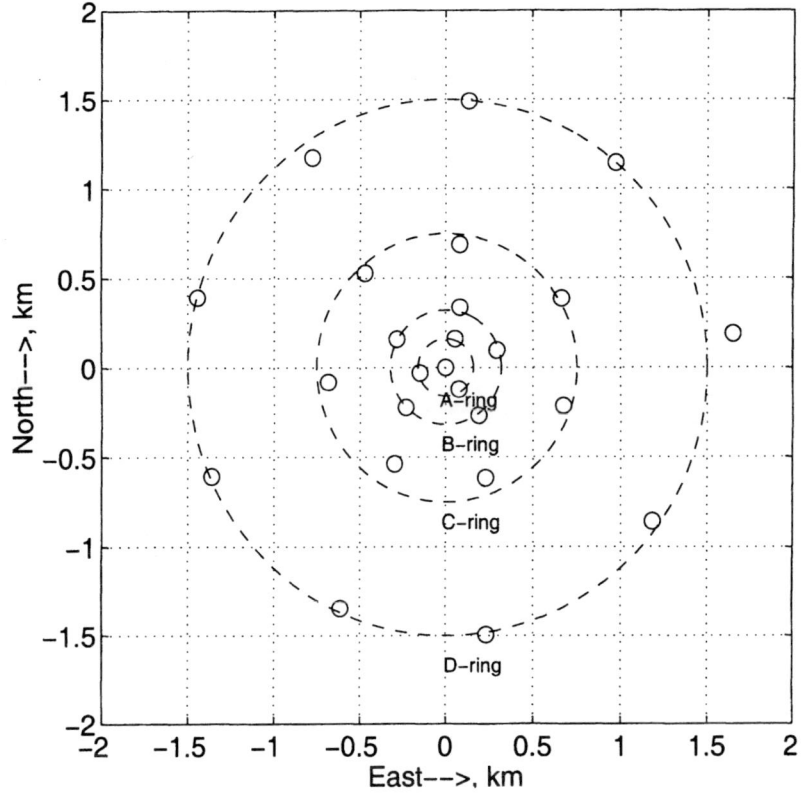

Figure 3
Geometry of the array ARCES. The open circles represent sensor locations related to the reference station (0, 0). There are four concentric rings in this array.

of the lengths of the long and short half-axes. For the array YKA, the R_3dB is 0.10 radian/km.

The resolvable minimum wavenumber starting from zero for the array is also defined by R_3dB:

$$k_{min} = R\_3dB \ ,$$

The aliasing frequency in the wavenumber domain, $R\_grating$, is determined by the half distance between centers of the main lobe and the closest grating lobe. It defines the maximum wavenumber (aliasing frequency in the wavenumber domain) of the array:

$$k_{max} = R\_grating \ .$$

For the array YKA, $R\_grating$ is 1.27 radian/km.

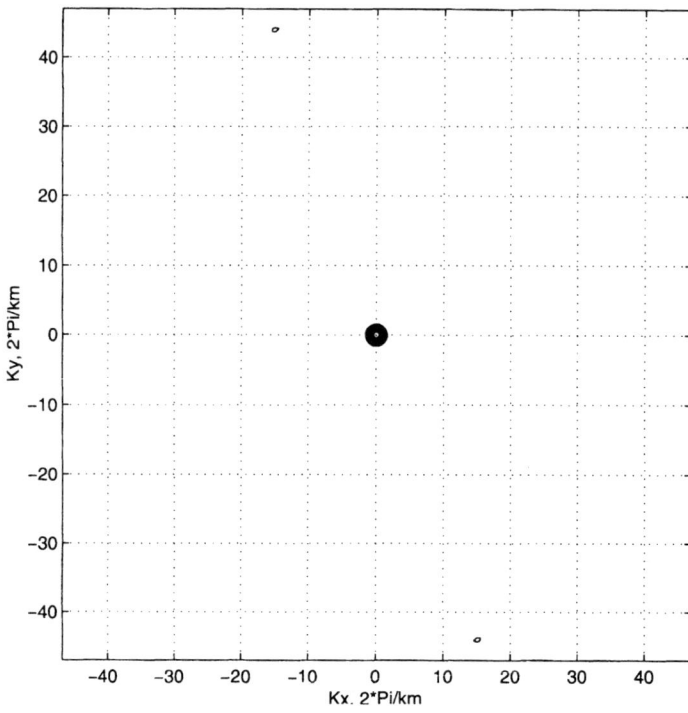

Figure 4

Contour map of the array pattern for ARCES. The main lobe is located in the center. There is no grating lobe in this case.

The relation of wavenumber k, wavelength λ, frequency f, and slowness s is given by

$$k = 2\pi/\lambda = 2\pi f s$$

the two constants $k_{\min}$ and $k_{\max}$ can be represented by

$$\log(s) = \log(k_{\min}/2\pi) - \log(f)$$

and

$$\log(s) = \log(k_{\max}/2\pi) - \log(f) \ .$$

They are two parallel straight lines with a slope of -1 on a log-log scale in the frequency-slowness coordinates. Figure 5 shows these two lines for the array YKA as an example. The area between the two lines is defined as the wavenumber passband of the array.

The basic assumption of seismic array processing is that a seismic signal is a coherent plane wave, and noises between any two sensors are incoherent. Ideally, a coherent plane wave produces identical signals from sensor to sensor except from a

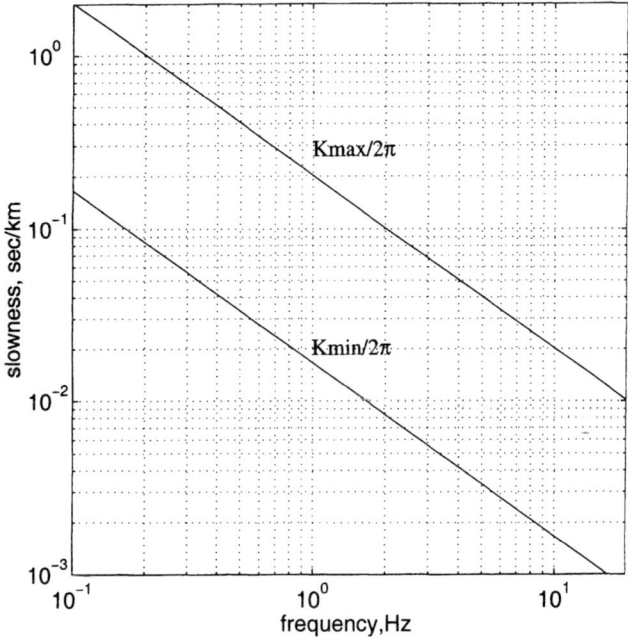

Figure 5
Wavenumber passband for the array YKA. The area between the two lines is defined as the wavenumber passband for the array YKA.

predictable propagation delay. The delays between sensors are determined by the relative sensor locations and by the speed and direction of the propagating wave. The delay-and-sum beamforming is a primary algorithm for signal detection processing at arrays. By assuming a slowness vector of the plane wave as it propagates across the array, a beam signal is formed by combining waveform data received at individual sensors of the array:

$$b(t) = \frac{1}{N} \sum_{i=1}^{N} g((t - \tau_i), x_i, y_i)$$

where $g(.)$ is the seismic sensor records that can be expressed as the sum of a signal component $u_i(t)$ and noise component $n_i(t)$:

$$g(t, x_i, y_i) = u_i(t) + n_i(t) \ .$$

(x_i, y_i) are the spatial coordinates of the i-th sensor, and τ_i represents the propagating delay of the signal at the i-th sensor with respect to a reference sensor:

$$\tau_i = (x_i s_x + y_i s_y) = \mathbf{r_i s} \ .$$

Because the algorithm adds the desired signal only constructively, the beam signal is composed mostly of the signal with slowness vector **s**, which, in turn, corresponds to a slowness amplitude and azimuth pair ($|s|$, θ). The beam signal corresponding to the direction and slowness of the wavefield will have the maximum energy, which can therefore be used to determine the propagation parameters. To detect all possible signals propagating from different directions and velocities (slowness), a detection statistic, usually the ratio of a short-term average to a long-term average (STA/LTA), is calculated for all possible steered beam positions, that is, a set of polar-coordinates in terms of slowness amplitude and arrival azimuth. However, it is impractical to calculate too many steered beam positions in the slowness plane in a real-time operation. Furthermore, the geometry configuration of an array constrains its detection capability. This has prompted the development of a strategy for beam deployment for various arrays in a rational way.

3. Beam Deployment Scheme Used in PIDC

Seismic signals of interest cover a range of slownesses and frequencies; they could come from all directions of the array, and the coherency of seismic signals and noise depends on phase-type, frequency, and sensor separation. To properly deploy detection beams for the variety of IMS seismic arrays, a scheme that takes into account array geometry, slowness and frequency distribution, and coherency properties at each array has been developed at the PIDC. With this approach, slowness and azimuth of steered beams are selected consistently with even resolution coverage in each frequency band. The following steps describe the scheme for the initial beam deployment:

1. Compute the array pattern based on the array geometry by using equation (1) to determine the wavenumber resolution, R_3dB, and aliasing frequency in the wavenumber domain, $R\_grating$, if it exists. Figures 2 and 4 show the array patterns for the two arrays YKA and ARCES, respectively. Table 1 summarizes the computed R_3dB and $R\_grating$ for the current IMS seismic arrays for which data are processed at the PIDC. The values of R_3db and $R\_grating$ in Table 1 were derived by assuming that all sensors (full aperture) would be used. Subarray, which are combinations of selected sensors of a given array, will have different wavenumber passbands than the full array, and are not included in the table. Table 1 also includes the aperture and number of sensors of the IMS seismic arrays.

2. Plot the k_{min} and k_{max} values, derived from R_3dB and $R\_grating$, respectively, in the frequency-slowness plane on a log-log scale as illustrated for array YKA in Figure 5. The two straight lines define the lower and upper boundaries of steered coherent beam positions in the frequency-slowness plane.

3. Identify slowness ranges of the seismic phases, such as teleseismic phases and regional seismic *Pn*, *Pg*, *Sn*, *Lg*, and *Rg* phases, to be detected. Figure 6 shows a

Table 1

Basic parameters for current IMS seismic arrays

ARRAY	Aperture (km)	Number of sensors	R_3dB (2π/km)	R_grating (2π/km)
KVAR	0.26	4	7.02	24.18
SPITS	0.98	9	2.02	16.57
HFS	1.19	8	1.90	23.39
FINES	2.02	16	1.20	
ARCES	3.10	25	0.85	
BRAR	3.35	7	1.09	5.77
PDAR	3.63	13	0.61	
GERES	3.87	25	0.66	
NVAR	4.25	10	0.48	13.75
TXAR	4.39	10	0.61	5.68
EKA	8.85	20	0.29	3.47
ESDC	9.57	19	0.25	
CMAR	9.74	18	0.23	
ASAR	9.82	19	0.24	
KSAR	10.09	19	0.22	
ILAR	10.20	19	0.24	
MJAR	11.09	7	0.17	0.75
YKA	22.63	18	0.10	1.27
WRA	26.40	20	0.09	

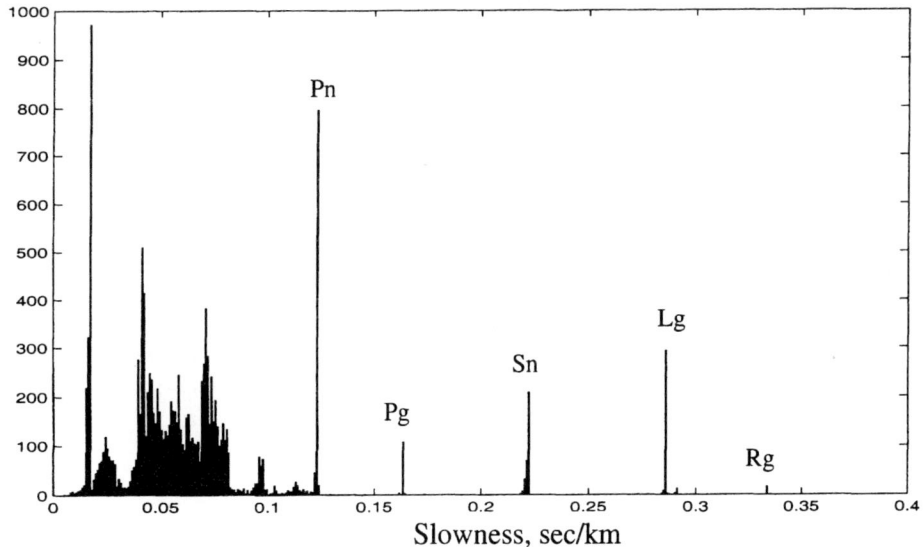

Figure 6

Histogram of slowness distribution of detected seismic phases at the PIDC between January 1 and April
20, 1997. Teleseismic phases have continuous observed slowness values smaller than the *Pn* slowness. *Pn,
Pg, Sn, Lg,* and *Rg* exhibit discrete slownesses.

slowness histogram of seismic phases detected at the PIDC from January 1 to April 20, 1997. It shows that the slowness values are more or less distributed continuously from 0 to 0.1237 s/km for teleseismic phases, and that five distinct peaks correspond to the regional phases Pn (0.1237 s/km), Pg (0.1640 s/km), Sn (0.2225 s/km), Lg (0.2857 s/km), and Rg (0.3334 s/km). The steered beams should cover the slowness values in range of teleseismic phases in a uniform interval, and include five individual slowness values of the regional phases.

4. Choose a set of overlapping bandpass filters that cover the frequency range of seismic signals of interest. This range is usually between 0.5 and 20 Hz. The number of frequency bands is not limited but is constrained by the number of beams so that the real-time processing is not too time-consuming. At the PIDC a set of octave frequency bands from 0.5 to 8 Hz with 50% overlap have been chosen. The upper limit is dictated by the drop in signal coherency at high frequencies for most arrays.

5. Determine slowness positions of detection beams in each frequency band by incorporating seismic slowness distribution. Five fixed slowness values that correspond to the regional primary and secondary phases (Pn: 0.1237 s/km, Pg: 0.1640 s/km, Sn: 0.2225 s/km, Lg: 0.2857 s/km, and Rg: 0.3334 s/km) are selected. For teleseismic phases, which are the dominant detections by the IMS seismic network, the slowness positions of steered beams are determined by the minimum slowness, *min_slow*, the maximum anti-aliasing slowness, *max_slow*, and the slowness interval, $\Delta slow$, in each frequency band.

The minimum slowness is defined by the minimum wavenumber k_{min}, or R_3dB, and the center frequency $f_c(i)$ of the frequency band, i:

$$min\_slow(i) = R\_3dB/(2\pi f_c(i)) \ .$$

The slowness interval is defined by the wavenumber resolution, R_3dB and the center frequency $f_c(i)$ of the frequency band, i:

$$\Delta slow(i) = R\_3dB/(2\pi f_c(i)) \ .$$

The maximum anti-aliasing slowness is defined by the maximum wavenumber k_{max}, or $R\_grating$, and the center frequency $f_c(i)$ of the frequency band, i:

$$max\_slow(i) = R\_grating/(2\pi f_c(i)) \ .$$

If the *max_slow(i)* is greater than the slowness value of the regional phase Pn, 0.1237 sec/km, or the k_{max} does not exist (no grating lobe existing) for a given array, the slowness upper bound of teleseismic phases will be the slowness value of the regional seismic phase Pn.

6. Derive the azimuthal distribution of steered beams by using the azimuthal resolution, which is a function of frequency, slowness and the wavenumber resolution:

$$\Delta Azi(i,j) = \Delta slow(i)/slow(i,j)(180/\pi)(\text{degrees}) \ ,$$

where $\Delta slow(i)$ is the slowness interval at the i-th frequency band and $slow(i, j)$ is the j-th slowness value at the i-th frequency band. The azimuthal distribution defined by this equation, which is an approximation of the theoretical angular resolution, gives a uniform 3_dB resolution coverage in the slowness plane for all detection beams with different slowness. For the sake of equal-spacing for the azimuth distribution, azimuth interval of each slowness is rounded to a factor of 360. For example, if the computed $\Delta Azi(i, j)$ is 32 degrees, 30 degrees is used.

7. Add one vertical beam (zero slowness) in each frequency band to detect possible signals coming from directly underneath the array. This beam actually performs data averaging for all channels.

8. Add incoherent beams for some teleseismic arrays because the slowness of regional seismic phases is outside of the wavenumber passband, or signals between large-spacing sensors lack coherency. The incoherent beam is calculated by averaging waveform powers over sensors with certain delays.

9. Repeat steps 1–7 if subarrays are configured in the given array for the beamforming detection. Subarray configurations are often used in beamforming to achieve optimum signal-to-noise ratio enhancement (gain) because the correlations of seismic signals and of noise among sensors are frequency- and distance-dependent. For example, MYKKELTVEIT et al. (1983) found that the noise correlation curves consistently had negative minima at certain sensor separations before tending to zero at larger separations, while the signal correlations degraded continuously as the distance increased. KVAERNA (1989) proposed the optimum subarrays for the small-aperture array NORESS based on the coherency features of signals and noise for different frequencies. CLAASSEN (1992) discussed how to choose the appropriate subarray at a given frequency to obtain optimal beam gain. For a large array WRA, the average coherency distances where the normalized correlation values drop to 0.7 were determined by WANG et al. (1998). Table 2 shows the average coherency distances of teleseismic P phases (T), regional P phases (P), regional S phase, and noise in different frequency bands for the array WRA. Based on these results, different subarrays were configured in their studies.

Table 2

Average coherency distances of different phase types at WRA

Frequency band (Hz)	T (km)	P (km)	S (km)	N (km)
0.50–1.0	26	26	4	2
0.75–1.5	26	26	3	0
1.0–2.0	26	26	2	0
1.5–3.0	26	15	0	0
2.0–4.0	18	3	0	0
3.0–6.0	2	0	0	0
4.0–8.0	0	0	0	0
5.0–10.0	0	0	0	0

As an initial beam deployment for IMS seismic arrays at the PIDC, subarray configuration is not used unless the given array has been carefully studied for its coherence properties among sensors.

4. Example of Initial Beam Deployment at ARCES Array

This section uses the initial beam deployment at ARCES array as an example to demonstrate the previously described scheme. Figure 3 shows the geometry of ARCES. It is an irregular planar array, which consists of four concentric rings with 25 sensors (MYKKELTVEIT et al., 1990). The A-ring, the innermost ring, has three sensors; the B-ring has five sensors; the C-ring has seven sensors; and the D-ring has nine sensors. The array pattern for all 25 elements (full array) has been computed using equation (1) and is shown in Figure 4. Two subarrays are used in the initial beam deployment for ARCES at the PIDC, based on previous coherence studies and operational experience. The first subarray consists of the center sensor a0, the C-ring, and the D-ring; the second subarray consists of a0, the B-ring, the C-ring, and the D-ring. The first subarray has larger inter-sensor spacing than the second subarray. Lower frequency bands (0.75–2.25 and 1.0–3.0 Hz) are used for the first one, and higher frequency bands (2.0–4.0 and 3.0–6.0) are used for the second one. Operational experience at the PIDC shows that few detections by the beams at frequencies higher than 6.0 Hz were associated in the analyst reviewed event bulletin.

Figure 7 shows beam positions in the frequency-slowness plane. The "plus" signs (+) represent all possible beam positions based on 3dB slowness resolution, and the open circles represent the selected beam positions. The slowness values of Pn, Pg, Sn, Lg, and Rg are marked in the figure. Because of small azimuth interval, detection beams for regional seismic phases are limited to a single frequency band to compromise computation time. Multiple frequency bands are used for teleseismic phases. Figure 8 shows the steered beam distribution of the designed detector in the slowness-azimuth plane. The different symbols represent the slowness positions in different frequency bands. The selection of slowness-azimuth positions is based on the slowness and frequency distribution of seismic phases and on a compromise with real-time processing speed for the operational systems. This beam deployment has full coverage for teleseismic phases with a 3dB resolution in the slowness-azimuth plane for all frequency bands, and has a reasonably good azimuth coverage for regional seismic phases.

The scheme described in this paper has been applied to all seismic IMS arrays processed at the PIDC except the very large aperture array NOA, which needs a special beamforming configuration (FYEN, 1997). As an initial implementation, the geometry constraints on detection capability of arrays were emphasized in beam deployment because no detailed information of spatial coherency of signals and noises among sensors are available for each array. If the coherence structure of

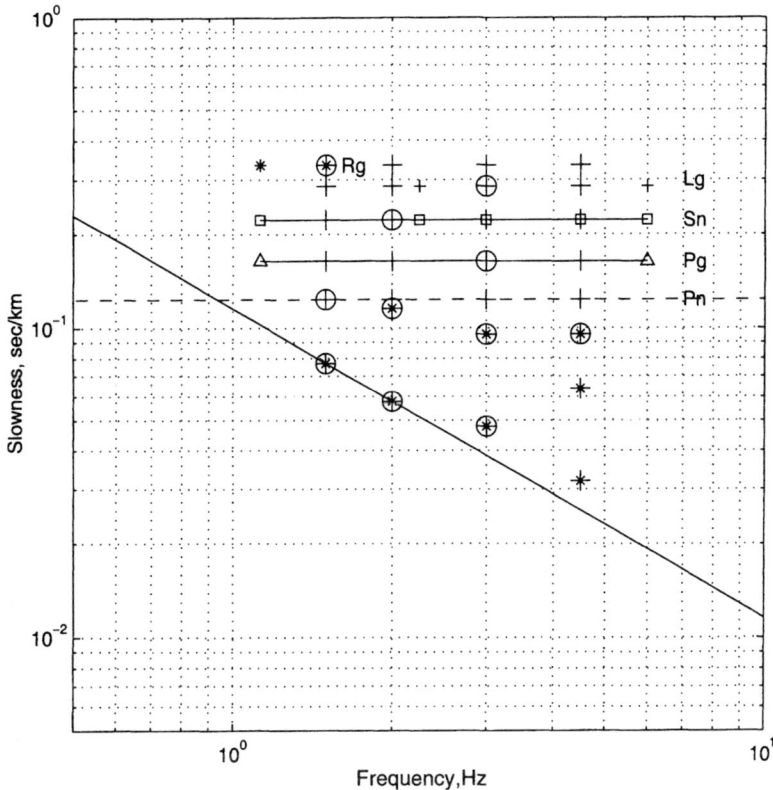

Figure 7
Beam positions in the frequency-slowness plane for ARCES. Open circles represent the frequency-slowness coordinates of selected beams, and the "plus" signs (+) represent all possible beam positions based on 3dB slowness resolution and the distinct regional slowness values. This selection is a compromise between the slowness and frequency distribution of seismic phases and processing speed requirements.

signals and noises among the array sensors is carefully studied in the future, different subarrays/frequency bands for each array can be reconfigured accordingly, therefore, the beam deployment can be modified using the framework of the proposed scheme.

5. Summary

This paper presents a scheme for initial beam deployment for a variety of seismic arrays. Beamforming is equivalent to a spatiotemporal bandpass filter, with a passband defined by the minimum and maximum wavenumbers, which are functions of the geometry configuration of the array. Different arrays may have different signal and noise coherence properties, which is an important factor for array signal processing. Deployment of coherent beams for array signal detection is based on the

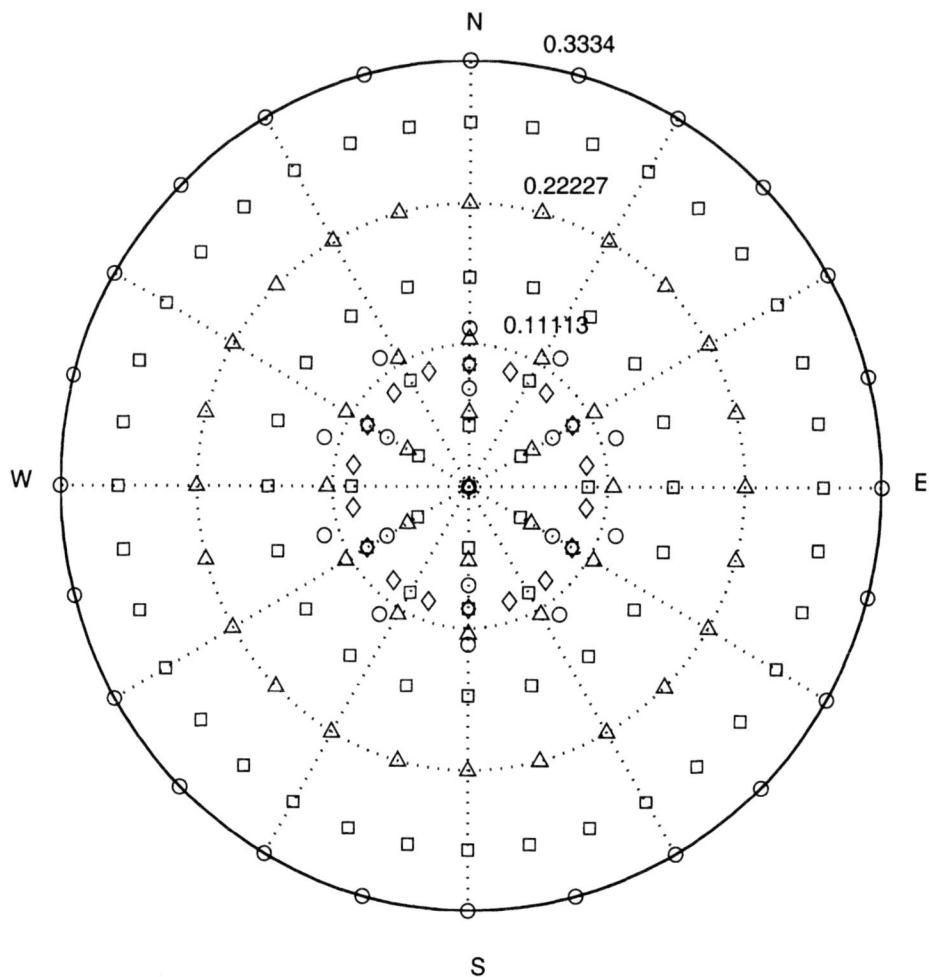

Figure 8
Beam positions in the slowness-azimuth polar plane for ARCES. Different symbols represent the slowness positions in different frequency bands. Open circles represent the beams in the frequency band of 0.75–2.25 Hz; triangles represent the beams in the frequency band of 1.0 to 3.0 Hz; the squares represent the beams in the frequency band of 2.0–4.0 Hz, and diamonds represent the beams in the frequency band of 3.0–6.0 Hz.

wavenumber resolution of the array (or subarray if necessary), slowness and frequency distribution of seismic phases, and coherence properties of seismic signals and noises among sensors. Within the wavenumber passband, all possible slowness values are determined by the resolution for each frequency band, and those that are outside the range of seismological interest are excluded. The appropriate azimuthal distribution for each selected slowness is determined from the azimuthal resolution. Using this scheme, detection beams at each array are rationally deployed in the slowness-azimuth and frequency domain. Beamforming detectors designed with this scheme have been operated at the PIDC with a satisfactory performance. However, there is often room for improvement for beam deployment, at least for regional and local events. After seismological experience is gained following a period of operation, additional considerations may be taken into account. Among these are:

—*The background noise field and local sources*. Are there frequency ranges that must be avoided due to high noise levels?

—*Regional wave propagation effects*. Are there unusual wave propagation effects that must be considered such as presence/absence of seismic phases?

—*Three-component observations*. Some of the arrays have additional three-component sensors. If several three-component sensors are available, beamforming can be applied to the rotated components. For given phases and distance ranges, improved detection capability can be gained by rotation into longitudinal, radial and transverse components.

The scheme presented in this paper describes one way to systematically design beamforming detectors. It can incorporate more sophisticated studies on individual arrays to optimize the detector performance. This scheme, in principle, can be applied to other data types of arrays, such as infrasonic and hydroacoustic arrays deployed in the International Monitoring System.

Acknowledgements

I am indebted to Hans Israelsson, Robert G. North, Keith Mclaughlin, Jerry Carter, and Chris Ferraro for their remarkable effort in finishing and improving this manuscript. I am also grateful to Robert Blandford, Raymond Willeman, Tom Sereno, Greg Beall, Roger Bowman, Richard Stead, and Henry Swanger for their valuable suggestions and discussions. Editor Zoltan A. Der and anonymous reviewers made valuable suggestions to improve this article. This work was sponsored by U.S. Department of Defense, Defense Threat Reduction Agency, under Contract No. DTRA01-99-C-0025.

REFERENCES

ANGLIN, F. M. (1971), *Detection Capabilities of the Yellowknife Seismic Array and Regional Seismicity*, Bull. Seismol. Soc. Am. *61*, 993–1008.

BACHE, T. C., MARSHALL, P. D., and YOUNG, J. B. (1986), *High-frequency Seismic Noise Characteristics at the Four United Kingdom-type Arrays*, Bull. Seismol. Soc. Am. *76*, 601–616.

CLAASSEN, J. P. (1992), *The Application of Multiply-constrained Minimum-variance Adaptive Beamforming to Regional Monitoring*, Bull. Seismol. Soc. Am. *82*, 2191–2212.

DER, Z. A., SHUMWAY, R. H., and LEES, A. C. (1988), *Frequency Domain Coherent Processing of Regional Seismic Signals at Small Arrays*, Bull. Seismol. Soc. Am. *78*, 326–338.

HAYKIN, S., *Advances in Spectrum Analysis and Array Processing* (Prentice-Hall, Inc., Englewood Cliffs, New Jersey 1991).

FYEN, J. (1997), *NORSAR Large Array Processing at the IDC Testbed*, Semiannual Tech. Summary, 1 April–30 September 1997, NORSAR Sci. Rep. 1-97/98, Kjeller, Norway.

JOHNSON, D. H. and DUDGEON, D. E., *Array Signal Processing* (Prentice-Hall, Inc., Englewood Cliffs, New Jersey 1993).

KVAERNA, T. (1989), *On Exploitation of Small-aperture NORESS Type Arrays for Enhanced P-wave Detectability*, Bull. Seismol. Soc. Am. *79*, 888–900.

MYKKELTVEIT, S., ASTEBOL, K., DOORNBOS, D. J., and HUSEBYE, E. S. (1983), *Seismic Array Configuration Optimization*, Bull. Seismol. Soc. Am. *73*, 173–186.

MYKKELTVEIT, S., RINGDAL, F., KVAERNA, T., and ALEWINE, W. R. (1990), *Application of Regional Arrays in Seismic Verification Research*, Bull. Seismol. Soc. Am. *80*, 1777–1800.

MYKKELTVEIT, S., BAADSHAUG, U., and KVAERNA, T. (1995), *Processing of Spitsbergen Array Data*, Semiannual Tech. Summary, NORSAR Sci. Rep. 2-94/95, Kjeller Norway.

RINGDAL, F. and HUSEBYE, E. S. (1982), *Application of Arrays in the Detection, Location, and Identification of Seismic Events*, Bull. Seismol. Soc. Am. *72*, S201–S224.

RINGDAL, F. (1990), *Teleseismic Event Detection Using the NORESS Array, with Special Reference to Low-Yield Semipalatinsk Explosions*, Bull. Seismol. Soc. Am. *80*, 2127–2142.

WANG, J., ISRAELSSON, H., and CARTER, J. (1996), *A Systematic Approach to Designing Detectors for Diverse Seismic Arrays*, Seism. Res. Lett. *60*(2), p 60.

WANG, J., ISRAELSSON, H., and NORTH, R. G. (1998), *Optimum Subarray Configuration Using Genetic Algorithms*, Proc. IEEE International Conference on Acoustic, Speech, and Signal Processing, 2129–2132, Seattle, May 1998.

WEICHERT, D. H. and HENGER, M. (1976), *The Canadian Seismic Array Monitor Processing System (CANSAM)*, Bull. Seismol. Soc. Am. *66*, 1381–1403.

WHALEN, A. D., *Detection of Signals in Noise* (John Wiley & Sons, New York 1970).

(Received March 20, 1999, revised November 5, 2000, accepted November 12, 2000)

 To access this journal online:
http://www.birkhauser.ch

Pure appl. geophys. 159 (2002) 1021–1041
0033–4553/02/051021–21 $ 1.50 + 0.20/0

© Birkhäuser Verlag, Basel, 2002

┃Pure and Applied Geophysics

Adaptive Training of Neural Networks for Automatic Seismic Phase Identification

JIN WANG[1]

Abstract — A neural network module has been implemented in the Prototype International Data Centre (PIDC) for automated identification of the initial phase type of seismic detections. Initial training of the neural networks for stations of the International Monitoring System (IMS) requires considerable effort. While there are many seismic phases in the analyst-reviewed database that can be assumed as the ground-truth resource of the initial phase type of Teleseism (T), Regional P (P), and Regional S (S), no ground-truth database of noise (N) is available. To reduce analyst effort required in building a ground-truth database, an "Adaptive Training Approach" is proposed in this paper. This approach automatically selects training patterns to take advantage of the learning ability of neural networks and information on the accumulated observation database. Using this approach, neural networks were trained on the data provided by station STKA, Australia. The performance of automated phase identification has been improved significantly by the retrained neural networks. This approach is also validated by comparison with the performance using the ground-truth noise database.

Key words: Artificial Intelligence, Neural Networks, CTBT, Seismic Phase Identification.

1. Introduction

Phase identification is a major task in operating an automated seismic monitoring system. The accuracy of the phase identification significantly impacts the final automated bulletin. The availability of digital three-component seismograms improves the chance of automatically identifying the *P*- and *S*-phase arrivals of seismic events. A number of algorithms have been developed to automatically identify seismic phase type, and research continues in improving the reliability of these algorithms and in reducing the overhead time. For example, JURKEVICS (1988) presented a technique for polarization analysis of three-component seismic data. ROBERTS *et al.* (1989) presented a technique for phase identification based on the auto- and cross-correlations of the three-component seismic data. BACHE *et al.* (1990) developed the first operational version of the

[1] Center for Monitoring Research, Science Applications International Corporation, 1300 N. 17th Street, Suite 1450, Arlington, VA 22209, U.S.A. E-mail: wangjin2000@yahoo.com

Intelligent Monitoring System, which now has been employed as the Prototype International Data Centre (PIDC) in Arlington, Virginia, USA, and as the International Data Centre for the Comprehensive Nuclear-Test-Ban Treaty Organization in Vienna, Austria. A rule-based expert system for initial phase identification was applied in that system. SUTEAU-HENSON (1991) presented a study on phase identification based on polarization analysis for three-component data. SERENO and PATNAIK (1993) developed a neural network module for automated identification of the initial phase type of seismic detections and implemented that module at the PIDC. CICHOWICZ (1993) developed an algorithm for automatically picking the S phase from three-component seismograms. WANG and TENG (1995, 1997) designed two artificial neural network based detectors for picking and identifying regional P and S phases. ANANT and DOWLA (1997) applied wavelet transform methods to phase identification for three-component seismograms.

While most of these examples have not been tested extensively in the operational environment, the neural network module developed by SERENO and PATNAIK (1993) has been implemented in the PIDC operational system since 1995. In the initial implementation, a set of average weights trained from a limited number of stations was used as a set of default weights for three-component stations. All detections at three-component stations were automatically classified into different phase types by the neural networks in routine operations. Like any other automatic monitoring system, the automatic network solutions were reviewed, and corrected if necessary, by skilled human analysts to generate final solutions. The corrections by analysts include changing signal onset time, changing assigned phase identifier, or changing which detections were grouped together and associated with a single seismic event. The reviewed results were called the "analyst reviewed bulletin" in the PIDC. Evaluations of phase identification with the default weights at the PIDC showed that the performance varies at different stations. The correct rate of the initial phase-type identification in the automated event bulletin compared to the analyst reviewed database ranged from 25.6% to 73.4%. The PIDC has accumulated millions of seismic phase readings for the current three-component stations, which were not available at the time of the original neural network implementation. Therefore, it is worthwhile to train neural networks for each specific station to improve the performance of automatic phase identification, which, in turn, will improve the quality of the automatic event bulletin. Although many seismic phases that were reviewed by analysts in the PIDC database can be assumed to be the ground-truth resource, no ground-truth database of noise is available. This is a major obstacle for updating the neural networks. This paper proposes an approach to training neural networks that takes advantage of accumulated seismic phases and the adaptive learning ability of neural networks.

2. Artificial Neural Networks

Artificial neural networks are inspired by biological systems in which large numbers of neurons, which individually function rather slowly and imperfectly, collectively perform tasks that even the fastest computers have not been able to match. This field is among the most rapidly developing scientific research today and is interdisciplinary in nature. Its potential applications include speech and image recognition, linear and nonlinear optimization, automatic control, and seismology.

As in a biological nervous system, fundamental elements of an artificial neural network are artificial neurons (also called cells, units, or nodes). The function of artificial neurons is identical to that of real neurons: they integrate input from other neurons and communicate the integrated signal to a decision making center. Computational models of a neural network try to emulate the physiology of real neurons. Artificial neural networks have two principal functions: one is the input-output mapping or feature extraction. The other is the pattern association or generalization. The mapping of input and output patterns is estimated or "learned" by neural networks with representative samples of the input and output pattern (training set). The generalization of neural networks is an output pattern in response to an input pattern, based on the neural network memories that were learned in the training process.

Among many different types of neural networks, a particular type, the Multi-Layered Perceptron (MLP), is used in the routine operation in the PIDC. A perceptron may be viewed as a neuron that computes activation with a sigmoidal activation function, which takes the following form (ZURADA, 1992):

$$o_i = F(d_i) = \frac{1}{1 + \exp(-\beta d_i)}$$

where

$$d_i = \sum_i^N w_{ij} x_j - \theta_i \ ,$$

where o_i is the activity output at neuron i, constant β controls the slope of the semi-linear region, d_i is the weighted sum of neuron i for the input from the last lower layer with N neurons, x_j is the input from neuron j, w_{ij} is the weight between neurons i and j, and θ_i is the intrinsic threshold that can be treated as an individual weight with a negative sign.

The structure of an artificial neuron is shown in Figure 1. The one-to-one correspondence between an artificial neuron and a biological neuron is commonly drawn: the inputs correspond to the dendrites of a biological neuron; the weights correspond to the synapses; the summation and the transfer function correspond to the cell body; and the output to other neurons corresponds to axons. Artificial neural

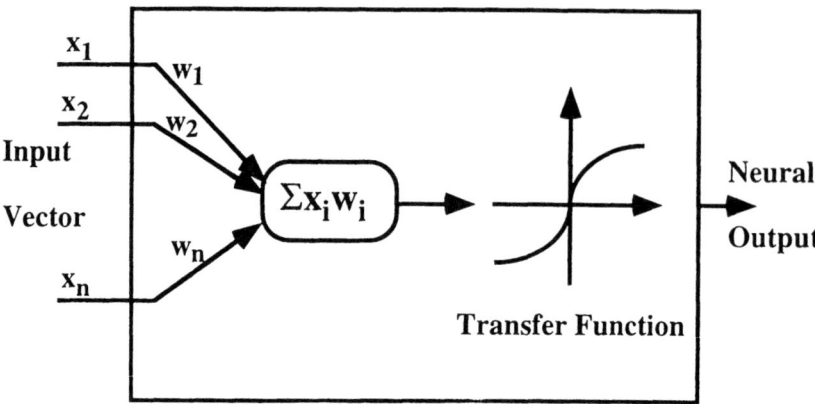

Figure 1
Architecture of single artificial neuron consisting of N input units. The input vector fans in from the left and fans out to the right. Interconnection of artificial neurons forms a complex architecture called an artificial neural network.

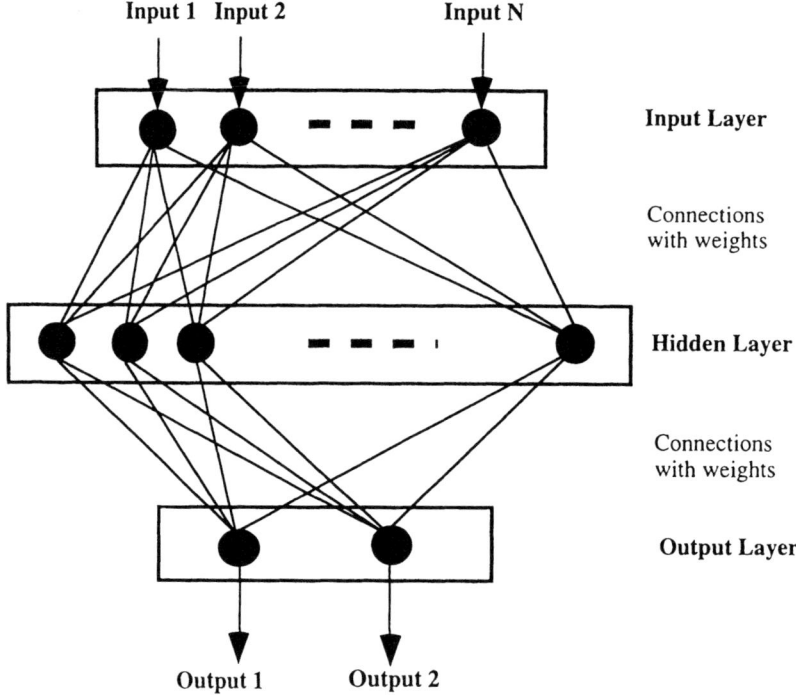

Figure 2
Structure of multi-layered perceptron neural network. It is a feedforward and partially connected network. Neurons are only connected to the adjacent layer neurons.

networks can be constructed by linking the neurons through the weights. A multi-layer feed-forward network is shown in Figure 2. The input data are propagated or feedforwarded through neurons and their connections in the hidden layer to the output neurons.

The weights of a neural network are obtained by learning through the training data set. The most widely used algorithm for a multi-layer feedforward neural network is the backward error propagation algorithm for supervised learning (RUMELHART et al., 1986). An objective function in the output layer is defined by

$$J_p = \frac{1}{2M} \sum_{k=1}^{M} [T_k(p) - O_k(p)]^2 \ ,$$

and the normalized sum of objective function over total training set is defined by

$$J = \frac{1}{P} \sum_{p=1}^{P} J_p \ ,$$

where M is the number of outputs of the neural network, $T_k(p)$ is the desired (or called target) output of the k-th neuron for the training pattern p, and $O_k(p)$ is the actual neural output of the k-th neuron. The back propagation learning algorithm uses a least-squares error minimization criterion to minimize J by adjusting the weights. That is, the learning process is supervised by the given input-output pattern set, and it will end if the J has converged to a desired value.

The trained neural network will be used to process and classify unknown input data. Successful application of the neural networks for pattern recognition is highly dependent on the training process. For supervised learning, the performance of a trained neural network is related to the quality of the selected training patterns.

3. Automatic Phase Identification at the PIDC

Phase identification is a major task in operating an automated seismic monitoring system. In the PIDC software system, signals detected in seismic stations first are categorized as one of four initial wave types: noise (N), teleseismic waves (T), regional P wave (P), or regional S wave (S). Signals are characterized based on their features and qualities extracted from the waveforms near the arrival time. For detections from three-component stations, a neural network is applied to determine the initial wave type.

In the software developed by Science International Applications Corporation (SAIC), the four initial wave-type classification problem is solved by three cascaded neural networks of the MLP (SERENO and PATNAIK, 1993). Each of the three neural networks consists of three layers: an input layer, a hidden layer, and an output layer. The input layer consists of 15 nodes, corresponding to 15 features; the hidden layer consists of 6 nodes, and the output layer consists of two nodes that correspond to the two output classes. The first neural network distinguishes between noise and signal, the second distinguishes between teleseismic and regional phases, and the last determines if the regional phases are P or S. The three types of initial phase (T, P, S) are further divided into final phases such as P, PKP, Pn, Pg, Sn, and Lg by other subprograms, and are used for automatic event formation and phase association. The noise detections, however, are not further processed by automatic operations.

The 15 attributes used for the input of each of the three neural networks include dominant period, polarization attributes, contextual attributes, and a spectral representation of the horizontal-to-vertical power ratio. Polarization is the main source of information, which can be derived from three-component seismic data. The method used for polarization analysis was developed by JURKEVICS (1988), and its implementation in the IMS was described by BACHE et al. (1990).

For polarization measurement, the convariance matrix of three-component seismic data, S, is evaluated by:

$$S_{jk} = \frac{1}{N} \sum_{i=1}^{N} x_{ij} x_{ik} \; ,$$

where x_{ij}, x_{ik} are the waveform data in a time window, i is the index of the sample in the time window, N is the number of samples, and j or k is the index of the component (z, n, e).

The terms of the 3×3 convariance matrix are the auto- and cross-variances of the three components of the ground motion:

$$S = \begin{bmatrix} S_{zz} & S_{nz} & S_{ez} \\ S_{zn} & S_{nn} & S_{en} \\ S_{ze} & S_{ne} & S_{ee} \end{bmatrix} .$$

The polarization ellipsoid is computed by solving the eigenvalue problem for the covariance matrix, which yields the eigenvalues $(\lambda_1, \lambda_2, \lambda_3)$ and eigenvectors (u_1, u_2, u_3) of the matrix.

Feature extraction (attribute selection) and pattern classification are two basic steps for a pattern recognition. The two steps are closely related; a better feature extraction will result in an easier classification. The success of a neural work system for pattern classification primarily depends on the designer's physical understanding of the problem. Fifteen attributes used in the neural networks for initial wave-type classification are based on previous studies and are proven successful (SERENO and PATNAIK, 1993). Other different feature combinations that have better performance might be found by further research. However, that discussion is beyond the scope of this paper. The goal of this paper is to tune the neural networks under the same structure that has been used for the PIDC. The following 15 attributes are used in the neural networks:

(1) *period*: Dominant period of the detected phase.

(2) *rect*: Signal rectilinearity: $rect = 1 - \frac{\lambda_2 + \lambda_3}{2\lambda_1}$, where $\lambda_1 > \lambda_2 > \lambda_3$ are eigenvalues obtained by solving the 3-D eigensystem.

(3) *plans*: Signal planarity: $plans = 1 - \frac{2\lambda_3}{\lambda_1 + \lambda_2}$, which is a measure of the planar characteristic of the polarization ellipsoid.

(4) *inang1*: Long-axis incidence angle: $inang1 = \frac{a\cos(|u_{11}|)}{90}$, where *inang1* is a normalized angle measured clockwise from the vertical axis, and u_{11} is the direction cosine of the eigenvector associated with the largest eigenvalue.

(5) *inang3*: Short-axis incidence angle: $inang3 = \frac{a\cos(|u_{31}|)}{90}$, where *inang3* is a normalized angle measured clockwise from the vertical axis, and u_{31} is the direction cosine of the eigenvector associated with the smallest eigenvalue.

(6) *hmxmn*: Ratio of the maximum to minimum horizontal amplitude: $hmxmn = \sqrt{\frac{\lambda_1}{\lambda_2}}$, where λ_1 and λ_2 are the maximum and minimum eigenvalues obtained by solving the 2-D eigensystem using only the horizontal components.

(7) *hvratp*: Ratio of horizontal-to-vertical power: $hvratp = \frac{s_{nn} + s_{ee}}{2s_{zz}}$, where s_{zz}, s_{nn}, and s_{ee} are the diagonal elements of the covariance matrix measured at the time of maximum rectilinearity.

(8) *hvrat*: Similar to *hvratp*, however measured at the time of the maximum three-component amplitude.

(9) $N_{after} - N_{before}$: Difference between the number of arrivals before and after the arrival in question within a fixed time window (60 seconds by default). The value of the number difference is scaled to a small range near ± 1 by dividing by 10.

(10) $T_{after} - T_{before}$: Mean-time difference between the arrival in question and those arrivals before and after it within a fixed time window (60 seconds by default). The value of the mean time difference is scaled to a small range near ± 1 by dividing by 100.

(11) *htov1*: Horizontal-to-vertical power ratio in an octave frequency band centered at 0.25 Hz.

(12) *htov2*: Horizontal-to-vertical power ratio in an octave frequency band centered at 0.5 Hz.

(13) *htov3*: Horizontal-to-vertical power ratio in an octave frequency band centered at 1.0 Hz.

(14) *htov4*: Horizontal-to-vertical power ratio in an octave frequency band centered at 2.0 Hz.

(15) *htov5*: Horizontal-to-vertical power ratio in an octave frequency band centered at 4.0 Hz.

This paper uses the three-component station STKA, which is located at Stephens Creek, Australia (31.88 S, 141.60 E), as a demonstration station. Figures 3–5 show histograms of the 15 attributes for the selected *T*-, *P*-, *S*- and *N*-type phases for station STKA from January 1 to October 26, 1996. In Figure 3, the distribution of signal periods for the four types is similar; the rectilinearity distributions can be

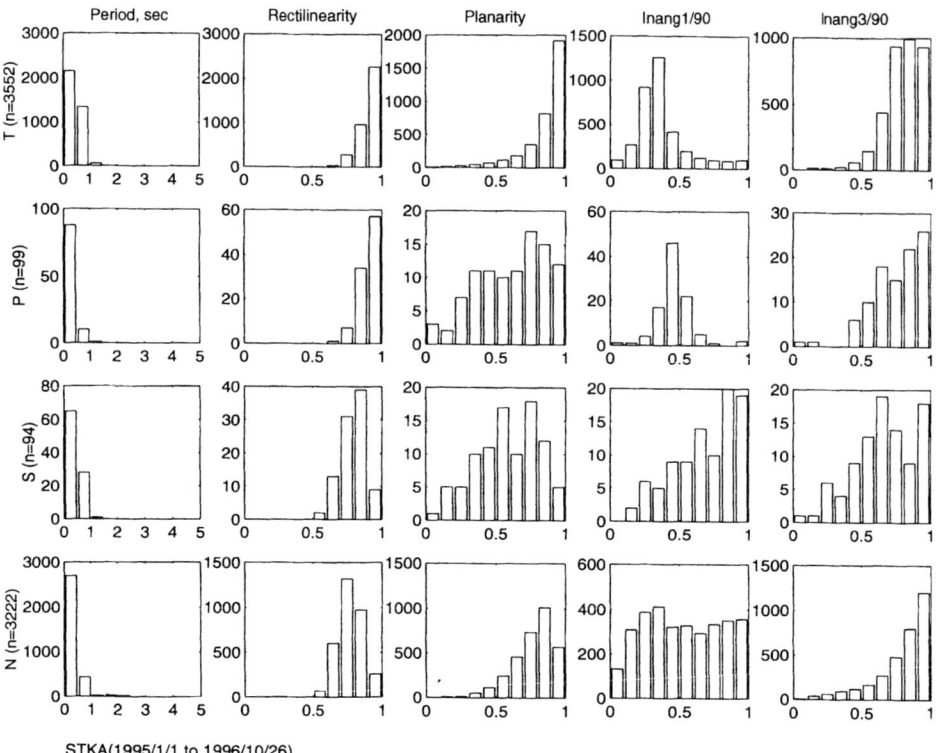

STKA(1995/1/1 to 1996/10/26)

Figure 3

Histograms of period, rectilinearity, planarity, long-axis incidence angle and short-axis incidence angle grouped by four initial wave types for station STKA. The distribution of signal periods for the four types are almost the same; the rectilinearity distributions can be divided into two groups, one for teleseismic and regional *P* phases (*T* and *P*), and one for regional *S* wave and noise (*S* and *N*); the planarity distributions are different for each type, and *Inang*1 distributions are also different for each type; the *Inang*3 distributions are similar for the *T* and *P* types, but are distinguished for the *S* and *N* types.

divided into two groups; one for teleseismic and regional *P* phases (*T* and *P*), and one for regional *S* wave and noise (*S* and *N*); the planarity distributions are different for each type, and the *Inang*1 are also different for each type; the *Inang*3 distribution is similar for the *T* and *P* types, but is different for the *S* and *N* types. In Figure 4, the five attribute distributions are quite different but exhibit some overlapping. In Figure 5, the horizontal-to-vertical power ratios in five frequency bands show some differences although with no significance. Therefore, the 15 attribute distributions in Figures 3–5 demonstrate considerable overlapping among the four initial phase types. Neural networks are well-suited for this condition because they are capable of constructing nonlinear decision surfaces across complex class boundaries from high-dimensional input data.

As the initial implementation, a set of average weights trained from a limited number of stations was used as a set of default weights for three-component stations

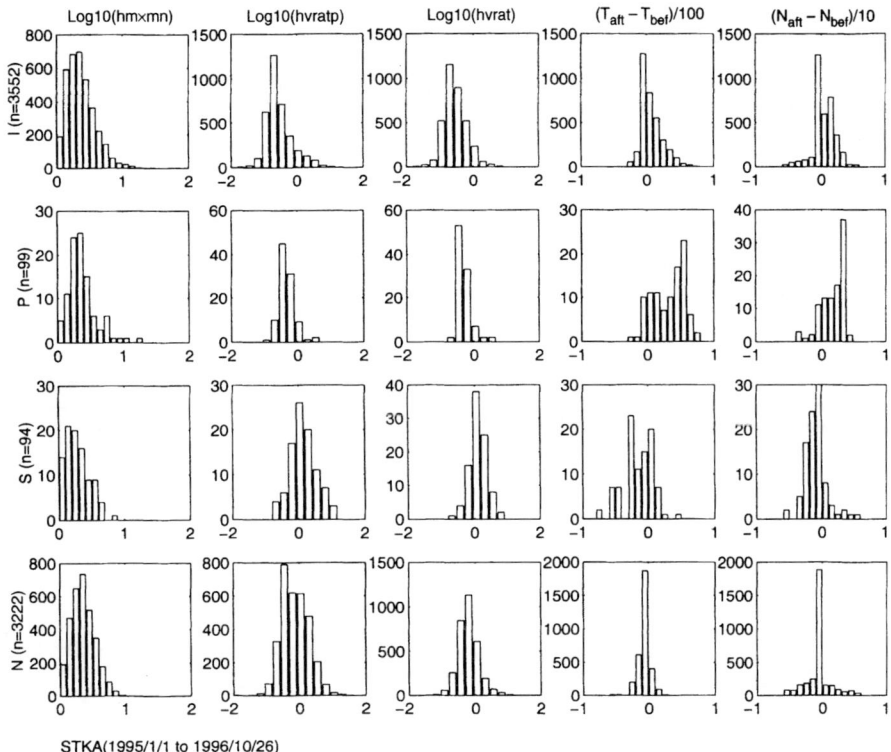

STKA(1995/1/1 to 1996/10/26)

Figure 4

Histograms of maximum-to-minimum horizontal amplitude ratio, horizontal-to-vertical power ratio at the time of maximum rectilinearity and at the time of maximum three-component amplitude, the difference of arrival times $((T_{aft} - T_{bef})/100)$ and detection numbers $((N_{aft} - N_{bef})/10)$ between that after and before the detection in question within a fixed time window of 60 seconds grouped by four initial wave types for station STKA. In the figure, five attribute distributions are quite different but have some overlapping.

in the PIDC. To evaluate how well the default weights performed for the automatic initial wave-type identification, the initial wave type of detections in the automatic bulletin was compared to those in the final analyst reviewed bulletin for station STKA. Table 1 indicates a "confusion matrix" of initial wave types from July 11 to November 25, 1996. The matrix indicates how the phase types classified by the neural networks were finally classified by analysts. Each row in the table is the number of initial wave types determined by the automatic system, and each column is the number of the types confirmed by analysts. Ideal performance of the automated system should result in zero for off-diagonal elements of the matrix. The "correct rate" is defined as the ratio of the sum of the diagonal elements to the total number of the phases in the final analyst reviewed bulletin; for STKA it was 704/ 2755 * 100% = 25.6%. Table 1 shows that many teleseismic phases in the final analyst reviewed bulletin were incorrectly identified as regional P-type phases in the automatic bulletin. In addition, 22.8% (629/2755) of phases in the final analyst

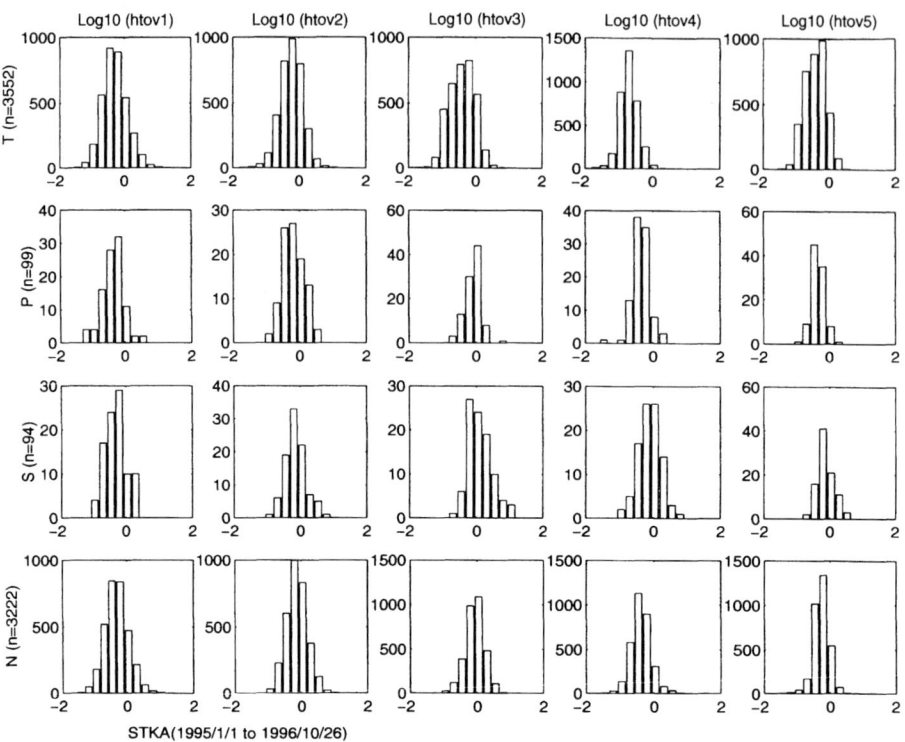

Figure 5
Histograms of horizontal-to-vertical power ratio at center frequencies of 0.25, 0.5, 1.0, 2.0, and 4.0 Hz
grouped by four initial wave types for station STKA. They show some differences but with no significance.

Table 1

*Confusion matrix of initial phase classification for STKA using default weights of neural networks (1996/07/
19–1996/11/25). Each row in the table is the result by the automatic system, and each column is the final result
confirmed by analysts. Off-diagonal elements are misclassification numbers*

		Results by analyst			Sum
		T	P	S	
Results by system	T	681	3	0	684
	P	1381	14	3	1398
	S	35	0	9	44
	N	626	1	2	629
	Sum	2723	18	14	2755

reviewed bulletin were incorrectly identified as N-type phases in the automatic
bulletin. This percentage, defined as the "N-phase rate", is another measure of the
automatic system's performance. Because the N phases in the automatic bulletin
would not be associated to form and locate events by the automatic system, reducing

the number of N phases that must be renamed and associated in the final analyst bulletin would reduce the work load of analysts.

4. Adaptive Training Procedure for Neural Networks

The evaluation described in the previous section shows that the initial default weights do not work well for station STKA. The neural networks for initial wave-type classification need to be retrained. Selection of a training data set is an important process in the application of neural networks. For supervised learning, the performance of the trained neural networks is related to the quality of the selected training patterns. The initial phase types reviewed by analysts in the analyst reviewed bulletin can be assumed to be accurate (ground truth). This means that many seismic phases are available in the final analyst reviewed bulletin as a ground-truth resource of the initial phase type of T, P, and S. For noise, however, no corresponding ground-truth resource is available. Building a ground-truth database with the help of analysts requires considerable effort. This is a major obstacle for updating the neural networks. This paper proposes a training approach that automatically and adaptively selects training patterns of seismic phases to take advantage of the learning ability of neural networks and information regarding the accumulated operational database. This "Adaptive Training Approach" is described in the following sections.

For teleseismic phases (T), training samples are randomly selected from the analyst reviewed database, of which the corresponding arrivals have no onset-time corrections between the automatic and the analyst reviewed databases. This restriction ensures that the 15 attributes of signals were calculated accurately from proper time windows.

Because the analyst reviewed database has fewer regional phases, the training samples for regional P- and S-type phases (P and S) are selected from the analyst reviewed database for those arrivals with onset-time corrections less than two seconds between the automatic and the analyst reviewed databases. This looser restriction compared with the T-type phase selection allows the use of more regional phases at the limited risk of less accurate signal attributes.

Because no ground-truth information is available for noise phases, training samples are randomly selected from N phases that were classified to noise detections in the automatic database but were not renamed/associated in the analyst reviewed bulletin. This initial selection assumes that the existing neural network classified the noise detections reasonably well after excluding phases that were renamed by analysts. Most of the selected N phases in the initial training set are likely to be true noise detections, although some could be seismic phases that were not associated in the analyst reviewed bulletin.

During the training process, a subset of the selected data is used for the supervised learning. A separate subset of data is used for testing after each training iteration. The comparison between input and output of the trained neural network can be written as a matrix:

	Output Signals	Output Noise
Input Signals	N_{11}	N_{12}
Input Noise	N_{21}	N_{22}

N_{11} is the number of "true" signal detections correctly classified by the neural network; and N_{22} is the number of "true" input noise detections correctly classified. The off-diagonal numbers of the matrix, N_{12} and N_{21}, reflect phase confusion by the trained neural network at each cycle of the training iterations. Ideally, for a training data set, output of the neural network that is being trained would converge to a perfect "correct rate" (100%) after a certain number of iterations. In reality, however, training patterns of two classes may not be selected perfectly correct; the neural network cannot distinguish two classes, therefore the "correct rate" of the neural network output only converges to a percentage value.

As a starting point of training the neural network for classification of signals and noise (stage 1) in this experiment, the noise training patterns were selected from the automatic database under some assumptions. The initial training set of noise phases might be contaminated by unassociated seismic signals, so that the neural network cannot distinguish them from the training patterns of the seismic signals during the training process. The initial training set for station STKA consists of 905 patterns (two thirds of the total), and the testing set consists of 453 patterns (one third of the total). Figure 6 shows the correct rate of training and test results versus the number of iterations for the initial training data set of the station STKA. Correct rates of neural network output converge after about 10,000 iterations: 87% of the training patterns was correctly identified, while 77% of the testing patterns was correctly identified. Specifically, the output of the trained neural network is listed in Table 2. Ninety-four patterns (group N_{21}) in the noise training class were classified to signal class by the trained neural network. Twenty-four patterns (group N_{12}) in the signal training class were classified to noise class by the trained neural network. The confusion matrix for the training result infers that the initial training set had some misclassified patterns.

As an adaptive step, the confused training patterns in group N_{21} were deleted from the initial training set. This group of detections may have similar features with the ground-truth signals and is likely to be unassociated seismic phases. The noise training data set will be purified after this group of detections is removed. The confused training patterns in group N_{12} are still kept as these confusions might be caused by the existence of group N_{21}. The remaining training data is called the

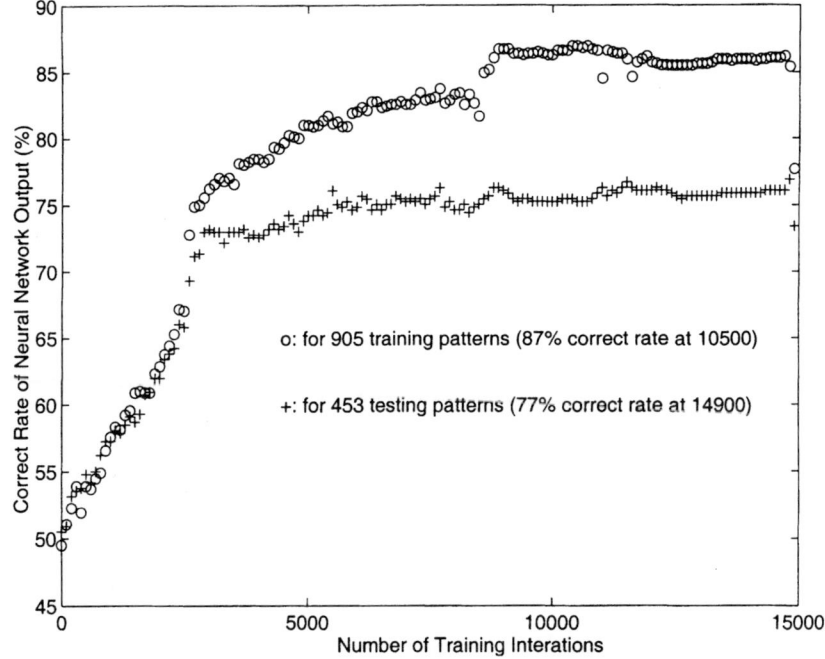

Figure 6

Training and test results of trained neural network (signals versus noise) with initial training set. Output of trained neural network converges after a certain number of learning iterations, and can only correctly classify 87% of the training data set and 77% of the testing data set.

Table 2

Confusion matrix of stage-1 neural network with initial training data set. Each row in the table is the known class of the training data result, and each column is the output of the trained neural network. Off-diagonal elements are misclassification numbers

	Output signal class	Output noise class	Total	Correct rate (%)
Input signal class	428	24	452	94.7
Input noise class	94	359	453	79.2
Total	522	383	905	87.0

"adaptive-1" training data set. In this experiment the size of the training data set is reduced from 905 to 811. The neural network is retrained with the adaptive-1 training set. Figure 7 shows the new results of the neural network for the adaptive-1 training data set. The output of the neural network converges after about 3000 iterations. The correct rate for the training data set increased from 87% to 99% in the initial training step. The neural network must classify training patterns correctly after the learning process. More importantly, the trained neural network can classify

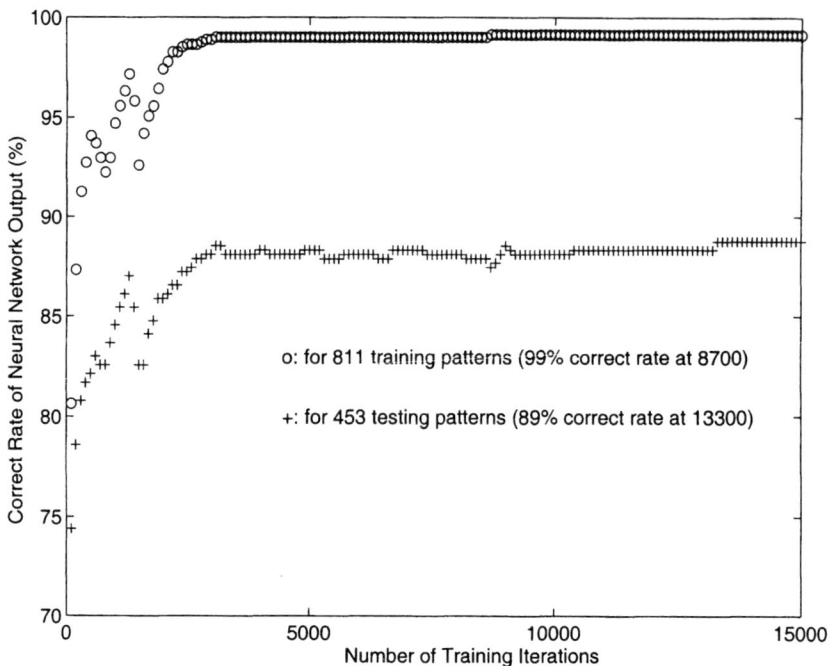

Figure 7
Training and test results of trained neural network (signals versus noise) with adaptive-1 training set. Output of the trained neural network now converges to 99% correct rate for the training data set and 89% for the testing data set.

the unlearned patterns correctly. The retrained neural network was tested by using the same testing data set as in the previous step. The correct rate for the testing data set increased from 77% to 89%. This increase was a significant improvement for classification. This improvement in classification with the adaptive-1 training data set does not indicate that better performance can be achieved with less information (fewer training patterns). In contrast, the proposed multi-step adaptive training process uses all of the information more efficiently.

Figure 7 shows that the adaptive-1 training data set still contains confused output. The next adaptive step is to delete the confused samples in both off-diagonal groups, N_{21} and N_{12}, from the adaptive-1 training data set. The group N_{12} is deleted in this step because the contaminating effect of the initial noise data set is assumed to have been minimized, and thus those small number of patterns is outliers of the signal class. The number of training data is reduced from 811 to 804. This reduced data set is called "adaptive-2" training data set. The neural network is retrained with the adaptive-2 training data set and tested with the same testing data set. The correct rate for the training data set increased to 100%, as shown in Figure 8. The correct rate for the same testing data set, however, is the same as in the previous adaptive

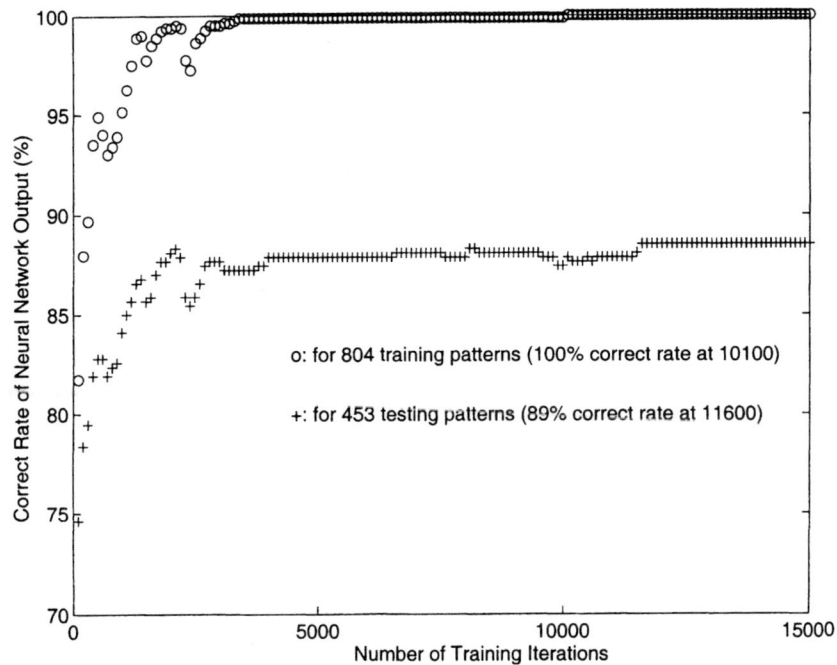

Figure 8
Training and test results of the trained neural network (signals versus noise) with adaptive-2 training set. Output of the trained neural network now converges to 100% correct rate for the training data set and 89% for the testing data set.

step, adaptive-1. This means that the second adaptive step has only a minor impact on the training results.

For stages 2 and 3 of initial phase identification (*T* and *P* versus *S*, and *T* versus *P*), a standard training procedure is used because all selected training patterns derive from the "ground-truth" signal database.

After training all three stages of neural networks, three sets of new weights of neural networks for initial phase identification were generated for station STKA. The new weights were applied to 130 days of data of station STKA in the analyst reviewed bulletin. This is the same data for which the performance of default weights were evaluated in Table 1. The initial wave type confusion matrix of the new output is shown in Table 3.

A comparison of the matrixes in Table 1 and 3 shows that a significant improvement has been obtained by the retrained neural networks. The "correct rate" of the output of the newly trained neural network compared to the default neural network has increased from 25.6% to 81.7%. The "*N*-phase rate" has decreased from 22.8% to 14.8%. While the number of *T*-type phases confused as *P*-type phases has been reduced from 1381 to 50, the number of *P*-type phases confused as *T*-type

Table 3

Confusion matrix of initial phase classification using new weights of neural networks with adaptive training set (1996/07/19–1996/11/25). Each row in the table is the result of the automatic system, and each column is the final result confirmed by analysts. Off-diagonal elements are misclassification numbers

		Results by analyst			Sum
		T	P	S	
Results by system	T	2236	6	1	2243
	P	50	8	0	58
	S	37	1	8	46
	N	400	3	5	408
	Sum	2723	18	14	2755

phases has increased marginally from 3 to 6. That is, the major decrease in confused *T*-type phases has been achieved at the cost of a small increase in confused *P*-type phases.

In summary, the "Adaptive Training Approach" has efficiently achieved the goal of updating default neural networks for initial wave-type classification with little human effort in constructing ground-truth databases.

5. Validation by the Ground-truth Noise Database

The neural networks that were retrained by the proposed "Adaptive Training Approach" outperform the default weights of neural networks for initial phase identification. This section will further validate that this approach performs against an analyst-reviewed ground-truth noise database constructed by analysts.

At the Center for Monitoring Research, SAIC, a ground-truth noise database for station STKA was constructed for validating the proposed "Adaptive Training Approach." When constructing the ground-truth noise database, the objective of the analyses was to classify automatic system detections as "noise" or "not-noise." To minimize ambiguity in noise selection, detections that had some "not-noise" characteristics were considered as being "not-noise," and therefore were not included in the noise database, even though the detections would not have passed a "real signal" threshold for inclusion in the final analyst reviewed bulletin. Numerous detections classified as "noise" by the automatic system were in fact considered to be "not-noise" of one variety or another. Other detections that were classified as "not-noise" by the automatic system were judged to be noise, and were reclassified as such during analysis. The noise detections included activity judged to be spikes, detections with little apparent distinction from preceding background in either amplitude or period content, and detections with very low vertical/horizontal coherency. These measurements were analyzed across at least three filter bands (0.5–2.0, 1.0–2.5 or 1.5–

3.0, and 3.0–6.0 Hz) that are commonly used to differentiate signal activity from noise during routine reviewing analysis. For station STKA, 898 noise phases were identified by the analysts after reviewing the waveform data of a five-day period during March 19–22, and March 27, 1997.

From this ground-truth noise database and the analyst reviewed bulletin, a training data set for the stage signals (*TPS*) versus noise (*N*) was constructed by random selection from the ground-truth databases. The number of training patterns and the ratio of signal detections to noise phases in the training set were selected intentionally to be about the same as those of the training data set that were automatically selected using the adaptive approach. The training set consists of 829 patterns, and the testing set consists of 426 patterns. The neural network for classification of signals and noise was trained with the ground-truth training data set. Figure 9 shows correct rates of output versus the number of training iterations for the neural network, which was trained by the ground-truth databases. For the given training data set and testing data set, the trained neural network converges to 95% and 84%, respectively, which are worse than the correct rates of the adaptively trained neural network.

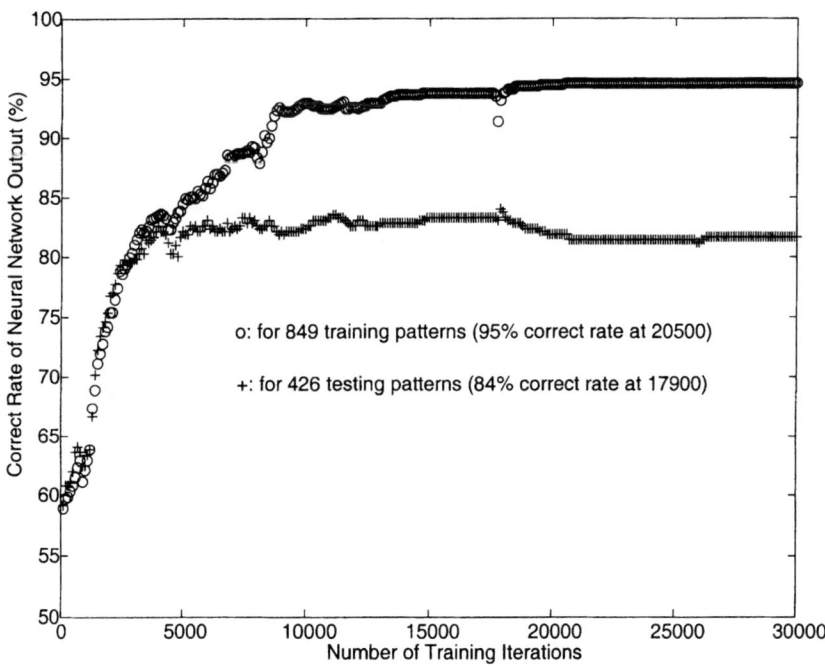

o: for 849 training patterns (95% correct rate at 20500)

+: for 426 testing patterns (84% correct rate at 17900)

Figure 9
Training and test results of trained neural network (signals versus noise) with ground-truth training set. Output of the trained neural network converges to 95% correct rate for the training data set and 84% for the testing data set.

Using the new weights of the neural network trained by the ground-truth data set, the same 130 days of data of STKA were tested again. The initial phase confusion matrix of the new output is listed in Table 4. Combined with Tables 1, 3, and 4, Table 5 gives a summary of the performance for different weights of neural networks. In Table 5, the Default column represents the performance of the default neural networks, New-1 represents the performance of the neural networks trained by the adaptive training approach, and New-2 represents the performance of the neural networks trained by the ground-truth training set. From the table we can see that the performance of initial phase identification is similar for both sets of the new weight of neural networks. However, for the output of the New-2 neural networks, more seismic phases were incorrectly identified as noise phases than for the New-1 neural networks. There are two possible explanations to these results. Firstly, the ground-truth noise database was selected from only a few days, while the adaptive training approach randomly selected training patterns over a considerably longer time period. The latter would better reflect the seasonal variation of noise detections. Secondly, the consistency with which human beings analyze noisy data is limited by analyst skills and experience.

To further illustrate the effect of the Adaptive Training Approach, the Mahalanobis distances (DUDA and HART, 1973) between the four wave types for

Table 4

Confusion matrix of initial phase classification using new weights of neural networks with ground-truth training set (1996/07/19–1996/11/25). Each row in the table is the result of the automatic system, and each column is the final result confirmed by analysts. Off-diagonal elements are misclassification numbers

		Results by analyst			Sum
		T	P	S	
Results by system	T	2214	5	1	2220
	P	42	7	0	49
	S	22	0	3	25
	N	445	6	10	461
	Sum	2723	18	14	2755

Table 5

Summary of testing using different NNET-weights for STKA. The 'Default' columns represent the performance of the initial default neural networks, 'New-1' and 'New-2' represent the performance of the neural networks trained by the adaptive training approach, and by the 'ground-truth' data set, respectively

Number of phases	Correct rate (%)			N-phase rate (%)		
	Default	New-1	New-2	Default	New-1	New-2
2755	25.6	81.7	80.7	22.8	14.8	16.7

Table 6

Mahalanobis distances between four wave types. The final noise data set selected using the adaptive approach gives better distinction, i.e., longer distance, with signals

	P	S	Initial noise data set	Final noise data set	Analyst-selected noise data set
T	4.2932	11.5904	4.6344	6.1566	5.4520
P		9.1040	4.9523	6.2631	5.5160
S			2.6378	2.6539	2.3352
Signals (T, P, S)			2.5044	3.5499	2.9467

each training set are listed in Table 6. The Mahalanobis distance between two N-dimension populations is defined as follows:

$$MD = \sum_{j=1}^{N} \frac{(\mu_{j1} - \mu_{j2})^2}{\sigma_{j1}^2 + \sigma_{j2}^2} .$$

It is a measure of the statistical distance between two classes based on their normal probability densities. In Table 6, the number in each cell represents Mahalanobis distance between the wave types in the corresponding row and column. Column 3 represents the Mahalanobis distances between signal wave types and the initial noise data set, which was automatically selected before training neural networks. Column 4 represents the Mahalanobis distances between signal wave types and the final noise data set, which was adaptive selected during the training process. Column 5 represents the Mahalanobis distances between the noise data selected by analysts and signal wave types. A comparison of column 5 with columns 3 and 4 indicates that the initial noise training data set has a closer distance to the signals than that of the analyst-selected noise data set, and the final noise data set has the further distance than that of the analyst-selected noise data set. This geometrical presentation demonstrates the adaptive characteristic of the proposed training approach.

6. Summary

The neural network applications developed for IMS stations have the ability to automatically identify for high-dimensional input data. The adaptive learning ability of neural networks makes retraining possible and easy for continuous operational systems.

The selection of training patterns is a key factor for the successful application of neural networks. The training of neural networks must accommodate the imperfect knowledge of the noise because no noise detections have been reviewed by analysts in the routine operation. Compared to seismic wave types (Teleseism, Regional P, and Regional S), the selection of a noise training set is considerably more difficult and requires considerable analyst effort. An adaptive training approach to automatically select training data sets has therefore been proposed in this paper. This approach uses the adaptive learning ability of neural networks and information stored in the automatic operational database.

The neural networks for initial wave-type classification have been trained using the proposed approach for station STKA, and thus specific neural network weights have been generated. Testing result of the new weights of the retrained neural networks shows significant improvements with the "correct rate" increasing from 25.6% to 81.7%.

The proposed "Adaptive Training Approach" has been validated by the ground-truth noise database, which was constructed manually by analysts. Comparisons of the performance of phase identification and the statistical distribution of training patterns demonstrate that the proposed approach is efficient and reliable.

Acknowledgements

I am indebted to Hans Israelsson, Robert G. North, Keith Mclaughlin, and Chris Ferraro for their remarkable effort to enhance the manuscript of this paper. I especially thank Michael Clark and Richard A. Reed for their work in constructing the ground-truth noise database used in this work. I am also grateful to Tom Bache, Greg Beall, Roger Bowman, Rick Jenkins, Tom Sereno, and Henry Swanger for their invaluable assistance and suggestions. Dr. Zontan A. Der and two anonymous reviewers made valuable suggestions to improve this article. This work was supported by the U. S. Department of Defense, Defense Threat Reduction Agency (Contract Number DTRA-99-C-0025).

REFERENCES

ANANT, K. S. and DOWLA, F. U. (1997), *Waveform Transform Methods for Phase Identification in Three-component Seismograms*, Bull. Seismol. Soc. Am. *87*, 1598–1612.

BACHE, T. C., BRATT, S. R., WANG, J., FUNG, R. M., KOBRYN, C., and GIVEN, J. W. (1990), *The Intelligent Monitoring System*, Bull. Seismol. Soc. Am. *80*, 1833–1851.

CICHOWICZ, A. (1993), *An Automatic S-phase Picker*, Bull. Seismol. Soc. Am. *81*, 180–189.

DUDA, R. O. and HART, P. E. *Pattern Recognition and Classification*, (John Wiley, New York, 1973) 482 pp.

JURKEVICS, A. (1988), *Polarization Analysis of Three-component Array Data*, Bull. Seismol. Soc. Am. *78*, 1725–1743.

ROBERTS, R. G., CHRISTOFFERSSON, A., and CASSIDY, F. (1989), *Real-time Event Detection, Phase Identification and Source Location Estimation Using Single-station Three-component Seismic Data*, Geophys. J. *97*, 471–480.

RUMELHART, D. E., MCCLELLAND, J. L., and the PDP RESEARCH GROUP (1986), *Learning Representations by Back-propagating Errors*, Nature *332*, 533–536.

SERENO, T. and PATNAIK, G. (1993), *Initial Wave-type Identification with Neural Networks and its Contribution to Automated Processing in IMS Version 3.0*, Technical Report, SAIC-93/1219.

SUTEAU-HENSON, A, (1991), *Three-component Analysis of Regional Phases at NORESS and ARCESS: Polarization and Phase Identification*, Bull. Seismol. Soc. Am. *81*, 2419–2440.

WANG, J. and TENG, T. L. (1995), *Artificial Neural Network-based Seismic Detector*, Bull. Seismol. Soc. Am. *85*, 308–319.

WANG, J. and TENG, T. L. (1997), *Identification and Picking of S Phase using an Artificial Neural Network*, Bull. Seismol. Soc. Am. *87*, 1140–1149.

ZURADA, J. M. *Introduction to Artificial Neural Systems* (West Publishing Company, St. Paul, 1992) pp. 163–250.

(Received May 20, 1999, revised May 15, 2000, accepted May 29, 2000)

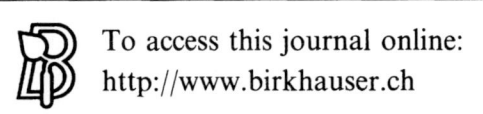

To access this journal online:
http://www.birkhauser.ch

Infrasound

Pure appl. geophys. 159 (2002) 1045–1062
0033–4553/02/051045–18 $ 1.50 + 0.20/0

❙ Pure and Applied Geophysics

Constraints on Infrasound Scaling and Attenuation Relations from Soviet Explosion Data

J. L. Stevens,[1] I. I. Divnov,[2] D. A. Adams,[1] J. R. Murphy,[1] and V. N. Bourchik[2]

Abstract—The Institute for the Dynamics of the Geospheres (IDG) in Moscow, Russia, contains an archive of infrasound recordings from Soviet atmospheric nuclear tests that were conducted in 1957 and 1961, and has digitized the highest quality records from this data set. We have measured the infrasound signals from these records and compared them with previously developed scaling and attenuation relations. We find that the data are in best agreement with a scaling and attenuation relation developed by the Los Alamos National Laboratory (LANL) which can be written as $\log P = 3.37 + 0.68 \log W - 1.36 \log R$ where P is zero to peak pressure amplitude in Pascals, W is the yield in kilotons, and R is the source to receiver distance in kilometers. We use the scaling relations to define an infrasound magnitude, and to estimate the detection capability of the International Monitoring System (IMS) being developed as part of the Comprehensive Nuclear-Test-Ban Treaty (CTBT). The detection threshold for the proposed 60-station IMS network is estimated to be slightly higher than the CTBT design goal of 1 kiloton in some locations.

Key words: Infrasound, scaling relation, attenuation, atmospheric explosion, nuclear explosion.

Introduction

The International Monitoring System (IMS) required by the Comprehensive Nuclear-Test-Ban Treaty (CTBT) will include sixty infrasound stations designed to detect atmospheric nuclear explosions. A design goal is that the system be able to detect and locate explosions as small as one kiloton anywhere in the world. In order to estimate the capability of the infrasound network, it is necessary to be able to predict the amplitude of an infrasound signal at any location, and to evaluate whether the signal would be detectable above noise levels at the recording stations. Scaling and attenuation relations are empirical and/or theoretical equations that relate the amplitude and period of infrasound signals to the explosion yield and source to receiver distance. Several different relations have been developed, based on theoretical infrasound modeling, and on recordings of atmospheric nuclear and chemical explosions, however the different scaling relations imply very different

---

[1] Science Applications International Corp., 10260 Campus Point Drive, San Diego, CA 92121, USA.
[2] Institute for the Dynamics of the Geospheres, Moscow, Russia.

detection threshold levels. In this paper, we use a data set of infrasound recordings from Soviet explosions to put constraints on these scaling relations and to estimate the detection capability of the IMS.

Scaling and Attenuation Relations

PIERCE and POSEY (1971) developed a solution for the excitation of the Lamb edge mode as an approximation for infrasound signals generated by atmospheric tests. They showed that this relatively simple approximation provided waveforms that agreed well with the first few cycles of observed waveforms from multi-megaton nuclear tests. From this they developed the relation between the yield W in kilotons, zero to peak pressure P in Pascals, the period of the first cycle of the waveform T, and the angular distance Δ given by:

$$W = 0.494P(\sin \Delta)^{1/2}H_s(cT)^{3/2} \ , \tag{1}$$

where c is the sound speed, H_s is the atmospheric scale height $c^2/\gamma g$, where γ is the adiabatic expansion constant, approximately 1.4 for air, and g is the acceleration of gravity. POSEY and PIERCE (1971) showed that this relation agreed well with a data set of observations of pressure and period measurements from large atmospheric nuclear explosions.

The derivation of equation (1) uses the approximation that the period being measured is much longer than the source duration. Pierce and Posey estimated the source duration to be approximately $T_s = 0.33W^{1/3}$ sec, which is only about 13 sec even for a 58 megaton explosion, and therefore always considerably less than the measured infrasound period. This approximation effectively makes the period independent of the yield and proportional to the cube root of distance. For the typical sound velocity (0.318 km/sec) and scale height (8 km) referenced in the paper, the period of the first cycle of the Lamb wave can be written as

$$T = 13.4R^{1/3} \ , \tag{2}$$

where R is the source to receiver distance in kilometers and T is in seconds. Equation (2) can then be combined with equation (1) to give

$$W = 34.8PR^{1/2}(\sin \Delta)^{1/2} \ . \tag{3}$$

The pressure is therefore predicted to be proportional to yield at a fixed distance and to decrease approximately inversely with distance. As can be seen from the data in the following section, these predictions are inconsistent with the data which scales decidedly more slowly than linearly with yield and exhibits a clear frequency dependence with yield.

Equation (1) has been widely used (MCKISIC, 1997) and proves to be in good agreement with the data when the observed period is used, even though the usage is inconsistent with the derivation of equation (1). BLANDFORD and CLAUTER (1995) modified equation (1) by assuming that the period T is proportional to the cube root of yield and independent of distance. From this it follows that pressure is proportional to the square root of yield and decays as the square root of distance from the source. BLANDFORD and CLAUTER (1995) used the square root of distance decay at distances less than 20 degrees, but modified the attenuation relation to have an exponential form at distances greater than 20 degrees to match the data set of WEXLER and HASS (1962). Using a larger data set, keeping the same yield scaling, but modifying the attenuation relation, CLAUTER and BLANDFORD (1998) derived the relation:

$$\log P = 0.92 + 0.5 \log W - 1.47 \log \Delta \ , \tag{4}$$

where Δ is the source to receiver distance in degrees, W is the yield in kilotons, and P is the zero to peak pressure amplitude in Pascals. They showed that this relation was consistent with a historical data set of nuclear and chemical explosions.

WHITAKER (1995) derived the following relation based on wind-corrected infrasound measurements from Los Alamos chemical explosions:

$$P_c = 2.35 \times 10^3 (R/W^{1/2})^{-1.36} \ , \tag{5}$$

where P_c is the wind corrected zero to peak pressure in Pascals, R is the source to receiver distance in kilometers, and W is the yield in kilotons (the original reference stated that yield is in tons, but should be kilotons (Whitaker, personal communication)). The wind correction applied in equation (5) is

$$P_c = 10^{-0.019v} P \ , \tag{6}$$

where v is the component of the stratospheric wind velocity in meters/second in the direction of wave motion and P is the measured pressure.

Russian scientists at IDG have used the following relationships to make approximate yield estimates:

$$P_m = k_p W^{1/3}/R \tag{7}$$

$$T = k_T W^{1/3}/R \tag{8}$$

$$\omega_m = k_\omega/W^{1/3} \tag{9}$$

where P_m is the zero to peak pressure in Pascals, T is the signal duration from the first arrival to the moment when the signal degrades to noise level, and 90% of the signal energy is contained in the frequency band below angular frequency ω_m. The three constants are empirically determined and are given in Table 1.

Table 1

Russian scaling constants

	k_p Pa-km/kt$^{1/3}$	k_T s/kt$^{1/3}$km	k_m kt$^{1/3}$/s
Downwind	2000	0.050	1.6
Perpendicular to wind	1000	0.025	1.2

The equations above can be written in a consistent format as:

$$\log P = -1.54 + \log W - 0.5 \log(R \sin \Delta) \quad \text{PIERCE and POSEY (1971)} \quad (10)$$

$$\log P = 0.92 + 0.5 \log W - 1.47 \log \Delta \quad \text{AFTAC(CLAUTER and BLANDFORD, 1998)} \quad (11)$$

$$\log P = 3.37 + 0.68 \log W - 1.36 \log R \quad \text{LANL(WHITAKER, 1995)} \quad (12)$$

$$\log P = 3.00 + 0.33 \log W - \log R \quad \text{Russian–Crosswind} \quad (13)$$

$$\log P = 3.30 + 0.33 \log W - \log R \quad \text{Russian–Downwind} \quad (14)$$

where P is zero to peak pressure in Pascals, W is yield in kilotons, R is distance in kilometers, and Δ is distance in degrees.

These scaling relations have very different implications for the infrasound detection threshold. Yield estimates for the threshold pressure level differ by several orders of magnitude, even though each relation was constrained by some infrasound data set. Table 2 shows the calculated yield at a nominal detection threshold of 0.1 Pascal for each of the scaling relations. The most important factor is the exponent in the pressure/yield relation. With the Pierce/Posey relation, which has a yield exponent of 1, pressure drops off far more rapidly with yield than with the other relations, leading to a very high threshold level. The 0.33 yield exponent in the Russian relations, however, implies a very slow decrease in pressure with yield and leads to very low threshold levels. The Whitaker and Clauter/Blandford relations, which have yield exponents of 0.68 and 0.5, respectively, predict intermediate threshold levels.

Table 2

Detection capability for a nominal detection threshold of 0.1 Pascal. The table shows yield in kilotons for each scaling relation and source to receiver distances from 10–60 degrees

	10	20	30	40	50	60
Clauter/Blandford	0.13	0.97	3.2	7.4	14	24
Whitaker	0.46	1.8	4.2	7.4	11	17
Russia Crosswind	0.0014	0.01	0.037	0.088	0.17	0.30
Russia Downwind	0.0002	0.0014	0.0046	0.011	0.02	0.037
Pierce/Posey	47	93	140	180	220	260

Soviet Infrasound Data

The Institute for the Dynamics of the Geospheres (IDG) in Moscow, Russia, contains an archive of approximately 300 recordings from 34 Soviet atmospheric nuclear tests that were conducted in 1957 and 1961. 20 of these explosions were located at the Novaya Zemlya test site, 12 at Semipalatinsk, and 2 at Kapoustin Yar (see Fig. 1). Of these, 233 recordings from 27 of the tests recorded at stations from 1000 to 5000 km were found to be of adequate quality for digitization. The yields of these tests range from 0.4 KT to 58 MT. The data set includes two high altitude explosions and the largest (58 megaton) atmospheric explosion ever detonated. The explosions corresponding to this data set are listed in Table 3. The number of records listed in the table is the total number of records digitized for each event.

Seventeen stations recorded data from these tests. These stations are listed in Table 4, and a map showing the paths from each event to each station in the current data set is shown in Figure 1. Data were recorded on instruments with varying low- and high-pass filters. The amplitude responses of three representative instrument types are shown in Figure 2. IDG has put considerable effort into identifying the instrument parameters that were in use at each of the recording stations. Some information regarding the data is not available. Absolute times are not known for any of the waveforms. For two of the stations, the source to receiver distance is known, but only the general area is known for the station location.

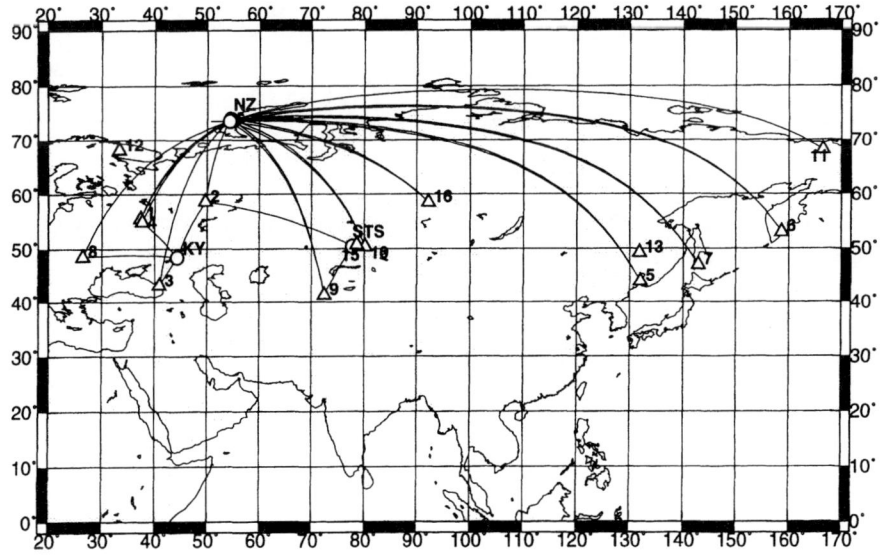

Figure 1
Stations recording infrasound signals from atmospheric explosions at three Soviet nuclear test sites, Semipalatinsk (STS), Kapoustin Yar (KY), and Novaya Zemlya (NZ).

J. L. Stevens *et al.*

Pure appl. geophys.,

Table 3

Soviet atmospheric nuclear explosions and the number of records for each event that have been digitized by IDG. STS is the Semipalatinsk test site, KY is Kapoustin Yar, and NZ is Novaya Zemlya

Test site	Explosion number	Date	Time (Moscow)	Latitude	Longitude	Height of burst (m)	Yield (KT)	Number of records
STS	088	1961/09/04	08:00:27	50.45	77.74	725	9	2
STS	089	1961/09/05	09:00:05	50.45	77.74	500	16	3
KY	091	1961/09/06		48.45	44.30	22700	11	6
STS	094	1961/09/10				390	12	2
NZ	095	1961/09/10	12:00:14	73.52	54.30	2000	2700	23
NZ	099	1961/09/12	13:08:00	73.52	54.30	1190	1150	23
STS	100	1961/09/13	08:01:55.8	50.45	77.75	710	14	1
NZ	102	1961/09/14	12:56:16	73.52	54.3	1700	1200	17
STS	103	1961/09/14	08:59:59.4	50.35	77.82	0.5	0.4	1
STS	105	1961/09/17	10:00:46.6	50.45	77.75	695	21	1
NZ	111	1961/09/20	11:12:12	73.52	54.30	1600	2000	6
STS	112	1961/09/21	17:01:01.6	50.33	77.70	110	0.8	2
NZ	113	1961/09/22	11:11:00	73.52	54.30	1300	260	4
STS	114	1961/09/26	10:01:19.8	50.45	77.75	665	1.2	2
NZ	116	1961/10/02	13:30:50	73.92	54.55	1500	250	13
STS	117	1961/10/04	10:01:19.9	50.44	77.76	605	13	1
NZ	118	1961/10/04	10:30:55	73.52	54.30	2100	3000	13
KY	119	1961/10/06		48.45	44.30	41300	40	4
NZ	120	1961/10/06	10:00:08	73.52	54.30	2700	4000	20
STS	122	1961/10/12	08:31:03.6	50.45	77.75	670	15	1
STS	123	1961/10/17	10:00:00.8	50.45	77.75	505	6.6	2
STS	124	1961/10/19	08:30:42.6	50.45	77.73	710	10	2
NZ	125	1961/10/20	11:07:03	73.52	54.30		1450	17
NZ	126	1961/10/23	11:31:22	73.5	54.3	3500	12500	21
NZ	128	1961/10/25	11:31:05	73.52	54.3	1450	300	10
NZ	133	1961/10/30	11:33:27	73.52	54.30	4000	58000	21
NZ	147	1961/11/04	10:20:23	73.5	54.3	1750	150–1500	20

Some examples of infrasound records are shown in Figure 3. These records are from tests that range from 8 kilotons to 58 megatons in yield. The frequency content of the signals changes dramatically over this range.

Measurement of Russian Data

All of the Russian data were carefully measured in a consistent manner. The data were first filtered to remove long- and short-period noise outside the frequency band of the data. A Butterworth filter was used with corner frequencies of 0.01 and 0.2 Hz for events with yield less than 100 kilotons, 0.002 and 0.1 Hz for events with yields between 100 kilotons and 2 megatons, and 0.001 and 0.1 Hz for events with yield greater than 2 megatons. The amplitude and period were measured as half the maximum peak to peak amplitude and twice the time difference between the peak

Table 4

Stations recording Soviet atmospheric tests

Station num.	Station name	Latitude	Longitude	Novaya Zemlya distance (km)	Semipalatinsk distance (km)	Kapoustin Yar distance (km)
1	Troitskoe	55.5	37.3	2200	2700	900
2	Kirov	58.6	49.7	1700	2100	1100
3	Esheri	43.1	40.9	3500	2900	800
4	Mikhnevo	54.9	37.7	2200	2700	900
5	Oussouriysk	43.9	132.0	5000	4100	6300
6	Petropavlovsk	53.0	158.7	4850	5300	7100
7	Yuzno-Sakhalinsk	47.0	142.8	5000	4600	6700
8	Zvanets	48.5	26.4	3150	3650	1400
9	Mayly-say	41.3	72.5	3800	1140	2250
10	Semipalatinsk	50.4	80.2	2850	100	2500
11	Bilibino	68.1	166.5	3450	4900	6150
12	Olenegorsk	68.1	33.3	1000	3100	2300
13	Kooldour	49.2	131.8	4430	3800	5900
14	Kazakhstan			2950		
15	Semipalatinsk Reg.	50.7	78.6	2800	100	2450
16	Yeniseysk	58.5	92.2	2350	1300	3300
17	Leningrad Reg.			1850		

and trough, respectively. Measurements were made on both the acoustic wave and the low frequency Lamb wave if possible. Only data with known instrument responses were measured, and a digital correction for the instrument response was made at the observed period. Some of the data consisted of measurements of the same arrival recorded on different instruments. Measurements were made on each of

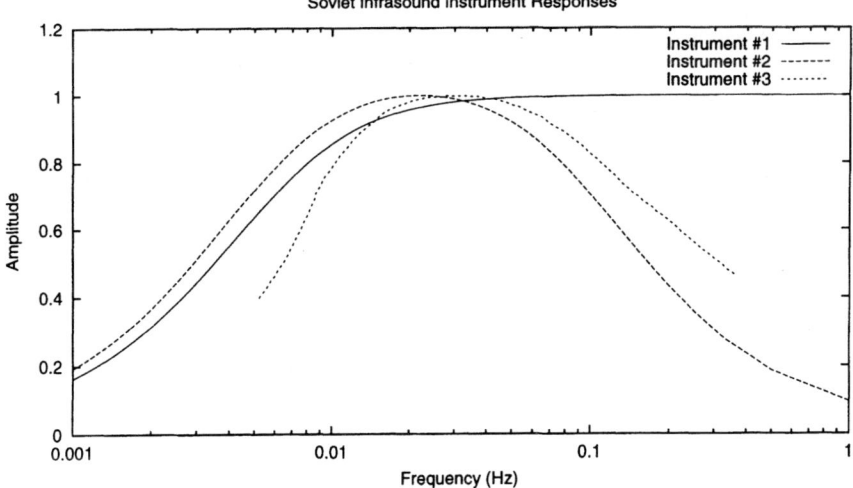

Figure 2
Instrument response curves for Russian infrasound recordings.

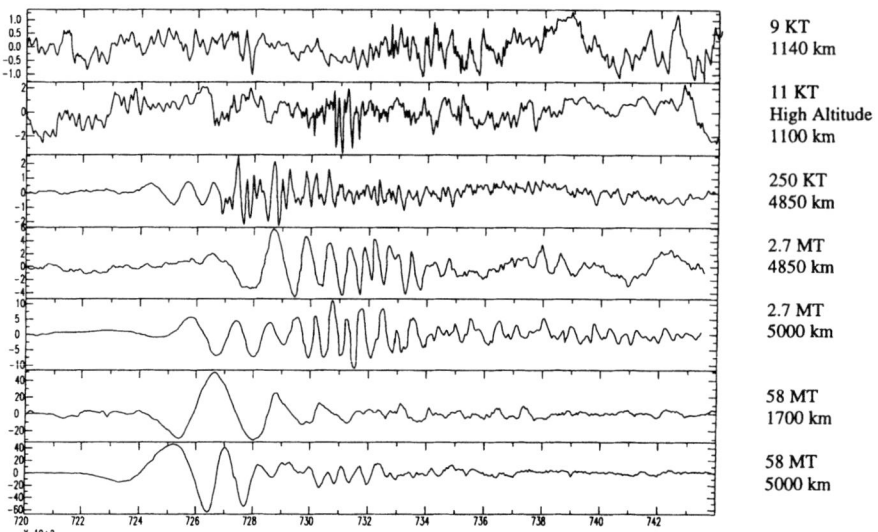

Figure 3
Infrasound signals from Soviet atmospheric tests. The top record is from test 088 at station 9, the second is from the high altitude test 091 at station 2, the third is from the test 116 at station 6, the fourth and fifth are from test 095 at stations 6 and 7, and the last two records are from the test 133 at stations 2 and 7. Amplitudes are in Pascals. All of these traces were recorded on instrument #2 except those for event 133, which were recorded on instrument #1.

these multiple records when the quality was comparable, and there are significant differences between the measurements, particularly for the measured period. A few signals with apparent calibration errors were not used. The final result is that measurements were made on a total of 133 waveforms. 107 acoustic waves from 17 events and 96 Lamb waves from 9 events were measured.

Figure 4 presents a comparison between pressure measurements made from the Russian data and equations (10)–(14). It is not possible to put them all on the same plot because the scaling relations have different functional forms. All pressures are zero to peak amplitudes in Pascals. The AFTAC relation (equation (11)) fits the lower yield data quite well. The Pierce/Posey relation (equation (10)) is a fairly good fit to the high yield Lamb wave data, but does not fit the acoustic wave data, particularly for the lower yield events. The LANL relation (equation (12)) appears to fit the data very well over the entire scaled range, although there is considerable scatter about the line. The pressure measurements have not been wind corrected since we have no information about wind conditions at the time of the tests. The Russian relations (equations (13)–(14)) also fit the data reasonably well with the crosswind equations matching the lower amplitude data and the downwind equations matching the higher amplitude data, however the observed data falls below the predicted curves for larger scaled ranges, and the data points for the largest yield events are well above the curve, while the lower yield events lie below the

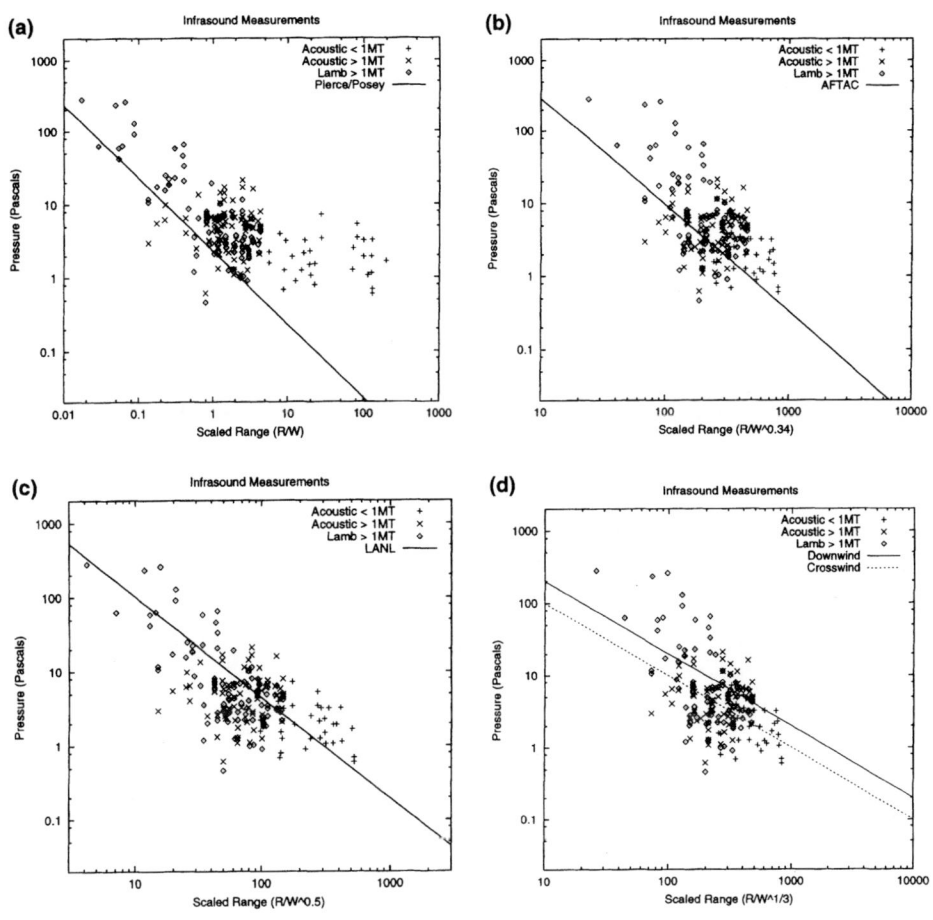

Figure 4
Comparison of Russian data with scaling relations. Top left is the Pierce/Posey relation equation (10), top right the AFTAC relation (Clauter and Blandford) equation (11), bottom left is the LANL relation (Whitaker) equation (12), and bottom right are the Russian relations equations (13) and (14).

curve. We conclude from this that the pressure/yield slope of 0.33 used in the Russian relation is too small, and the pressure/yield slope of 1.0 used in the Pierce/Posey relation is too large. The data is consistent with the intermediate slopes of 0.5–0.68 of the AFTAC and LANL relations, with the LANL relation fitting the data over the widest range.

Figure 5 shows the measured period plotted vs. yield for all data, together with lines showing cube root scaling of period with yield with the best fit to the acoustic and Lamb data taken as independent data sets. The cube root scaling approximately fits the data, however there is substantial scatter and the slow increase of period with yield would cause considerable uncertainty in any yield estimate based on measured period.

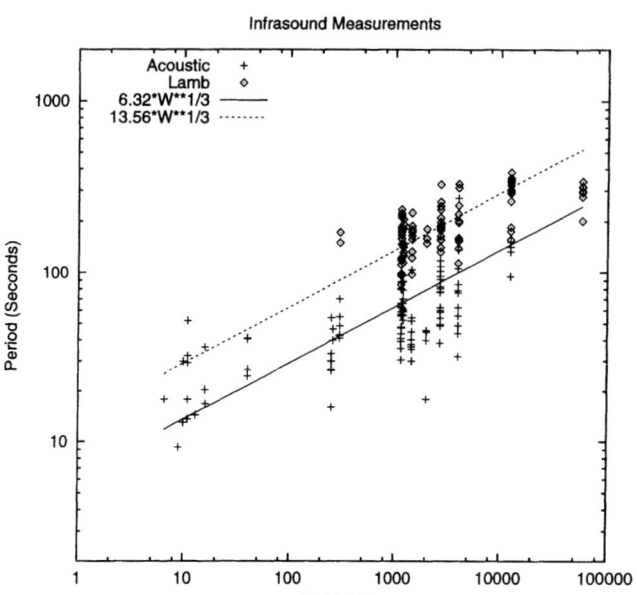

Figure 5
Period vs. explosion yield for all data.

Spectral Measurements

Additional insight into the scaling laws can be obtained by examining the spectra of arrivals with different yields at the same station. Station 7, at Yuzno-Sakhalinsk, recorded four atmospheric explosions with yields of 250, 1450, 2700, and 58,000 kilotons. The spectra of these four arrivals are shown in Figure 6. The shape of the spectra changes dramatically over this yield range, with considerable low frequency energy at higher yields.

Figure 7 indicates the change in amplitude for three frequencies plotted as a function of yield, showing that the slope of the log amplitude vs. log yield curve changes as a function of frequency. Figure 8 shows the slope of the amplitude vs. yield curve (as shown in Fig. 7) plotted as a function of frequency. This figure shows that the slope of the amplitude/yield curve is strongly frequency-dependent, and that the slope is close to 1, as in the Pierce/Posey model, at very low frequencies, but declines to approximately 1/3, as in the Russian scaling laws, at higher frequencies.

The strong dependence of the infrasound spectra on yield suggests that the explosion source function is much longer than the duration of the explosion and its associated near field nonlinear effects. As noted earlier, these effects last only a few seconds, while the spectral shape is affected at periods of hundreds of sceconds.

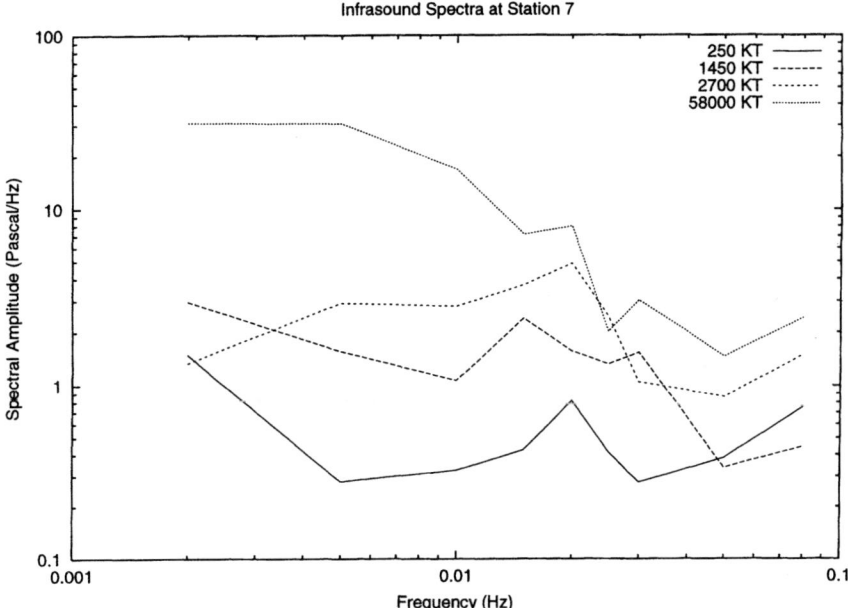

Figure 6
Spectra of four infrasound signals recorded at station 7 with yields of 250, 1450, 2700, and 58,000 kilotons.

Network Detection Simulations

In order to predict the performance of the proposed 60 station IMS infrasound network, we modified the network simulation program NetSim (SERENO et al., 1990) to include the models of infrasound propagation described in equations (10)–(14). NetSim uses these equations to calculate the pressure as a function of yield and range. These are used together with station locations, a noise model, a minimum signal-to-noise ratio for reliable measurement, and the number of stations required for a signal to be reported, to determine the network detection threshold as a function of position on the earth. We have calculated detection thresholds for the proposed IMS network for a constant noise level and for station-dependent noise levels. The calculations were performed using the following parameters:

1. For the constant noise simulations, noise estimates were taken from BLANDFORD and CLAUTER (1995). The amplitude of background noise is log normally distributed with log mean value of −1.14 (Pascals) and log standard deviation of 0.37. In this case, the noise distribution is taken to be geographically uniform.
2. For the simulation with station-dependent noise levels, we used noise estimates from BLANDFORD et al. (1995). The log noise levels varied from −1.37 to −0.09 (Pascals), with the highest levels occurring in oceanic regions. Log standard deviation was again taken to be 0.37.
3. The minimum signal-to-noise ratio for detection is 2.

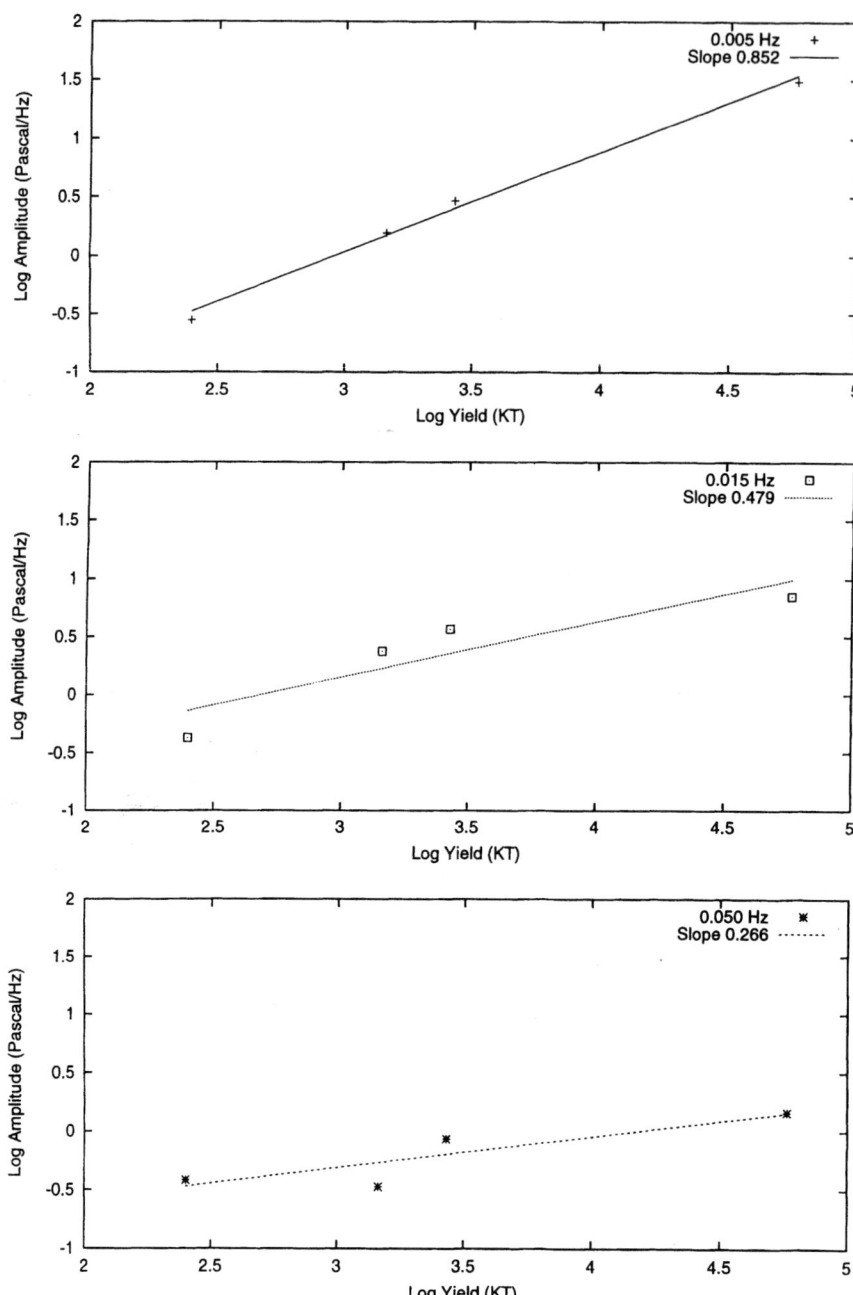

Figure 7
Spectral amplitude plotted vs. yield at frequencies of 0.005 Hz (top), 0.015 Hz (middle), and 0.050 Hz (bottom). The slope of the amplitude yield curve decreases with increasing frequency.

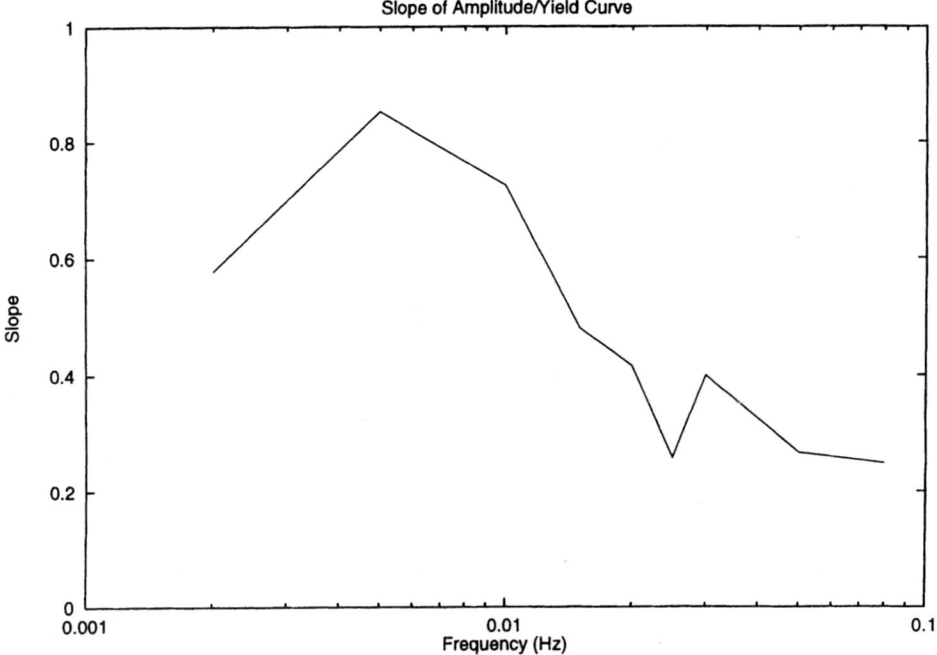

Figure 8
The slope of the log amplitude vs. log yield curve plotted as a function of frequency.

4. Two stations detect infrasound signals at a 90% confidence level.
5. Four element infrasound arrays increase signal-to-noise ratio by a factor of 2.
6. Propagation error has a log standard deviation in log signal of 0.3.
7. Station reliability is 95%.

Detection threshold maps were calculated for the AFTAC and LANL scaling relations. We did not calculate the detection threshold for the Pierce Posey or Russian models because they are clearly unrealistic for yields near the detection threshold. The results are shown in Figures 9–10 for the uniform noise model, and Figures 11–12 for the station-dependent noise models. For the AFTAC model, the detection threshold is less than a kiloton at all locations for the uniform model, and ranges from 0.3 KT to about 1 KT for the variable noise model. For the LANL models, the detection threshold is between 0.4 and 1.1 KT for the uniform model, and between 0.5 and 2 KT for the variable noise model. In the uniform noise simulations, the variations in threshold level are determined by station coverage, since other factors are the same for all stations. These scaling relations predict, therefore, that the infrasound detection threshold for the IMS network is somewhat higher than the design goal of one kiloton in some locations, particularly in broad ocean areas.

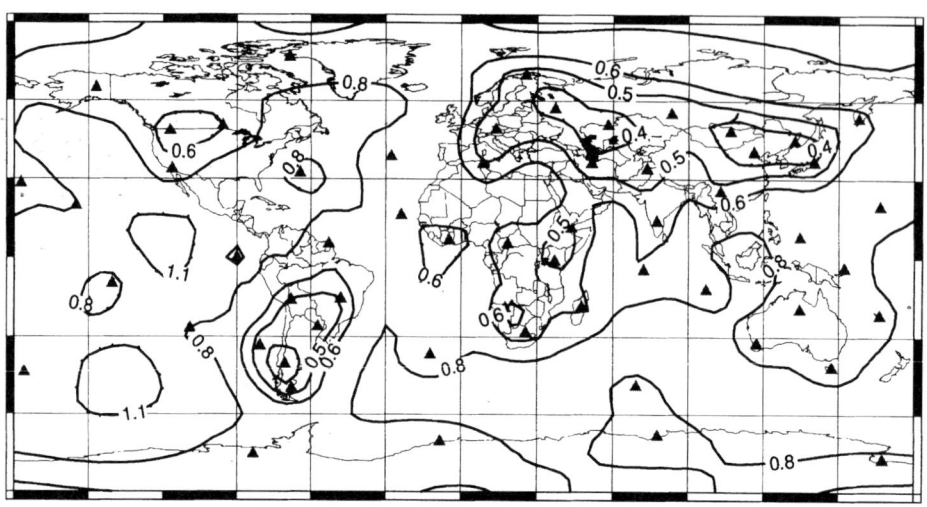

Figure 9
Contours showing detection thresholds with a 90% level of confidence for detection at 2 infrasound stations with uniform noise levels using the LANL scaling relation (equation (12)). The intervals are logarithmically spaced with labels in kilotons.

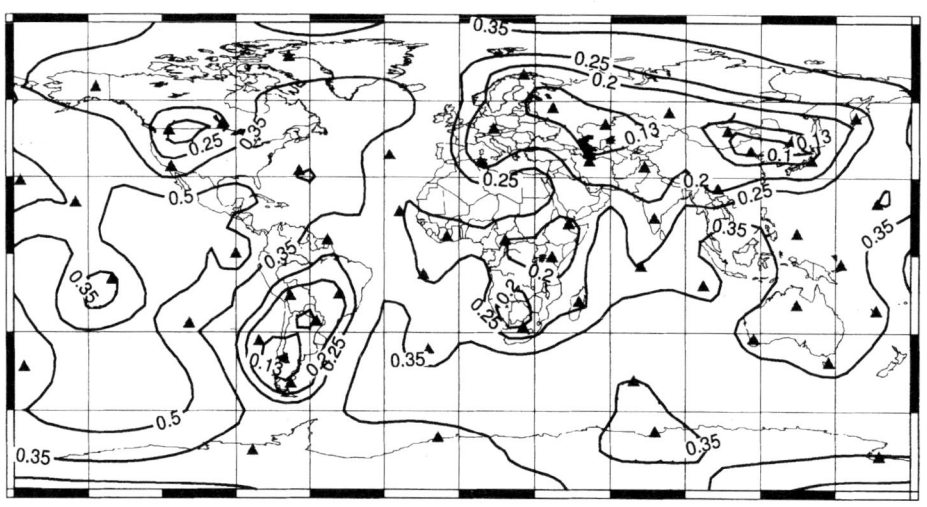

Figure 10
Contours showing detection thresholds with a 90% level of confidence for detection at 2 infrasound stations with uniform noise levels using the AFTAC scaling relation (equation (11)). The intervals are logarithmically spaced with labels in kilotons.

Discussion and Conclusions

In the analysis above, we used a data set of infrasound waveforms from Soviet atmospheric tests ranging in yield from 6 kilotons to 58 megatons to place constraints on

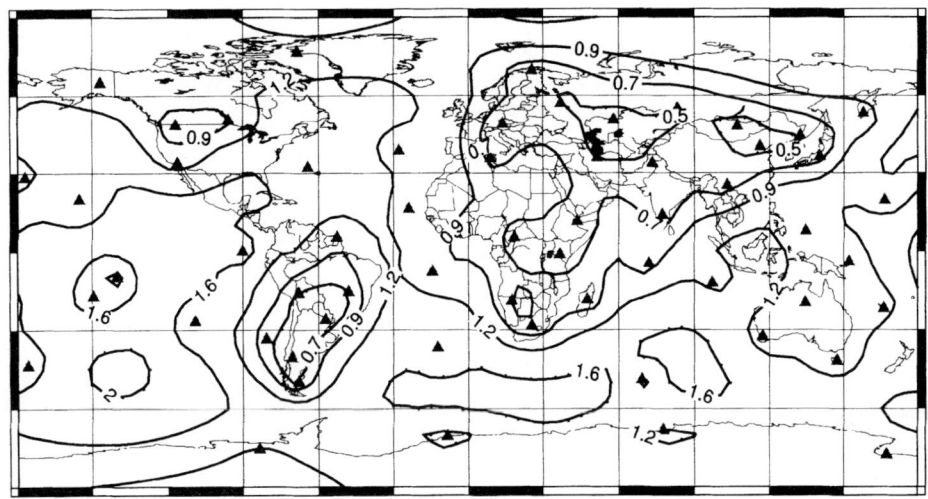

Figure 11
Contours showing detection thresholds with a 90% level of confidence for detection at 2 infrasound stations with station-dependent noise levels using the LANL scaling relation (equation (12)). The intervals are logarithmically spaced with labels in kilotons.

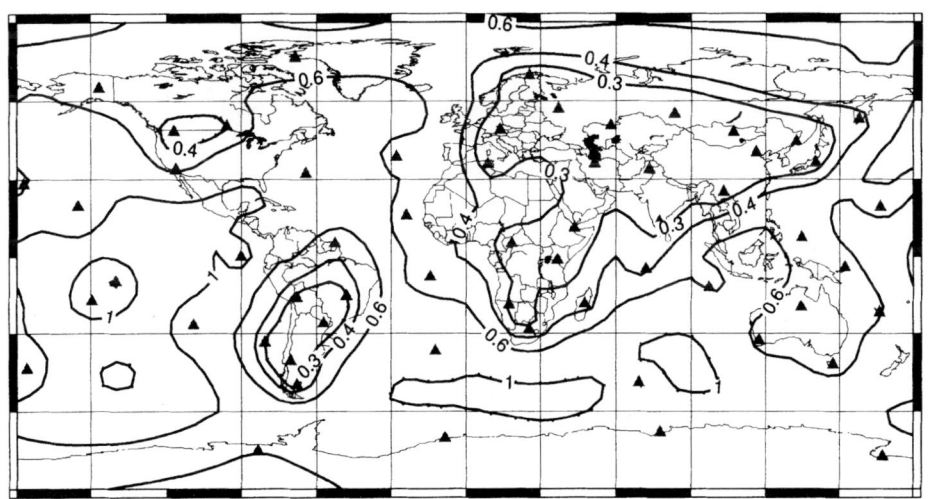

Figure 12
Contours showing detection thresholds with a 90% level of confidence for detection at 2 infrasound stations with station-dependent noise levels using the AFTAC scaling relation (equation (11)). The intervals are logarithmically spaced with labels in kilotons.

infrasound scaling relations and to estimate the detection threshold of the future International Monitoring System. Analysis of 133 waveforms reveals that measured pressures are consistent with yield and attenuation scaling relations developed at LANL for HE tests, and also somewhat consistent with a scaling relation developed by AFTAC. The scaling relations developed by Pierce and Posey are consistent with Lamb waves from very large explosions, however not with acoustic waves from smaller explosions.

Because the LANL relation is consistent with data over a very wide yield range, it has recently been adopted as the basis for an infrasound magnitude by the International Data Center (IDC) (BROWN, 1999). A magnitude is a useful quantity for providing an estimate of source size that is independent of the distance at which the signal is measured. The magnitude equation is:

$$M_i = \log_{10} P + 1.36 \log_{10} R - 0.019v \ , \tag{15}$$

where the last term corrects for wind as discussed earlier. Infrasound magnitudes for the Russian data set (without wind correction), are listed in Table 5.

Magnitudes for the acoustic wave and Lamb wave are listed separately. Figure 13 shows M_I plotted vs. yield for the seven Soviet explosions. Also shown is the LANL relation, equation (12), rewritten as a magnitude/yield relation

$$M_I = 0.68 \log W + 3.37 \ . \tag{16}$$

Table 5

Infrasound magnitude for eighteen Soviet explosions

Test site	Explosion	M_I Acoustic	$\sigma(M_I)$ Acoustic	Number Acoustic	M_I Lamb	$\sigma(M_I)$ Lamb	Number Lamb
STS	088	4.44		1			
STS	089	4.31	0.05	3			
KY	091	4.53	0.18	5			
NZ	095	5.52	0.26	16	5.50	0.25	16
NZ	099	5.29	0.46	18	5.26	0.37	20
NZ	102	5.29	0.34	13	5.13	0.31	9
NZ	111	5.56	0.40	5	5.42	0.13	3
NZ	113	4.94	0.37	2			
NZ	116	4.98	0.32	6			
STS	117	4.52		1			
KY	119	4.45	0.48	4			
NZ	120	5.38	0.48	11	5.36	0.53	12
STS	123	4.46		1			
STS	124	4.47	0.40	2			
NZ	125	5.43	0.56	9	5.24	0.41	10
NZ	126	5.33	0.35	4	6.15	0.43	14
NZ	128	5.04	0.38	6	4.92	0.38	2
NZ	133				6.69	0.40	10

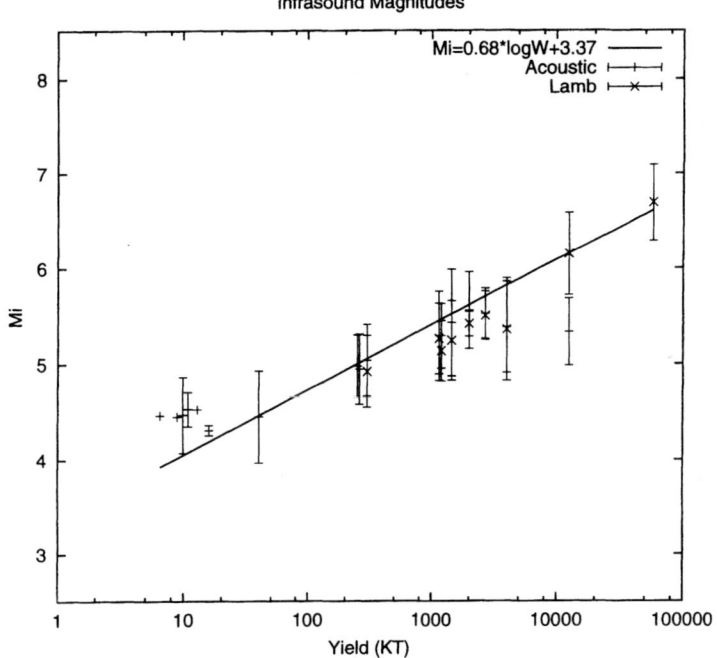

Figure 13
Infrasound magnitude plotted vs. yield for eighteen Soviet explosions.

As can be seen in the figure, equation (16) fits the data over this very wide yield range very well. An exception is the low yield, high altitude explosion at Kapoustin Yar which lies above the curve by about 0.5 magnitude units.

The network simulation results obtained using the best constraints which are currently available indicate that the detection threshold of the future IMS infrasound network may be somewhat higher than the one-kiloton design goal in some locations. This result depends, of course, on numerous assumptions that went into the simulations. We made the assumptions that a four element array leads to a factor of two improvement in signal/noise ratio, that the noise levels at all stations are independent, and that a signal will be identified with a signal to noise ratio of 2. In general, these assumptions are optimistic, although experienced analysts may be able to detect a signal at lower S/N ratios. Improvements could also be made in the signal modeling, particularly by including stratospheric winds, which will have the effect of improving detection in some directions and degrading it in others. Excitation and propagation, however, appear to be modeled quite well by the LANL relation, equation (12), over a wide range of explosion yields and distance ranges, and this relation can be used to predict infrasound amplitudes in network simulations.

Acknowledgement

We thank Bob Blandford and Dean Clauter for their comments and for providing the station dependent noise estimates. This work was supported by Defense Threat Reduction Agency contract DSWA01-97-C-0129.

REFERENCES

BLANDFORD, R. R. and CLAUTER, D. A. (1995), *Capability Estimation of Infrasound Networks*, AFTAC Report.

BLANDFORD, R. R., CLAUTER, D. A., WHITAKER, R. W., and ARMSTRONG, T. (1995), *Infrasound Network Options*, AFTAC/DOE White Paper.

BROWN, D. J. (1999), *Summary of Infrasound Source Location Meeting: San Diego, November 9–10, 1998*, Center for Monitoring Research Technical Report CMR-99/02, January.

CLAUTER, D. A. and BLANDFORD, R. R. (1998), *Capability Modeling of the Proposed International Monitoring System 60-Station Infrasonic Network*, Proceedings of the Infrasound Workshop for CTBT Monitoring, Santa Fe, New Mexico, August 25–28, 1997, LANL report number LA-UR-98-56.

MCKISIC, J. M. (1997), *Infrasound and the Infrasonic Monitoring of Atmospheric Nuclear Explosions*, Tracor Applied Sciences Final Report to Phillips Laboratory, PL-TR-97-2123, February.

PIERCE, A. D. and POSEY, J. W. (1971), *Theory of the Excitation and Propagation of Lamb's Atmospheric Edge Mode from Nuclear Explosions*, Geophys. J. R. Astr. Soc. *26*, 341–368.

POSEY, J. W. and PIERCE, A. D. (1971), *Estimation of Nuclear Explosion Energies from Microbarograph Records*, Nature *232*, 253.

SERENO, T. J., BRATT, S. R., and YEE G. (1990), *NetSim: A computer Program for Simulating Detection and Location Capability of Regional Seismic Networks*, SAIC Annual Technical Report to DARPA, SAIC 90/1163, March.

WEXLER, H. and HASS, W. A. (1962), *Global Atmospheric Pressure Effects of the October 30, 1961 Explosion*, J. Geophys. Res. *67*, 3875–3887.

WHITAKER, R. W. (1995), *Infrasonic Monitoring*, Proceedings of the 17th Annual Seismic Research Symposium in Scottsdale, AZ, September 12–155, 997–1000.

(Received May 24, 1999, revised July 10, 2000, accepted July 17, 2000)

To access this journal online:
http://www.birkhauser.ch

Pure appl. geophys. 159 (2002) 1063–1079
0033–4553/02/051063–17 $ 1.50 + 0.20/0

❚ Pure and Applied Geophysics

Infrasound Detection of Large Mining Blasts in Kazakstan

MIKE T. HAGERTY,[1] WON-YOUNG KIM,[1] and PAVEL MARTYSEVICH[2]

Abstract — We describe infrasonic observations recorded since October, 1997, at the Kurchatov Observatory in Kazakstan from large mining blasts in Kazakstan and Siberia. Seismic signals are regularly recorded on a 21-element cross-array from events located at the Ekibastuz mine, 250 km NW of Kurchatov. However, associated infrasonic detections are infrequent and appear to be seasonal, with maximum numbers of detections occurring during November to January. Raytracing through model atmosphere temperature and wind profiles predicts enhanced infrasound reception during the winter months, when the prevailing stratospheric winds blow towards Kurchatov. In addition, raytracing confirms that the first infrasound arrivals at Kurchatov propagate through the troposphere and are followed, some 50–70 s later, by a stratospheric arrival.

Key words: Infrasound, mining blasts, Kazakstan.

Introduction

In order to meet the monitoring requirements of the Comprehensive Test-Ban Treaty (CTBT), three technologies – seismic, hydroacoustic, and infrasonic – will be relied upon to detect acoustic waves produced by nuclear explosions detonated on land, in the sea, and in the air. While seismic and hydroacoustic technologies are sufficiently evolved to meet the monitoring needs of the CTBT, in its current state, infrasonic technology, is not. There are several reasons why development of the infrasound component lags behind the other two. Chief among these is the greater difficulty encountered in recording acoustic signals in the atmosphere versus in the ocean or in the earth. The greater levels of cultural and flow noise in the atmosphere hinder reliable infrasonic detection. Detection can be improved by utilizing inlet hose or pipe arrays, designed to decorrelate wind noise by spatially filtering the input signal, however, there is considerable debate about the best configuration and material to use, and today wind noise reduction remains an art. Furthermore, temperatures in the atmosphere vary on time scales of hours to months, and wind speeds (up to 100 m/s) are a significant fraction of the average sound speed (~330 m/s).

[1] Lamont-Doherty Earth Observatory, Columbia University, Palisades, NY 10964, U.S.A. E-mail: wykim@ldeo.columbia.edu
[2] Institute of Geophysical Research, National Nuclear Center, Semipalatinsk-21, Kazakstan.

The result is that the acoustic propagation channels in the atmosphere are highly variable. This variability must be quantified for infrasonic monitoring to meet the needs of the CTBT. There is currently an acute lack of reliable infrasonic observations that can be utilized to examine the effects of time-dependent variations in atmospheric properties on the detectability and characteristics of infrasound signals. This is particularly true for infrasound generated from smaller explosions, which the CTBT aims to monitor, but which are likely to be recorded by only a few (1–3) of the nearest IMS stations.

Since October, 1997 we have conducted infrasound observations at the Kurchatov Geophysical Observatory in Kazakstan using available microphones coupled with existing noise reduction systems in order to address some of the infrasound monitoring issues outlined above. The Kurchatov Observatory (Fig. 1) is an ideal site for research on infrasound and on the application of synergistic (seismic and acoustic) methods of event discrimination, as it operates both a 21-element short-period seismic cross-array (Fig. 2) and a three-component broadband seismic station, and because of its close proximity to several large (100+ ton) mining operations (Fig. 1). In addition, conditions appear to be favorable for long-range infrasound propagation in Kazakstan, where infrasound signals have been detected out to 2000 km distance (AL'PEROVICH *et al.*, 1985).

Available noise reduction pipe arrays at the site are depicted in Figure 2. Three types of noise reduction configuration were utilized in this study: 1) six, 70-m long underground pipes extending radially from a central chamber and referred to as "East-" and "West-" spiders; 2) a single, 30-m long pipe with an outlet at the center of the length; and 3) an "H-pipe" array consisting of two, 300-m long pipes joined at their centers by a 100-m long pipe (Fig. 2). The 300-m long pipes are raised 1 m above the ground and have sampling nozzles spaced at 3-m intervals located on their undersides, facing the ground. The internal diameter of the 300-m long pipes varies from 1/2 inch at the ends to 2 inches in the middle. The 30-m long pipe is raised 0.5 m above the ground and has sampling nozzles spaced at 0.5-m intervals. We presume that these systems, which were constructed during the Soviet era, were designed to be functional during the severe local winters, when snow covers the ground for five months each year and prohibits the use of conventional plastic hoses. Transfer functions for these systems are not yet determined.

Two types of capacitor microphone – Globe and Soviet K301 – have been utilized with the pipe arrays described above. Globe microphones have been used widely in infrasound research for many years and their broadband response is well known (e.g., DONN and POSMENTIER, 1968). The K301s were originally installed at Kurchatov in the early 1970s by the Russian Ministry of Defense and recorded on paper. It is noteworthy that the recording site is only about 1400 km from the Chinese nuclear test site at Lop Nor, where some 12–13 atmospheric nuclear tests were conducted in the 1970s and one (the last) in 1980. Since February, 1995, the analog signal from a K301 sensor has been digitized and recorded by a 16-bit A/D

Figure 1

Map of Kazakstan and central Siberia showing locations of broadband seismic stations in Kazakstan (solid traingles), active mining areas (diamonds), Kuzbass and Abakan mining areas (shaded squares), the Balapan nuclear test site (star), and the Kurchatov Geophysical Observatory (KUR), where the seismic and infrasonic observations were made.

system together with the seismic channels from the 21-element bore-hole, short-period (0.5–5 Hz) vertical seismic array. The K301 microphone has variable gain (mv/Pa) and has exceptionally good sensitivity to longer periods (0.01–0.3 Hz). Since October, 1997, 4–8 microphones connected to the various noise reduction systems have been simultaneously recorded.

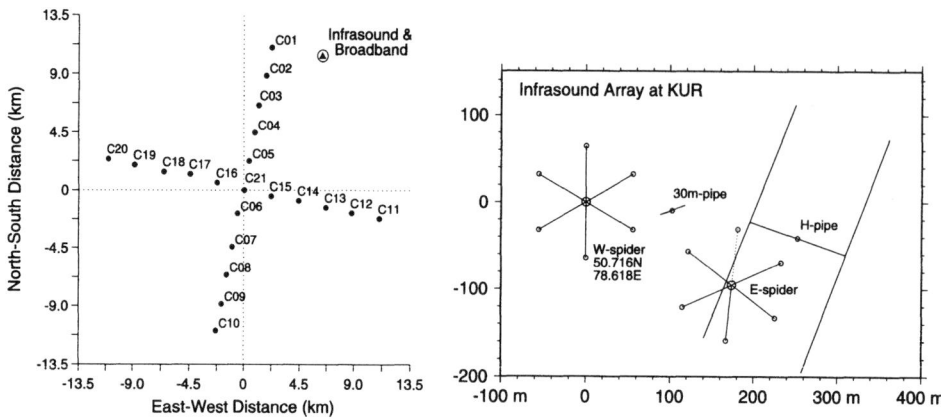

Figure 2

Left panel: Plan view of the 21-element seismic borehole array (cross array) at Kurchatov and location of infrasound sensors. Right panel: Infrasound noise reduction systems (pipe arrays) utilized in this study. All are located within the Kurchatov Geophysical Observatory compound, which also houses the central recording unit for the cross array and a three-component broadband seismometer installed in a 25 m shaft.

Observations

Several large mines in the region generate explosions that are routinely detected seismically and, in some cases, are also detected with infrasound. The mines range in distance from 80 to 750 km from the infrasound array. The Ekibastuz mine, 250 km NW of the array, regularly produces 4–6 seismic detections per day. However, associated infrasonic detections are found only for roughly 10% of the events. Between October 1997 and January 1998 we detected infrasound signals generated by 26 Ekibastuz events. The location and origin time of each event were determined from the seismic cross-array and are listed in Table 1. Ekibastuz comprises a number of coal mines centered about (51.67°N, 75.40°E) as determined by satellite photographs (THURBER *et al.*, 1989).

The infrasound wavetrain generated by Ekibastuz explosions can be classified into two different types. The first type, shown in Figure 3a, consists of 1 or 2 simple pulses, with travel times of approximately 740 and 810 s with respect to the seismically estimated origin time. The second arrival is observed in about 60% of the events (18) from Ekibastuz; when the second arrival is present, it generally follows the first by 50–70 s, though this can range from 24 to 85 s (Table 1). The travel time of the first arrival exhibits great variation and probably reflects varying atmospheric conditions such as transient propagation ducts. This is discussed further in the next section. Figure 3b shows a 20 s time window containing the first infrasound arrival from the Ekibastuz event shown in Figure 3a. An array beam has been obtained by slant stacking the traces with a 2 s moving window and locating the maximum power within each window. Maximum power occurs when the traces are aligned with a

Table 1

List of infrasound signals from mining blasts at Ekibastuz, Kazakstan

Id no.	Origin time		Magnitude	Phase id	Travel time (s)	Group velocity (m/s)	Amplitude (Pa)
	mm/dd/yy	hh:mm:sec					
01	10/16/1997	08:16:13	2.1	T1	820	303	0.22
02	10/16/1997	08:28:43	2.6	T1	810	307	0.23
03	10/16/1997	08:34:39	2.3	T1	825	301	0.11
04	10/17/1997	11:39:14	2.2	S1	646	325	0.13
				T1	731	287	0.06
05	11/19/1997	09:37:57	2.2	S1	730	341	0.09
06	11/19/1997	10:20:07	2.5	S1	793	313	0.13
07	11/27/1997	07:43:17	2.2	S1	763	326	0.29
08	11/27/1997	09:47:45	1.9	S1	788	315	0.16
09	11/27/1997	09:53:18	2.4	S1	765	325	0.54
10	11/28/1997	08:34:14	2.5	S1	763	326	0.13
11	11/28/1997	08:58:06	2.4	S1	761	327	0.19
12	11/28/1997	09:30:49	2.3	S1	757	328	0.45
13	12/02/1997	08:48:11	1.9	S1	734	339	0.29
				T1	793	313	0.17
14	12/02/1997	09:23:38	1.6	S1	756	329	0.08
				T1	794	313	0.05
15	12/02/1997	09:27:25	2.0	S1	746	333	0.41
				T1	786	316	0.22
16	12/02/1997	10:29:01	2.1	S1	743	335	0.13
				T1	779	319	0.08
17	12/11/1997	07:51:46	2.4	S1	739	336	0.23
				T1	809	307	0.29
18	12/11/1997	07:57:43	1.9	S1	718	346	0.04
				T1	799	311	0.03
19	12/11/1997	08:36:50	2.8	S1	747	333	0.09
				T1	809	307	0.05
20	12/11/1997	08:58:22	2.5	S1	750	332	0.19
				T1	812	306	0.17
21	12/11/1997	10:49:47	2.2	S1	737	337	0.09
				T1	798	312	0.06
22	01/14/1998	07:02:34	2.0	S1	796	312	0.26
				T1	827	301	0.49
23	01/14/1998	08:13:40	2.4	S1	795	313	0.26
				T1	829	300	0.31
24	01/14/1998	09:16:47	2.6	S1	815	305	0.17
				T1	839	296	0.25
25	01/28/1998	07:31:04	3.0	S1	753	330	0.15
				T1	777	320	0.08
26	01/28/1998	08:15:21	2.4	S1	764	325	0.25
				T1	788	315	0.27

horizontal phase velocity of 0.38 km/s and a backazimuth of 267°. The estimated backazimuth (267°) is 30° off the backazimuth predicted from the seismic location (297°). There are several possible reasons for this large discrepancy. First, the

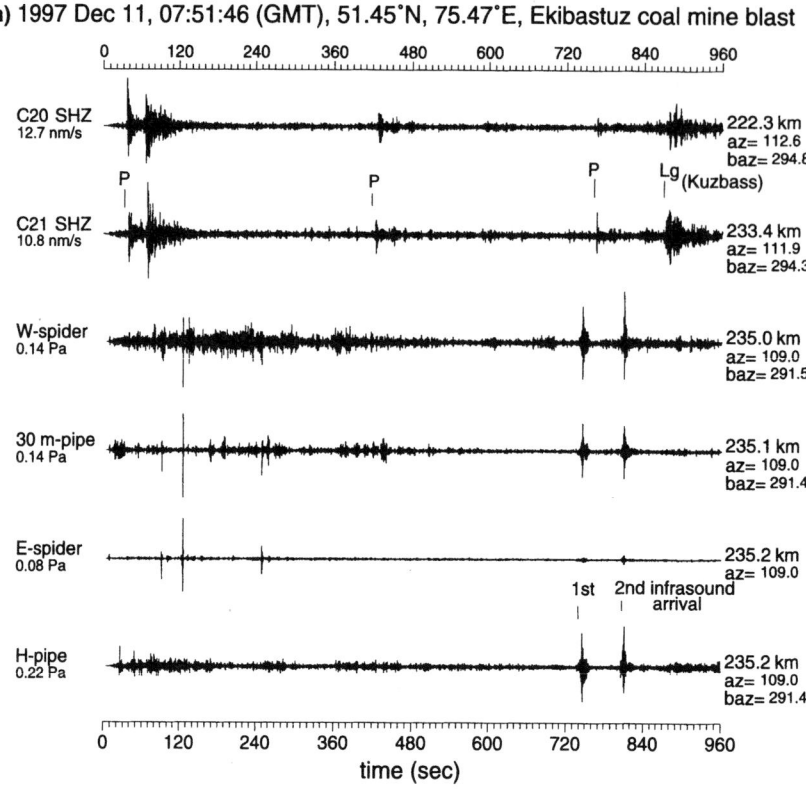

a) 1997 Dec 11, 07:51:46 (GMT), 51.45°N, 75.47°E, Ekibastuz coal mine blast

Figure 3

(a) Selected cross-array seismic channels (top two traces) and infrasound signals recorded with four different noise reduction systems (Fig. 2) for an Ekibastuz coal mine blast on Dec. 11, 1997. (b) Expanded view of first infrasound arrivals and contour plot of beam power obtained by slant-stacking infrasound traces at different slownesses. Maximum beam power is obtained for phase velocity = 0.38 km/s and backazimuth = 267°.

infrasound array has a very small aperture and, hence, very poor directional resolution of incoming signals. Second, cross-winds in the upper atmosphere can greatly alter the apparent azimuth of infrasound signals. Although poorly resolved, the phase velocity of the second arrival appears to be lower than that of the first.

The second type of infrasonic wavetrain associated with Ekibastuz events, shown in Figure 4a, consists of a series of pulses of growing amplitude lasting some 20–30 s. The event shown in Figure 4a is actually two explosions closely spaced in time. While the seismic waves from the two explosions overlap, making it difficult to resolve two events, the lower phase velocities and shorter durations of the infrasonic waves allow the two events to be distinguished (Fig. 4b). For this reason, infrasound observations may be crucial for detecting a shallow nuclear test hidden in the seismic coda of an earlier event. There is no evidence of multiple cast firing within either of the two

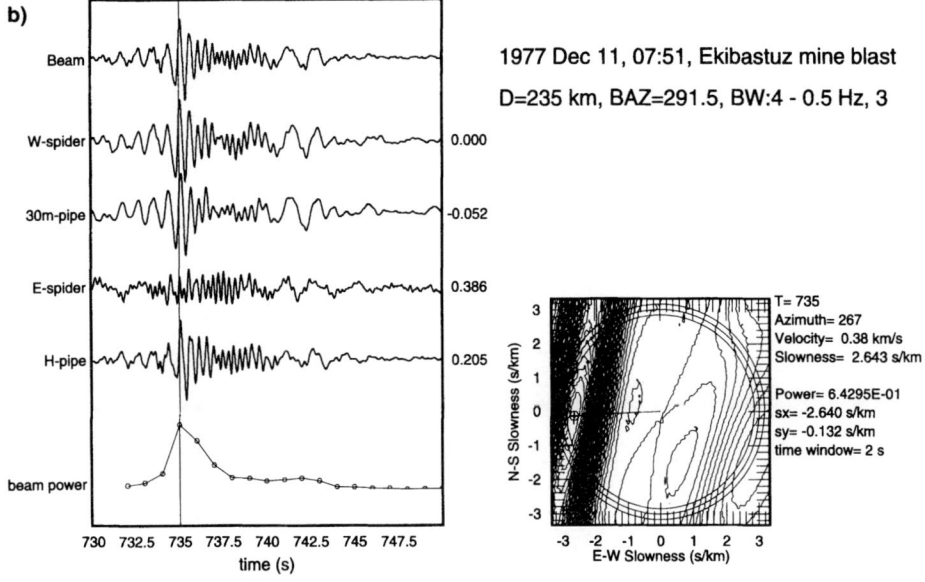

Figure 3 (b)

seismic events, hence, the multiple phases observed in the infrasound data must be produced by propagation effects. This is discussed further in the next section.

During the observation period, three infrasound signals from events in the Kara-Zhyra coal mine in the Balapan former Soviet nuclear test site were identified (Table 2). Figure 5 shows seismic and infrasonic signals originating from one such event, located 80 km S of Kurchatov. Peak infrasound pressure is about 0.35 Pa and there is evidence of dispersion; the period of the initial arrival is 0.85 s and gradually decreases to 0.37 s.

Modeling

In order to identify the infrasound arrivals, we raytraced through various suitable atmospheric models. Figure 6a shows the sound speed (c) as a function of height in the atmosphere derived from mid-latitude (45°N) temperature profiles for January and July (VALLEY, 1965) and the well-known theoretical relation $c(T) = 20.1\sqrt{T(K)}$, where T is the temperature in degrees Kelvin. Above 130 km, average sound speeds from the 1962 U.S. Standard Atmosphere have been used. In Figure 6b mid-latitude zonal wind models for summer and winter adapted from GEORGES and BEASLEY (1977) are shown, along with wind profiles measured by the high-resolution Doppler imager aboard the Upper Atmosphere Research Satellite (UARS) (FLEMING et al., 1996). While the details vary among the models, both display the prominent seasonal

a) 1997 Nov 27, 09:53:16 (GMT), 51.67°N, 75.40°E, Ekibastuz mine blasts

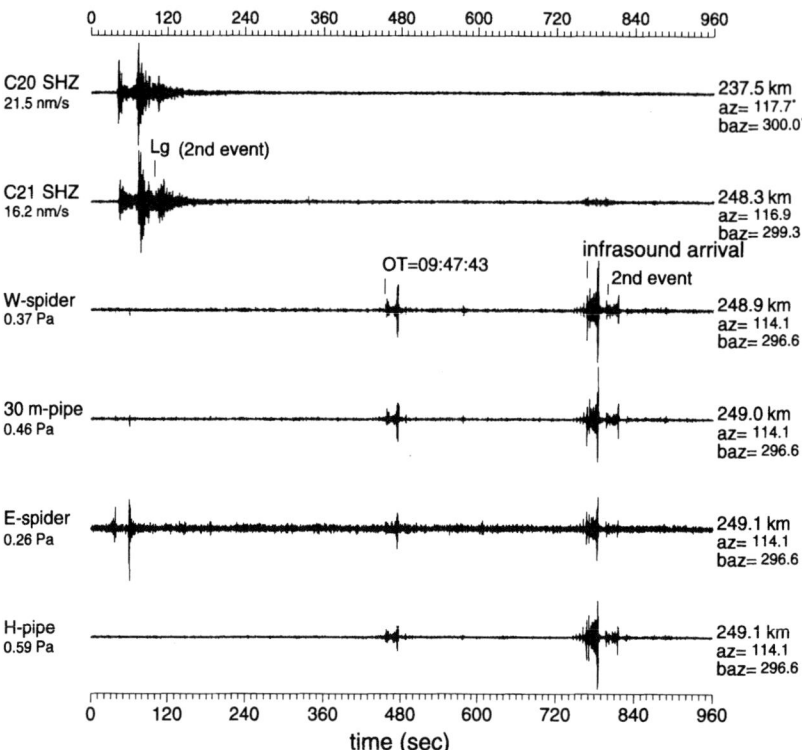

b) 1997 Nov 27, 09:53:16, Infrasound signals from two Ekibastuz mine blasts

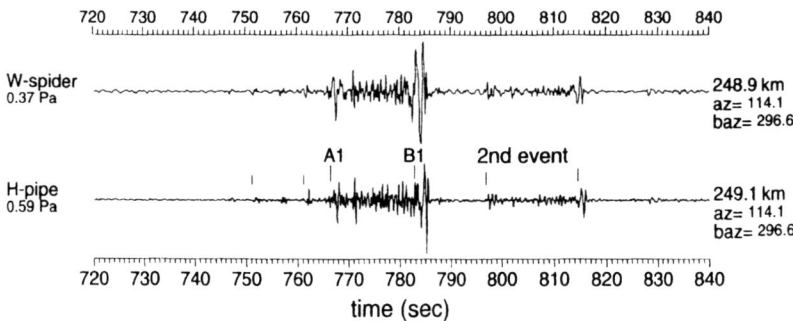

Figure 4

(a) Same as Figure 3a for Ekibastuz mine blast on Nov. 27, 1997. This event is actually two closely spaced explosions which produce overlapping *Lg* phases in the seismic channels but produce separate infrasound arrivals. An additional infrasound arrival at $t \sim 460$ s is from an earlier Ekibastuz event. (b) Blow-up of the infrasound signals shown in (a). Each event is seen to consist of a sequence of pulses of increasing amplitude lasting 20–30 s.

Table 2

List of infrasound signals from other mining blasts

Id	Origin time		Lat. (°N)	Long. (°E)	Mag.	Phase id	TT (s)	Group velocity (m/s)	Amp (Pa)
	mm/dd/yy	hh:mm:sec							
	Novotaubinka near Semipalatinsk								
1	11/19/1997	09:24:38	49.80	80.85	0.5	D	624	302	0.10
	Kara-Zhyra mine in Balapan								
1	11/19/1997	10:28:27	50.01	78.80	3.7	D	252	317	0.58
2	12/11/1997	11:58:10	49.99	78.59	1.8	D	232	234	0.39

Mag = magnitude, TT = travel time, Amp = amplitude.

shift in stratospheric wind direction; at 50–80 km, strong summer easterlies give way to winter westerlies. This has an important effect on the seasonal reception of infrasound. Figure 6c shows mid-latitude meridional wind models taken from GEORGES and BEASLEY (1977); the meridional winds are smaller than the zonal winds at altitudes less than 100 km and therefore have less impact on infrasound propagation. Figure 7 presents the results of raytracing through a sound-speed profile given by the mid-latitude winter temperature profile alone, neglecting the effects of wind. As can be seen, in this high-frequency ray approximation, no energy is predicted to return to the earth surface at distances less than 280 km. Beyond 280 km, the first arrivals turn in the thermosphere, at altitudes of $\sim$150 km, followed by secondary arrivals that turn at $\sim$125 km; no energy is returned from the stratosphere (40–80 km). Thus, in the absence of wind, no favorable propagation paths exist between Ekibastuz and Kurchatov ($\Delta = 250$ km). In order to correctly compute the predicted raypaths and travel times through the atmosphere, we must consider both the temperature and wind effects on the resulting sound speed. Several past studies have combined the temperature-derived sound speed, $c(z)$, and the component of horizontal wind velocity along the propagation path, $w'(z)$, into an effective sound-speed profile for a given propagation azimuth, $v(z) = c(z) + w'(z)$, that can be used, along with Snell's law, to compute ray paths and travel times. Here the effective wind speed along the propagation azimuth is $w'(z) = w \times \cos(W_d - az)$, where w is the wind magnitude, and W_d and az are the wind direction and the receiver azimuth measured in degrees clockwise from north. Thus, the horizontal wind enhances sound propagation in the wind direction (downwind) and retards it in the opposite direction (upwind).

However, as noted by THOMPSON (1972) and recently by GARCÉS et al. (1998), the above is an approximate treatment of sound propagation in a moving medium, and is not strictly correct. A more correct treatment is obtained by raytracing in a coordinate frame that is moving at the wind velocity. It is still convenient to rotate the wind vector $\vec{u}$ with respect to the receiver azimuth, $\vec{u} = u(z)\hat{i}_x + v(z)\hat{i}_y$, where $u(z)$ and $v(z)$

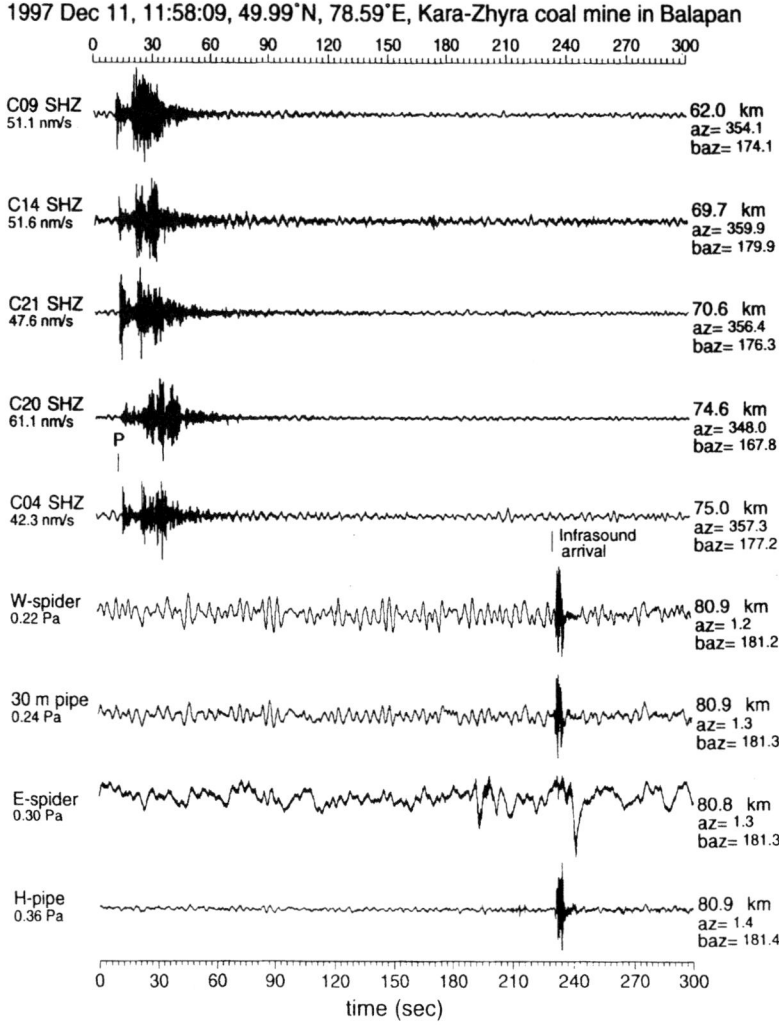

Figure 5
Selected cross-array seismic channels (top five traces) and infrasound signals recorded with four different noise reduction systems (Fig. 2) for an event located in the Kara-Zhyra coal mine in Balapan, some 80 km south of Kurchatov, on Dec. 11, 1997.

are the wind components parallel and transverse to the receiver azimuth, respectively. Raypaths computed using the effective speed and the 'correct' wind treatment are compared in Figure 8 for take-off angles of 45° and 85°. While the rays still turn at a height where the effective speed equals the horizontal phase velocity at the surface, the horizontal distance where this occurs is different for the two methods. Figure 9 compares the travel-time, phase velocity, and group velocity curves that result from using the two different methods to ray trace through the winter temperature and wind

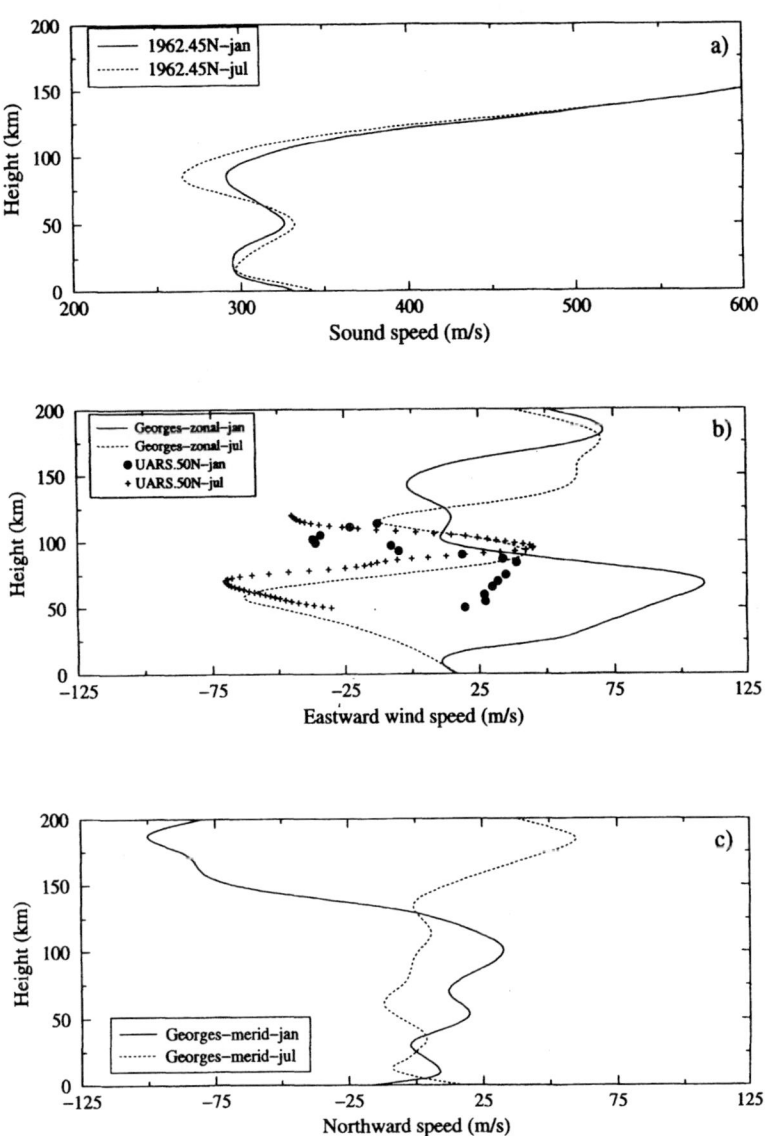

Figure 6
(a) January and July sound-speed profiles derived from the 1962 U.S. Standard Atmosphere mid-latitude (45°N) temperature profiles (VALLEY, 1965). (b) January and July mid-latitude zonal winds taken from GEORGES and BEASLEY (1977). Also shown for comparison are zonal winds measured by the high-resolution Doppler imager aboard the Upper Atmosphere Research Satellite (UARS) (FLEMING et al., 1996). (c) January and July mid-latitude meridional winds from GEORGES and BEASLEY (1977).

models (Fig. 6). While the travel-time and group velocity curves are indistinguishable, the phase velocity curves are offset horizontally. This will have important implications for CTBT monitoring, as the error in location derived from the phase velocity

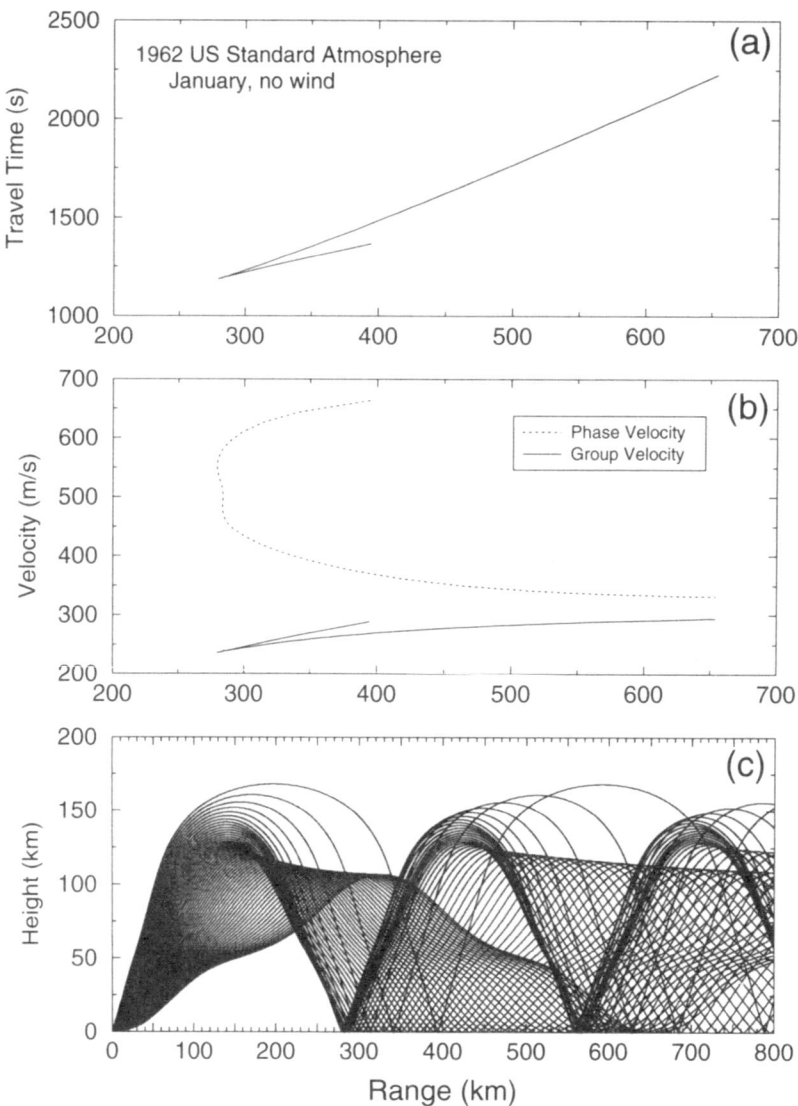

Figure 7

Results of raytracing through a sound speed profile given by the 1962 U.S. Standard Atmosphere mid-latitude temperature profile for January, neglecting wind. (a) Travel-time curve. (b) Group and phase velocity curves. (c) Raypaths for take-off angles of 30–90°.

obtained from raytracing with the effective speed may be as large as 25–40 km for the thermospheric returns, and will increase with increasing numbers of ray bounces (Fig. 8). However, for our present needs, which are to identify the observed infrasound phases, the effective sound speed method is adequate and is used hereafter. Figure 9c shows that the first rays arriving at 220–250 km have turned in the

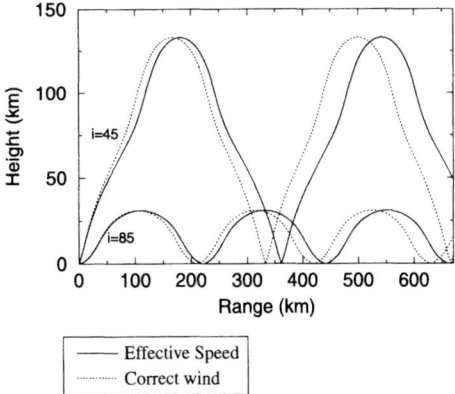

Figure 8
Comparison of raypaths computed using the effective speed method (solid line) and a method that includes the wind speed in the ray equations (dotted line). For both cases, the sound-speed model used is the same as in Figure 7, and the wind model used is the projection of the mid-latitude wind model of GEORGES and BEASLEY (1977) (Fig. 6) in the direction of the azimuth from Ekibastuz to Kurchatov. Rays are compared for take-off angles of 45° and 85° from vertical.

stratosphere (h ~ 38 km). The predicted travel time and phase velocity match the observed values at Ekibastuz reasonably well, however, the second observed arrival is not predicted by this model. In order to produce two arrivals with the observed time separation (~70 s) at $\Delta = 250$ km, a tropospheric duct must exist between Ekibastuz and Kurchatov. This is supported by the observation of a direct (tropospheric) arrival from Kara-Zhyra explosions, 80 km away and well within the shadow zone predicted by models with no tropospheric duct (Fig. 9). The existence of a tropospheric duct implies an increase in the effective sound speed between the surface and the tropopause (0–15 km); we cannot distinguish whether this is caused by a temperature inversion or a westerly jet, however, we prefer the former as the duct must exist in both the easterly and northerly directions. Support for a temperature inversion can be found in a study on infrasound reception of explosions detonated at 1 km height in the atmosphere in southern Kazakstan during winter (AL'PEROVICH et al., 1985); direct arrivals observed out to distances of 200 km were linked with meteorological measurements in the troposphere that confirmed a temperature inversion between the surface and 2–4 km.

A modified sound-speed profile which contains a favorable tropospheric duct is presented in Figure 10 along with the resulting travel-time and velocity curves. For this model, the first arrival at 250 km propagates through the tropospheric duct and is followed some 70 s later by a stratospheric arrival. The modified tropospheric profile results in good matches to the observed first arrival travel times at 80 and 250 km (Fig. 10c). It proved very difficult to find a model that would predict two arrivals at a distance of 250 km separated by such a large amount of time (70 s). While the model shown in Figure 10 is by no means unique, only models with a fast

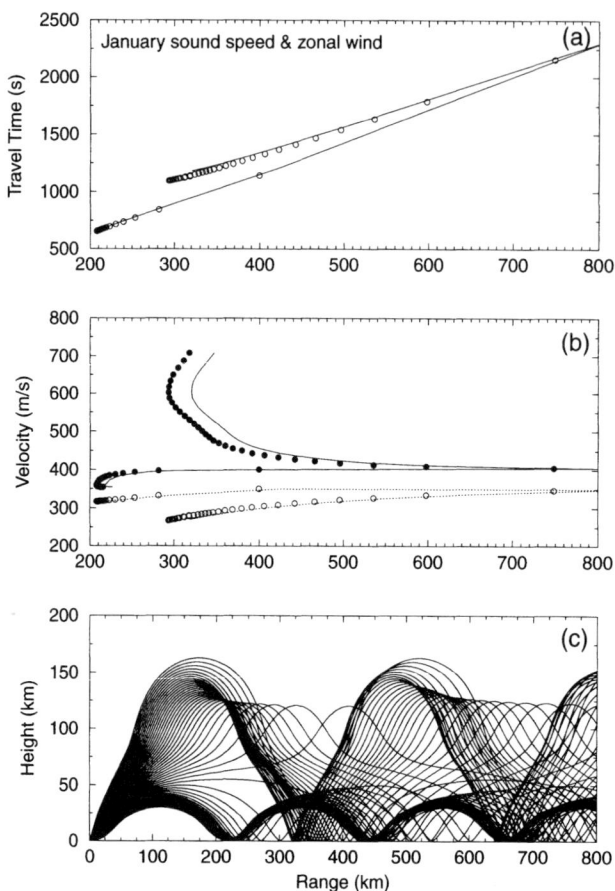

Figure 9
Results of raytracing through a sound speed profile given by the 1962 U.S. Standard Atmosphere mid-
latitude temperature profile and the January zonal wind model. Solid and dotted lines correspond to the
exact wind treatment; open and filled circles correspond to the approximate (effective speed) method.
a) Travel times. b) Phase (solid line and filled circle) and group (dotted line and open circle) velocities.
c) Raypaths computed for the exact method.

troposphere and an elevated stratospheric lid were able to match the observed arrival
times (Fig. 10c, inset). More conservative modifications to the winter models
(Fig. 9), which consisted of increasing the sound-speed gradient in the troposphere
and stratosphere, succeeded in producing several arrivals at a range of 250 km,
however, all arrived within a time window of about 20 s. This suggests that the
multiple arrivals observed for the second type of infrasound signals (Fig. 4b) likely
result from strong tropospheric and stratospheric gradients.

As shown by GARCÉS *et al.* (1998), the effect of the transverse wind component is to
translate the ray-coordinate frame a distance Δy perpendicular to the original wave
normal direction. This will not affect the ray parameter, however it will affect the

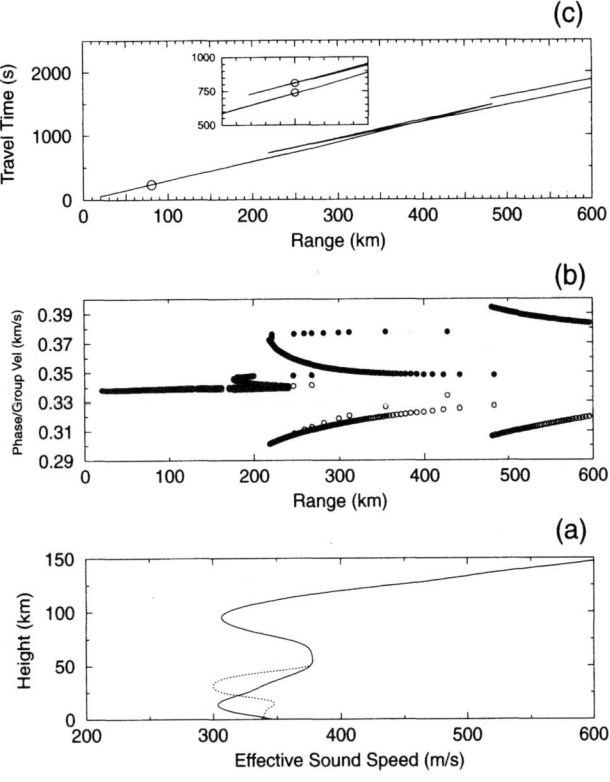

Figure 10
a) Effective sound-speed profiles computed for a winter model (solid line) and modified below 50 km (dotted line). b) Phase (filled circles) and group (open circles) velocity curves for the modified model shown in a. c) Travel-time curve for the modified model shown in a. Also shown are the infrasound arrival times observed at Kurchatov from Balapan ($\Delta = 80$ km) and Ekibastuz ($\Delta = 250$ km) (inset) events.

backazimuth measured at the array. However, both synthetic modeling (GARCÉS *et al.*, 1998) and observations (GEORGES and BEASLEY, 1977), suggest that this deviation will be less than 10° for realistic wind models. In particular, since the dominant winds are zonal (E–W) and since the backazimuth from Kurchatov to Ekibastuz is nearly E–W, the deviation should be quite small. Hence, the discrepancy between the observed and predicted backazimuths for Ekibastuz arrivals is likely due to the poor azimuthal resolution of the array, resulting from the small sensor separation.

Discussion

We have presented here an initial interpretation of the characteristics of infrasound propagation observed in northern Kazakstan. The infrasound signals associated with Ekibastuz events can be classified into two types. The first type consists of two pulses spaced 50–70 s apart, while the second type consists of multiple pulses arriving within about a 20–30 s window. Infrasound arrivals from both Ekibastuz ($\Delta = 250$ km) and Kara-Zhyra ($\Delta = 80$ km) explosions support the existence of a tropospheric duct, produced by a temperature inversion and/or a westerly jet in the troposphere. The multiple arrivals characteristic of the second type of infrasound wavetrains likely result from strong positive sound speed gradients in the troposphere and, especially, in the upper stratosphere. The large variability in the character of infrasound signals generated by Ekibastuz events over short-time scales (hours to days) is indicative of the rapidity of atmospheric fluctuations.

In the absence of wind, no favorable propagation ducts exist between Ekibastuz and Kurchatov. This may explain the low detection rate (10%) of infrasound signals from Ekibastuz events. Indeed, there is evidence that the infrasound detectability is seasonal, with greater numbers of detections occurring during the winter months (Table 1), when westerly winds create a strong stratospheric duct between Ekibastuz and Kurchatov. Since the dominant stratospheric winds are zonal (E–W), a seasonal dependence of infrasound detectability of Ekibastuz (Baz = 297°) explosions should be observed. However, it is also possible that periods of low detection coincide with high surface winds at the receiver, which mask infrasound reception. In fact, we do find evidence for low infrasound detectability when surface winds exceed about 1.7 m/s and we are investigating this possibility further. Finally, infrasound detection could be affected by variable coupling between the seismic and acoustic wavefields at the source.

Measurements of the backazimuth of different infrasound arrivals have the potential to resolve some of the ambiguity that exists in using infrasound to infer the horizontal wind structure in the atmosphere. However, far more accurate backazimuth estimates than can be obtained by a small aperture array are required. For this reason we have extended the small array at Kurchatov into a larger triangular array with 2 km sides. In addition, we intend to install a second three-element infrasound array at Borovoye, 377 km NW of Ekibastuz (Fig. 1), in order to examine the seasonal effects of the stratospheric winds both upwind and downwind from the source. We hope that these improved infrasound arrays will provide unique data that may be utilized to advance the infrasound technology used for CTBT monitoring.

Acknowledgments

We thank the personnel of the Institute of Geophysical Research, National Nuclear Center, Kazakstan, for their on-going efforts to maintain working instruments and to collect high-quality geophysical data. David Lentrichia provided valuable technical assistance.

Our work was sponsored by contract DSWA01-97-C-0156 and DTRA01-00-C-0077 with the Defense Threat Reduction Agency. Lamont-Doherty Earth Observatory contribution number 6292.

REFERENCES

AL'PEROVICH, L. S., AFRAYMOVICH, E. L., VUGMEYSTER, B. O., GOKHBERG, M. B., DROBZHEV, V. I., YERUSHCHENKOV, A. I., IVANOV, E. A., KALIKHMAN, A. D., KUDRYAVTSEV, V. P., KULICHKOV, S. N., KRASNOV, V. M., MORDUKHOVICH, M. I., MATVEYEV, A. K., MAGORSKIY, P. M., PONOMAREV, E. A., SALIKHOV, N. M., TARASHCHUK, YU, E., TROITSKAYA, V. A., and FEDOROVICH, G. V. (1985), *The Acoustic Wave of an Explosion*, Izvestiya Academy of Sciences USSR, Physics of the Solid Earth (USA) *21*(11), 835–842.

DONN, W. L., and POSMENTIER, E. *Infrasonic waves from natural and artificial sources*, In *Acoustic-gravity Waves in the Atmosphere* (ed. T. M. Georges) (Boulder, Co., 1968) pp. 195–208.

FLEMING, E. L., CHANDRA, S., BURRAGE, M. D., SKINNER, W. R., HAYS, P. B., SOLHEIM, B. H., and SHEPHERD, G. G. (1996), *Climatological Mean Wind Observations from the UARS High-resolution Doppler Imager and Wind Imaging Interferometer: Comparison with Current Reference Models*, J. Geophys. Res. *101*, 10,455–10,473.

GARCÉS, M. A., HANSEN, R. A., and LINDQUIST, K. G. (1998), *Traveltimes for Infrasonic Waves Propagating in a Stratified Atmosphere*, Geophys. J. Intl. *135*, 255–263.

GEORGES, T. M., and BEASLEY, W. H. (1977), *Infrasound Refraction by Upper-atmospheric Winds*, J. Acoust. Soc. Am. *61*(1), 28–34.

THOMPSON, R. J. (1972), *Ray Theory for an Inhomogeneous Moving Medium*, J. Acoust. Soc. Am. *51*(5), 1675–1682.

THURBER, C., GIVEN, H., and BERGER, J. (1989), *Regional Seismic Event Location with a Sparse Network: Application to Eastern Kazakstan, USSR*, J. Geophys. Res. *94*, 17,767–17,780.

VALLEY, S. L., *Handbook of Geophysics and Space Environments* (ed. S. L. Valley) (McGraw-Hill, New York, 1965).

(Received April 7, 1999, revised June 1, 2000, accepted June 8, 2000)

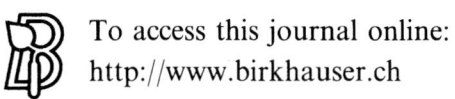

To access this journal online:
http://www.birkhauser.ch

Pure appl. geophys. 159 (2002) 1081–1125
0033–4553/02/051081–45 $ 1.50 + 0.20/0

▌Pure and Applied Geophysics

Infrasonic Signal Detection and Source Location at the Prototype International Data Centre

DAVID J. BROWN,[1,4] CHARLES N. KATZ,[2] RONAN LE BRAS,[2]
MEGAN P. FLANAGAN,[2,3] JIN WANG[1] and ANNA K. GAULT[1]

Abstract—This paper describes an automatic and interactive data processing system designed to locate impulsive atmospheric sources with a yield of at least one kiloton by detecting and characterizing the airborne infrasound radiated by the source. The infrasonic processing subsystem forms part of a larger system currently under development at the Prototype International Data Center (PIDC) in Arlington, Virginia where seismic, hydroacoustic, radionuclide and infrasonic methods are used to detect and locate impulsive sources in any terrestrial environment. Infrasonic signal detection is achieved via a coincidence detector which requires both the normalized cross correlation and the short-term-average/ long-term-average ratio of a beam in the direction of maximum correlation to exceed predetermined threshold values simultaneously before a detection is declared. The infrasound propagation model currently used to infer travel-time information assumes the horizontal sound speed across the ground to be 320.0 m/s. This crude model is currently being replaced by a model which predicts travel-time information through a ray-tracing algorithm for acoustic waves in an atmosphere with seasonal representations for temperature and wind. A novel feature of the source location process is the fusion of all available arrival information, whether it be seismic, hydroacoustic or infrasonic to locate a single source where it is reasonable to hypothesize a common source. In its final configuration the infrasonic subsystem will routinely process data from the global 60-station International Monitoring System (IMS) infrasonic network currently under development.

Key words: Infrasonic nuclear monitoring, CTBT, IMS, automatic and interactive processing, synergy.

1. Introduction

Interest in atmospheric infrasound can be divided into several well-defined epochs. Prior to the 1940s, the interest was purely scientific; a notable achievement of the early infrasonic research being the inference of the gross vertical atmospheric temperature variation up to a height of 50 km (MILNE, 1921; LINDEMANN and

---
[1] SAIC, Center for Monitoring Research, 1300 N. 17th Street, Suite 1450, Arlington, VA 22209, U.S.A.
[2] SAIC, 10260 Campus Point Dr., San Diego, CA 92121, U.S.A.
[3] Now at Lawrence Livermore National Laboratory.
[4] Corresponding author, E-mail: djb@cmr.gov

DOBSON, 1923; GUTENBERG, 1939).[1] The first time infrasonic recording equipment was deliberately deployed to measure the signal generated by a large atmospheric explosion was at Bikini Atoll in 1946.[2] For the next 20 years, research into infrasound was actively pursued, being promoted mainly by the military as an accurate tool for determining when nuclear weapons testing was occurring.

The next epoch commenced with the signing of the Limited Test-Ban Treaty in 1963. This treaty was a tri-lateral agreement between the Soviet Union, the United States, and Britain which banned the testing of all nuclear devices in space, and the oceans and atmosphere.[3] The testing of nuclear devices moved from the atmosphere to the underground environment and culminated in the last atmospheric test by China in October, 1980.

During the ensuing years, infrasonic research for purely scientific pursuit did occur, examples being the investigations by BOWMAN and BEDARD (1971), and by GEORGES (1973) of storm generated infrasound, and later of BEDARD *et al.* (1988) of avalanche generated infrasound. Nevertheless the major impetus given infrasonic research by the nuclear monitoring program disappeared. That being the case, it must be emphasized however that there was a Department of Energy (DoE) program in the 1980s at Los Alamos National Laboratory for infrasonic monitoring of the acoustic signals from underground nuclear tests (UGT). In support of TTBT, DoE explored several techniques that would provide additional means of UGT detection. Through DoE support, the Los Alamos program maintained infrasound technology for measurements, analysis, signal processing, source modeling, and propagation modeling, including wind effects. Of particular value for future CTBT use, was the long-range detections of large atmospheric chemical explosions at White Sands Missile Range (WHITAKER *et al.*, 1990).

Renewed interest in achieving a global comprehensive nuclear test-ban treaty (CTBT) in the latter part of the 1990s however saw a resurgence in interest in infrasonic research. This is based on the premise that if one is serious about maintaining a CTBT, which covers the testing of nuclear weapons in any environment, it becomes mandatory to monitor the atmosphere for such clandestine detonations.

In the early days of infrasonic signal detection, single sensor stations were deployed. The major objective here was to detect the low frequency acoustic-gravity waves radiated by large sources. Consequently, substantial effort went into the development, through empirical and theoretical methods, of accurate dispersion relationships for the possible acoustic-gravity modes (PFEFFER and ZARICHNY, 1962; DONN and EWING, 1962; PRESS and HARKRIDER, 1962; HARKRIDER, 1964; DONN and SHAW, 1967); armed

---

[1] See also McKISC (1997) for a summary of this work.

[2] The 21-kT air-dropped 'Able' Atom bomb test (see, DoE-1994).

[3] The Tri-lateral Test-Ban Treaty (TTBT).

with such knowledge one could deduce crude travel-time information based on observed period. Single sensor data was originally recorded on paper records, and eventually as an analogue signal onto magnetic tape. Signal detection for these single sensor records consisted of either analogue signal processing via an electronic sound spectrograph (DONN *et al.*, 1963), or, after converting to a digital signal, calculating the Fourier Transform of a portion of the wave record centered at successive times. Amplitude was then plotted simultaneously as a function of both period and group velocity. When the amplitude was contoured in the period-velocity plane, the low frequency acoustic-gravity modes became clearly discernible.

It became quite clear by the early 1960s that the only way to differentiate between the local meteorological pressure variations and the infrasonic signals from distant sources was to use an array of sensors with separations of at least several miles (BROWN, 1963). Initially, signal feature extraction was achieved by 'visual congruence', or correlation, where signals on paper records were visually superimposed using a transparent table top illuminated from below.

A significant development in infrasonic signal detection was the introduction of the automatic correlator; BROWN (1963) gives the first appraisal of such a device. The correlator in this case is a largely mechanical device.[4] A metal plate, which has rods attached to one face with a recording head attached to the other end of each rod, is allowed to tilt in any direction via a gimble mechanism. In this way a range of slowness and azimuth values can be explored; the correlator output being an average of the cross correlations between pairs of its inputs.

The increased capability of digital electronic computing devices in the 1970s meant that more sophisticated algorithms could be utilized in the signal detection process. Beyond the delay and sum beamformer, the *Fk* detector became popular, especially in light of the Fast Transform algorithms being developed. Refinements to the basic *Fk* detection algorithm were also discussed in the literature. For example, SMART and FLINN (1971) suggested using a Fisher Statistic (*F*-stat.) to assign a significance weighting to the arrival indicated via a prior *Fk* analysis.

Presently, there is a substantial effort underway at the Prototype International Data Center to create a near real-time automated infrasonic signal detection and source location system. This system is designed to monitor compliance with the Comprehensive Test-Ban Treaty, although the design goals differ somewhat from the monitoring efforts of the 1950s and 1960s. Previous efforts focused on detecting large sources at known test sites, and consequently large aperture arrays were used with preferred steering directions. The goal of detecting 1-kT sources at distances of 2500 km at unknown locations imposes additional constraints on the new system. The smaller yield implies that discrimination of signal types becomes an important issue because many environmental sources are capable of generating impulsive

---

[4] KERR (1971) presents a digital version of this correlator.

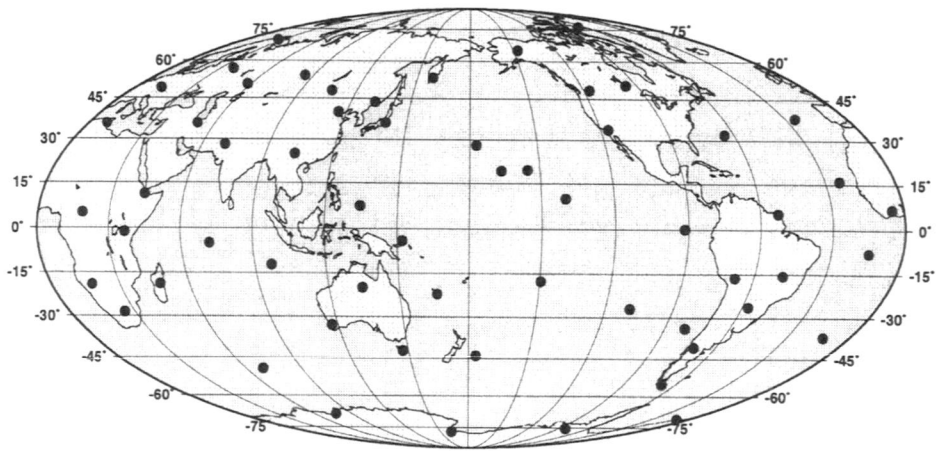

Figure 1
60-station IMS infrasonic network.

signals in the frequency range of interest for small explosions.[5] In addition, in most locations every direction from the array should be given equal consideration when seeking impulsive sources.

Complementing the existing hydroacoustic, seismic and radionuclide programs, the infrasonic data processing system gives the PIDC the ability to estimate the temporal and spatial source location for impulsive sources that couple in some part to the atmosphere. It is intended that the completed system will process continuously transmitted data from the full 60-station IMS infrasound network (see e.g., CHRISTIE, 1998; VELOSO, 1998) shown in Figure 1.

It is therefore necessary that reasonably accurate automatic event processing take place prior to any interactive analysis and production of a reviewed event bulletin.

The PIDC currently receives continuous infrasound data from the six arrays shown in Table 1.

The infrasonic subsystem can be divided into several separate and clearly-defined components as shown schematically in Figure 2.

The Continuous Data Subsystem

As in the case of seismic and hydroacoustic data, infrasonic data are sent to the PIDC by the individual states signatories according to the CD-1 protocol (IDC3.4.2). Although IMS specifications call for infrasonic data to be sampled at 20 Hz, data with sampling frequencies ranging from 10 Hz to 40 Hz are being received from the

---

[5] Microbaroms, for example, with periods around 5 s become a significant source of noise for the new system.

Table 1

Infrasonic arrays currently sending data to the PIDC. The total number of detections from Dec. 04 1998 to Jul. 01 1999 is also indicated

Array	Lat. (°)	Lon. (°)	Number of detections Dec. 04, 1998 to Jul. 01, 1999
LSAR	35.8667	−106.3342	10340
DLIAR*	35.8676	−106.3342	1823
SGAR	37.0153	−113.6153	4935
PDIAR	42.7663	−109.5939	790
TXIAR	29.3338	−103.6670	112
WRAI	−19.9402	134.2260	413

*IMS type array.

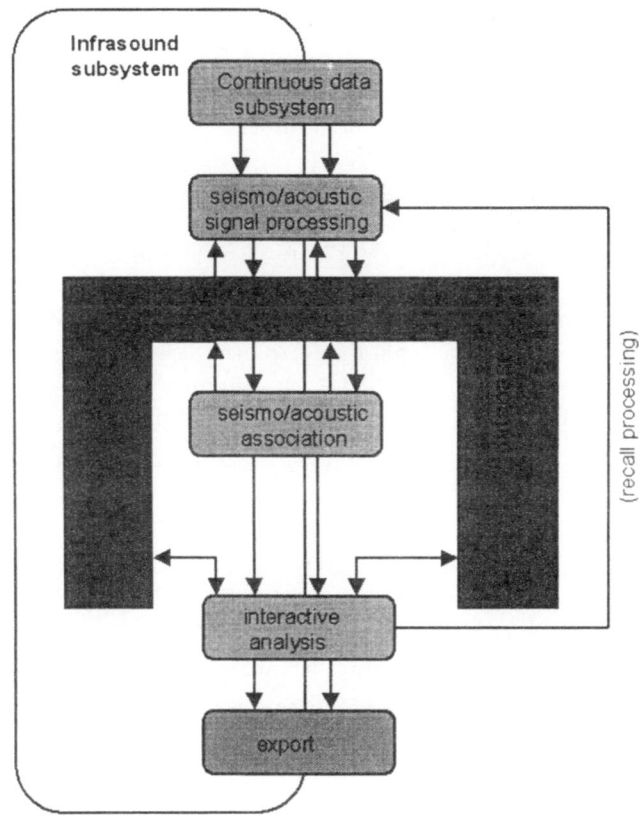

Figure 2
Schematic of PIDC automatic and interactive data processing system.

non-IMS infrasonic arrays. Additionally, data from both differential and absolute sensors are being received.

Seismo-acoustic Signal Processing

Numerous methods exist for extracting coherent signal information from array waveform data. Each of these makes certain suppositions concerning the spatial and temporal coherence of the signal, and indeed the background energy. Any infrasonic signal detection scheme chosen for use at the PIDC must operate well in the low signal-to-noise (SNR) environment that characterizes the events of interest and would have to operate at faster than real time if data from the full IMS network are to be routinely processed. Ability to steer the detector in a preferred direction to search for a possible signal is also a necessary feature. Signal detection is achieved via a coincidence detector which requires that two independent statistics achieve predetermined threshold values simultaneously. The first is a normalized cross correlation, and the second is a short-term-average/(lagged) long-term-average ratio on a beam formed in the direction of maximum coherence. The infrasonic detection algorithm is discussed more fully in later sections.

Seismo-acoustic Event Association and Location

At the stage of event association and location, extracted features from all three waveform technologies are processed together to form association sets and locate the potential sources of the arrivals. Other than the specific travel-time models used, no particular distinction is made between the waveform technology for one arrival and that of any other arrival within the same association set. For example, the same seismic source may generate a seismic *P* wave that could be assigned travel-time information from the IASPEI velocity model (KENNETT, 1991), as well as hydroacoustic *T* phases from which predicted travel times can be obtained from seasonal and path-dependent models. Similarly, predicted travel times for infrasound arrivals may be inferred using a model such as that proposed by PIERCE and POSEY (1970), GARCES *et al.* (1998) or DIGHE *et al.* (1998). The location algorithm makes use of arrival-dependent weights which are the inverses of the *a priori* measurement and modeling errors on each feature of the arrivals. For instance, in the case of infrasonic data, the present travel-time model is very basic and does not take into account the wind along the propagation path. A large *a priori* error is therefore assigned to the infrasonic travel-time, which in turn means that the travel-time component of infrasonic arrivals is given lower weight in the location process than the travel-time component of a seismic arrival. If an event is formed with seismic and infrasonic arrivals, the seismic travel-time component will contribute more to the temporal definition of the event's occurrence, whereas the infrasound will primarily influence the spatial location of the event. The azimuth of infrasonic arrivals is assigned a very small *a priori* error and therefore the azimuth has a comparatively higher weight in the locator for infrasound data than it has for seismic data. In the future, infrasonic travel-time models of higher accuracy will be developed and it is hoped that they will allow us to take full advantage of the travel-time picks from infrasound arrays.

Interactive Analysis

All automatically formed events are scrutinized by analysts to check the validity of the automatic association, thus ensuring the integrity of the Reviewed Event Bulletin (REB). In addition, analysts scan the data manually for events missed during automatic processing. A graphical user interface, known as the Analyst Review Station (ARS), has been developed which allows an analyst to display waveforms for a specified time period from individual receivers from any, or all of the seismo-acoustic technologies. Fundamental signal processing features built into ARS enhance the analysts' decision making ability when assessing a prospective event. In addition, 'event definition criteria' have been established that provide a measure against which all events are created. This sort of regular analyst review of infrasound data has never been performed on a global scale, therefore careful instruction regarding signal interpretation is required. The product of interactive analysis is the Reviewed Event Bulletin, which contains event location and magnitude estimates as well as parameters of signals used to define and locate each event.

Product Export

The REB is published in a timely fashion and contains event information, such as estimated spatial and temporal source location as well as an hypothesized magnitude.[6]

As suggested by Figure 2, connections to the surrounding database occur at many different levels.

At a basic design level, there is slight difference between the infrasonic processing subsystem and those of the other seismo-acoustic technologies (i.e., seismic and hydroacoustic); the parallel design was chosen to facilitate the data fusion concept. One area where the basic design symmetry is lost, though, is in the area of recall processing. For several reasons, some database entries may not have been filled during routine automatic processing of the waveform data. This may be due, for example, to the late arrival of some data from a station. Recall processing allows an analyst to reprocess data in certain time periods to better characterize signals

---

[6] Note that the much smaller propagation speed of infrasonic signals ($\sim$ 300 m/s) over those of seismic ($\sim$ 9000 m/s for P waves) and Hydroacoustic ($\sim$ 1000 m/s) prevents the inclusion of infrasonic information in various early automatic event bulletins. In addition, it is not PIDC policy to 'classify' sources in any way, hence any source magnitude measure is related to physical observables rather than an implied kiloton yield.

associated with a prospective source. Although infrasonic recall processing is a desirable feature, it has not been implemented at the present time. Furthermore, when implemented it may deviate in detail from the procedures used for automated infrasonic processing.

In section two of this paper, a summary of the detector design and operation is given. Section three discusses our approach to phase identification; section four summarizes our synergistic approach to the source location problem and presents several examples. Section five discusses interactive analysis, highlighting aspects of waveform analysis that are particular to infrasound. A brief summary of ARS will be given and preliminary analyst event definition criteria are discussed. In section six, an evaluation of the system performance over a six-month period is presented. Section seven briefly discusses proposed system enhancements, and a brief summary and conclusion is given in section eight.

2. Signal Detection

The paradigm for achieving automated detection of airborne infrasonic arrivals is coincidence detection. Coincidence detection is the recognition that for a traditional coherent ('delay and sum') beam, $B_s(t)$, of digitally sampled waveform data there is energy excess beyond a fixed threshold, B_0, at the same time (i.e., coincidence) that there is spatial coherence beyond a fixed threshold C_0, in a sample-by-sample 'beam' $C_s(t)$ of a suitable measure of spatial coherence.[7] Both $B_s(t)$ and $C_s(t)$ are formed to the same vector slowness value **s**, and are generated from the bandpass-filtered waveform data from the same spectral band **B**, and the same subgroup **G**, of sensors. The basis for forming these two beams in slowness **s**, is a 'contact' cluster analysis wherein for any given macroscopic time interval (typically of 20 to 60 minutes duration) of detection processing, a primary detection space is generated. This space may contain either no contacts, in which case no beams are formed and no coincidence detections can be made, or some contacts, in which case one or more 'clusters' are formed, and one or more pairs of beams, $B_s(t)$ and $C_s(t)$ are generated, a distinct pair for each cluster, and coincidence detections are made. The following sections identify and describe the nature of the algorithms for generating primary detection space and forming clusters where there are contacts.

2.1 Coincidence Detection

Given the beam $B_s(t)$, all time samples where $B_s(t) > B_0$ serve to define the temporal intervals of excess energy in the vector slowness value s sufficient to declare

---

[7] Section 2.1.2, below provides further detail on the nature of $C'_s(t)$.

energy detection. (The specific excess energy detection algorithm is described in further detail in section 2.1.1, below.) Similarly, where beam $C_s(t) > C_0$ there is sufficient spatial coherence to support the declaration of coherent detection. For those temporal intervals in which both criteria are satisfied there are one or more coincidence detections (the precise number is governed by the number of distinct[8] energy packets which are detected). Generally, it is not possible *a priori* to precisely predict the number of coincidence detections, if any, which will be realized for a given pair of beams.

At the time of the local peak energy of each distinct energy packet, which is also a coincidence detection, a 'vernier' *fk*-spectrum is invoked for estimating the precise value of vector slowness, **s**, which is to be attributed to this energy packet. (See section 2.1.2, below, for further details concerning the vernier *fk*-spectrum.) In addition to the foregoing parameter estimates, the vernier *Fk*-spectrum provides estimates of the variance in vector slowness including both measurement and modeling error. The measurement error estimate is formulated in a fashion analogous to a Cramer-Rao bound. The vector slowness s in which to form beams $B_s(t)$ and $C_s(t)$ is a slight modification[9] of the vector slowness, $\bar{s}_i$, of the 'seed' contact (i.e., the contact exhibiting the largest measure of spatial coherence) in the ith cluster which is being evaluated.

2.1.1 Energy detection via the STA/LTA detection statistic applied to $B_s(t)$

The traditional coherent ('delay and sum') beam $B_s(t)$ of bandpass-filtered waveform energy (beamformed for only those sensors in the sensor subgroup and for the spectral band of the 'seed' contact of the cluster) is evaluated for temporally abrupt increases of a sufficient magnitude in beam energy. $B_s(t)$ is defined as

$$B_s(t) = \sum_{j=1}^{J} x_j(t + \tau_j) \; , \tag{1}$$

where $x_j(t)$ are the bandpass-filtered waveform data sampled at time t for the jth sensor, and τ_j is the time alignment required to register the jth sensor's time-domain data for the given vector slowness value, s. The time-integrated beamformed data yield the quantity, $p_{sc}(s; T, t)$, as in

---

[8] Local energy maxima which are sufficiently separated in time with a sufficiently deep minimum between them are declared as distinct energy arrivals.

[9] In order to facilitate the accurate formation of beams $B_s(t)$ and $C_s(t)$, without the need to interpolate between time-domain samples, a small slowness region (approximately 10 degrees wide in azimuth by 10% wide in magnitude slowness) is searched in fine increments to find a value of vector slowness, s, near $\bar{s}$, which yields a list of time alignment lags (for beamforming) that are as close as possible to integer numbers of samples. The vernier *fk*-spectrum slowness estimate is ultimately added vectorially to s to estimate the net vector slowness of the arrival for the given energy packet.

$$p_{sc}(s;T,t) = \sum_{t'=t}^{t+T} \left| \left[\sum_{j=1}^{J} x_j(t' + \tau_j) \right] \right|^{Q}, \qquad (2)$$

where $t + \tau_j$ is the start time of the integration window for the jth sensor-channel and T is the time duration. The integer Q assumes the value of either 1 or 2 for *L1* and *L2* norms, respectively. Normally, for geophysical applications, both seismic and infrasonic, the algorithm is applied with a *L1* norm. A *L1* norm has better immunity to the effects of isolated (1 to 3 samples of time-domain duration) spikes in the waveform data. The detection statistic, $\rho(s;t)$, applied to accomplish the energy detection is the 'short-term average' (STA) compared with the time-lagged (by Δt) 'long-term average' (LTA), as in:

$$\rho(s;t) = \frac{\left[\frac{p_{sc}(s;T_S, t)}{T_S} \right]}{\left[\frac{p_{sc}(s;T_L, t-\Delta t)}{T_L} \right]} = \left[\frac{T_L}{T_S} \right] \cdot \left[\frac{p_{sc}(s; T_S, t)}{p_{sc}(s; T_L, t - \Delta t)} \right], \qquad (3)$$

where the short averaging time is T_S, the long averaging time is T_L, the time separation between the centers of the STA and LTA windows is $\Delta t - (T_L - T_S)/2$, the vector slowness of the beam is s and the nature of the norm (*L1* or *L2*) is governed by Q. In the absence of spatially coherent arrivals, and in the presence of white Gaussian noise, not the typical infrasonic ambient, the detection statistic $\rho(s;t)$ is approximately unity due to the normalization of each sum by the integration time. Where $\rho(s;t)$ exceeds a threshold, e.g., $\rho_0 > 1$, there is sufficient energy excess to allow for the declaration of coincidence detection (should there be a time-overlapped interval of sufficient spatial coherence exhibited by $C_s(t)$). The number of arrivals during unbroken intervals where $\rho(s;t) > \rho_0$ for all t satisfying $t_1 < t < t_2$, is governed by the number of distinct energy packets wherein peaks in $\rho(s;t)$ are sufficiently separated in time and have a sufficiently 'deep' minimum between neighboring peaks.

It is feasible to apply variations on the behavior of the STA/LTA detection statistic. One variation which may have merit for the peculiar nature of wind burst behavior at infrasonic sensor array sites is that it may be useful in formulating the LTA to replace the traditional coherent ('delay and sum') beam $B_s(t)$, with the 'minimum' beam $\underline{B}_s(t)$, defined as

$$\underline{B}_S(t) = x_j(t + \tau_j)$$
$$1 \leq j \leq J \ni |x_j(t + \tau_j)| < |x_k(t + \tau_k)|$$
$$1 \leq k \leq J$$
$$k \neq j.$$

Such a beam is likely to be dominated less by isolated wind bursts on a small subset of the sensors which are included in the sensor subgroup. Furthermore, the functional behavior of the STA/LTA detection statistic can also be altered to a form in which the LTA window is frozen in place at the instant $\rho(s;t) > \rho_0$, i.e., at $t = \bar{t}$, only after $\rho(s;t) < \rho_0$ and $t > \bar{t} + \Delta t$ is the LTA window advanced to its normal time alignment Δt earlier than the STA window position. This form of STA/LTA detector is more effective at sensing a succession of arrival energy packets because the LTA window does not move into the early energy packets.[10] (When the LTA window moves into the early arrival energy packets its value can be increased by the energy in these arrivals to the point of diminishing $\rho(s;t)$'s value when it encounters later arrivals which are distinct and should be detectable.)

2.1.2 Spatial coherence estimation and evaluation applied to $C_s(t)$

In order to estimate and evaluate the spatial coherence in the vector slowness, s, a 'beam' of coherence estimates is produced using the F-Statistic estimator generated in a sample-by-sample mode so as to produce a time-series $C_s(t)$ sampled at the same rate as the bandpass-filtered waveform data (from the sensor-channels, J in number). In fact, any of several different detection statistics could have served for generating the spatial coherence 'beam' $C_s(t)$. Historically, for long-range airborne infrasound processing, spatial coherence has been estimated by $\Gamma(s)$, the average of time-aligned normalized cross-correlation coefficients (see the definition of $\Gamma(s)$ in section 2.3, below). It has recently been shown (KATZ, 1999) that the F-Statistic, $C_s(t)$ in the notation used herein, another popular estimator of spatial coherence, can be calculated directly from an algebraic transformation of $\Gamma(s)$. As detectors of spatial coherence both $\Gamma(s)$ and $C_s(t)$ have identical minimum detectable levels (for fixed statistical criteria such as false alarm probability less than 'a' and detection probability in excess of 'b'); however, secondary properties of the F-Statistic favor its application for also estimating the onset time of spatial coherence in excess of a fixed threshold. Consequently, for the algorithm described herein, the spatial coherence 'beam' is a sample-by-sample value of the F-Statistic (BLANDFORD, 1974), $C_s(t)$, and can be expressed as:

$$C_S(t) = \left(\frac{J-1}{J}\right) \cdot \frac{\sum\limits_{t'=t}^{t+T}\left[\left|\sum\limits_{j=1}^{J} x_j(t'+\tau_j)\right|\right]^2}{\sum\limits_{t'=t}^{t+T}\left[\sum\limits_{j=1}^{J}\left|x_j(t'+\tau_j) - \left[\frac{1}{J}\cdot\sum\limits_{m=1}^{J} x_m(t'+\tau_m)\right]\right|^2\right]} \tag{4}$$

where the $x_j(t)$ are the bandpass-filtered waveform data for the jth sensor-channel at the time sample corresponding to time t, and the τ_j's are the time-domain time-lags

---

[10] Both types of variation to the usual STA/LTA detection statistic formulation have been incorporated into the algorithm, but have not been implemented at the time of writing.

required to achieve the vector slowness value s. In order to succeed, it is generally essential to reduce each sensor-channel's data set to an exactly zero-mean constant variance process. Otherwise, the exquisite cancellation called-for in the denominator of this last equation will fail. The denominator is simply the sum of the power in the residuals over the time-window from t to $t + T$ for the J sensor-channels which are in the requisite sensor subgroup. The residuals in each sensor-channel are generated by subtracting from the time-aligned waveform data in that sensor-channel a suitably scaled and time-aligned beam. If the scale of a given sensor-channel is too far from the scale of the normalized beam, the residuals will be too large. Although by definition bandpass-filtered waveform data are zero-mean in an ensemble sense, any particular time-window of such data may exhibit a mean slightly different from zero. Consequently, for best results the specific data samples to be used, e.g., for sensor-channel, m, the time samples corresponding to $t + \tau_m$ through those corresponding to $t + \tau_m + T$, are explicitly rendered exactly zero-mean by finding their mean and subtracting this value (divided by the number of sample in time duration T) from each sample. Similarly, to establish the J sensor-channels on a common scale, once they have been reduced to exactly zero-mean, their summed powers are generated, P_m for $m = 1, \ldots, J$, and the samples of each sensor-channel are divided by the square-root of the corresponding P_m. Thus, for sensor-channel, m, of bandpass-filtered waveform data, $y_m(t)$ we transform to $x_m(t)$ as follows:

create intermediate time-series $z_m(t)$ as in:

$$z_m(t) = y_m(t) - \mu_m , \tag{5}$$

where the sensor-channel dependent mean, μ_m, is given by:

$$\mu_m = \frac{1}{T} \cdot \sum_{t'=t}^{t+T} y_m(t' + \tau_m) . \tag{6}$$

Then form the power P_m and rescale the intermediate data $z_m(t)$ to form the time-series $x_m(t)$ to be input to the $C_s(t)$ calculation, as follows

$$P_m = \sum_{t'=t}^{t+T} |z_m(t' + \tau_m)|^2 \tag{7}$$

$$x_m(t) = \frac{z_m(t)}{\sqrt{P_m}} . \tag{8}$$

In order to avoid excessive arithmetic error, loss in coherence, and the excessive processing burden of cubic interpolation of every time-series sample in every sensor-channel, the beam $C_s(t)$ is formed[11] to a vector slowness value as close as possible to the desired vector slowness value $\bar{s}$. Here the actual steered vector slowness value s leads to a set of delays τ_m which are as close as possible to integer-valued delays with respect to the sampling interval of the time-series, $y_m(t)$.

Since T is generally much larger than the sampling interval of the bandpass-filtered waveform data, the 'beam' $C_s(t)$ is slowly varying in value (from sample to sample). To evaluate the entire processing time interval for sufficient levels of spatial coherence in vector slowness, s, $C_s(t)$ is simply compared to a fixed threshold value, e.g., C_0, for the entire time interval of the processing (typically a batch of 20 to 60 minutes duration). All time samples for which $C_s(t) > C_0$ are 'declared' to have detectable levels of spatial coherence in the vector slowness value s.

2.1.3 The vernier fk-spectrum

The vernier fk-spectrum uses an algorithm nearly identical to the traditional Fk-spectrum, and is defined in terms of the equivalent F-Statistic transformation (SMART and FLINN, 1971) as

$$FS(s; f_1, f_2; T, t) = [J - 1] \cdot \left\{ \frac{E(s; f_1, f_2; T, t)}{E_0(f_1, f_2; T, t) - E(s; f_1, f_2; T, t)} \right\} \qquad (9)$$

where the slowness-dependent term, $E(s; f_1, f_2, T, t)$, is given by

$$E(s; f_1, f_2; T, t) = \frac{1}{J} \cdot \sum_{f'=f_1}^{f_2} \left| \left[\sum_{j=1}^{J} \left[\sum_{t'=0}^{T} e^{2\pi\sqrt{-1}f't'} \cdot x_j(t' + t + \tau_j) \right] \cdot e_j^{-2\pi\sqrt{-1}f'(s \cdot r)} \right] \right|^2 \qquad (10)$$

where the τ_j are the pre-alignment lags to center the Fk-spectrum about the vector slowness value of interest; the total in-band energy for this time-window and spectral band $E_0(f_1, f_2, T, t)$ is given by

$$E_0(f_1, f_2; T, t) = \frac{1}{J} \cdot \sum_{j=1}^{J} \left\{ \sum_{f'=f_1}^{f_2} \left| \left[\sum_{t'=0}^{T} e^{-2\pi\sqrt{-1}f't'} \cdot x_j(t' + t + \tau_j) \right] \right|^2 \right\} . \qquad (11)$$

In the foregoing formulae, t is the start time of the processing window, T is the time duration of the processing window, and f_1 and f_2 ($> f_1$) are the frequency limits of the processing band.

---

[11] A fine-scale search centered on $\bar{s}$ is performed where for each alternative to s the list of precise delays is generated. The sum of squared discrepancies between the actual delay values in the list and the closest integer-valued delays is generated. For that alternative slowness, s having the least sum of squared discrepancies, the actual beams are formed. Rarely does the discrepancy between a required delay and the closest integer-valued delay exceed 0.1 of the sampling time interval.

This definition for the vernier *Fk*-spectrum differs from the traditional *Fk*-spectrum in the following ways:

(i) the time-domain waveform data are pre-aligned in time (via the τ_j) to the approximate vector slowness $\bar{s}$ of the plane wave arrival of interest.

(ii) the overall slowness coverage is limited to a small region of slowness space Δs sufficient to encompass the slowness, s, of the actual arrival which is sought.

(iii) the resolution δs in vector slowness is very fine, i.e., $\Delta s/\delta s \gg 1$.

By pre-aligning the waveform data which are input to the *Fk*-spectrum to nearly the vector slowness of interest close to the vector slowness of the beam pair $B_s(t)$ and $C_s(t)$ in this application, the resultant *Fk*-spectrum is centered in slowness space on the detection of interest. Furthermore, by limiting the overall vector slowness coverage to only a small region of slowness space, the resultant spectrum is dominated by the detection of interest and is not 'captured' by a potentially stronger arrival at some other azimuth and/or magnitude slowness value. Finally, by generating the vernier *fk*-spectrum with a very fine vector slowness quantization, the position of the peak value is well estimated and can be readily interpolated if this is required.

2.1.4 *Errors in the vector slowness estimates from the vernier Fk-spectrum*

The errors in the vector slowness parameters estimated by the vernier *Fk*-spectrum (as in those slowness coordinates for which the value of the *Fk*-spectrum is maximum over its vector slowness coverage, Δs) are given by

$$\delta|\mathbf{v}'| = \frac{1}{|s'|\sqrt{FS}} \cdot \frac{\delta|s'|}{|s'|} \tag{12}$$

$$\delta\theta' = a\sin\left(\frac{1}{2} \cdot \frac{\delta|s'|}{|s'|}\right) , \tag{13}$$

where $|\mathbf{v}'|$ is the magnitude of the apparent phase velocity, θ' is the azimuth, FS is the value of the *F*-Statistic of the peak in the vernier *fk*-spectrum, and $\delta|s'|$ is the estimated error in magnitude slowness. The prime symbol on the foregoing parameters represents a vernier value of the parameter. Since this is a vernier *fk*-spectrum, these errors must be transformed through the vector sum of the pre-alignment slowness, s, and the estimated (from the vernier *Fk*-spectrum) slowness s' to yield the net signal arrival slowness, s_0:

$$s_0 = s + s' . \tag{14}$$

The resultant forms of $\delta|\mathbf{v}|$ and $\delta\theta$ are fairly complicated. These forms are remarkably similar to the parametric forms for the Cramer-Rao (most optimistic) bounds on the estimation errors for an ideal plane wave, planar array matched in geometry to the wavelength of the arrival.

2.2 Cluster Formation

The basis for achieving automated detection of airborne infrasonic arrivals is multiple hypothesis testing of a suitably selected detection statistic evaluated in a six-dimensional space; $\Gamma = \Gamma(s : B, G, E)$. That is, the detection statistic Γ is compared with a slowness-dependent threshold in a space with two dimensions of vector slowness s in the planar projection of the sensor array for a given spectral band B, a particular subgroup of sensors at that array site G, and a specific processing time window or epoch E (defined by start time and duration). The coverage in vector slowness must be sufficiently extensive to include the entire slowness space for plane wave arrivals from distant sources, typically 360 degree coverage in azimuth and coverage in magnitude slowness from approximately 1.0 to 5.0 s/km. For a given sensor site and processing interval, there is a slowness plane for each combination of spectral band, sensor subgroup and processing time epoch. Those slowness planes wherein the detection statistic, $\Gamma(s : B, G, E)$, exceeds the slowness-dependent threshold,[12] $\Gamma_0(s : B, G)$, are said to contain one or more contacts. If there is more than a single contact for a given slowness plane, the contacts are rank ordered based upon their degree of spatial coherence, and the best contacts (up to six in number) are reported into the Primary Detection Space (PDS). Contacts must be spatially distinct to be reported. That is, given a slowness plane with the detection statistic exceeding its threshold at more than one value of vector slowness, with vector slowness for the maximum value of detection statistic, noted as $\bar{s}_0$, an exclusion region in slowness space (defined by an azimuth extent and a magnitude slowness extent)[13] is centered on $\bar{s}_0$. If there are values of the detection statistic exceeding the threshold for slowness coordinates outside the exclusion region, then the best above-threshold value of the detection statistic outside the exclusion region defines the next best contact, and its coordinates, $\bar{s}_1$, define the vector slowness of this contact. An exclusion region of the same extent is centered on $\bar{s}_1$ and the procedure is applied again to seek the next best above threshold contact. For each slowness plane this procedure is terminated when there are no longer values of the detection statistic which exceed threshold outside of any of the prior exclusion regions (which are allowed to overlap) or when six contacts have been reported.

Once the entire set of PDS slowness planes has been evaluated for evidence of contact, if there are one or more contacts among all of the slowness planes, then a master slowness plane is defined. It has the same extent in vector slowness as the individual slowness planes. The contact with the highest degree of spatial coherence

---

[12] The threshold typically varies with spectral band and sensor subgroup as well, however it is time invariant for a given B, G, and s.

[13] The extent in vector slowness of the exclusion region is defined by the slowness width of the sensor array's spatial impulse response function.

is 'transferred' to the master slowness plane and serves as the seed contact of the 0th cluster. All other contacts of the PDS are evaluated to determine whether they are 'close enough'[14] in vector slowness to the slowness coordinates, $\bar{s}_0$ of this seed contact; if so, they are transferred to this 0th cluster. Once a contact is transferred it is no longer available for inclusion in or the formation of other clusters. If after all contacts are considered for inclusion in the 0th cluster there remain one or more contacts not transferred, then the remaining contact with the highest degree of spatial coherence is 'transferred' to the master slowness plane and serves as the 'seed' contact of the 1st cluster. All other remaining contacts of the PDS are evaluated to determine whether they are 'close enough' in vector slowness to the slowness coordinates of this seed contact; if so, they are transferred to this 1st cluster. As with the 0th cluster, once a contact is transferred it is no longer available for inclusion in the formation of other clusters. This process is continued until all contacts are transferred to clusters in the PDS or a maximum of twelve clusters have been formed. The numbers of clusters formed for a given sensor site and processing interval cannot be precisely predicted. Where no clusters are formed no coincidence detections can be realized. Where one or more clusters are formed, their 'seed' contacts define the beam pairs, $B_s(t)$ and $C_s(t)$, which are used to seek coincidence detection.

2.3 Primary Detection Space Generation

In the foregoing section the concept of a Primary Detection Space (PDS) was introduced. The PDS provides the indications as to where in slowness space there may be sufficient spatial coherence to seek detection of airborne infrasonic waves. Cluster formation and coincidence detection complete the search for detection; PDS generation provides the input for cluster formation. Since both energy excess and spatial coherence beyond a statistically significant level are required to achieve coincidence detection, in principle, the PDS could have been a search for energy excess rather than a search for sufficient spatial coherence. In fact, the reason for populating the PDS with a detection statistic for sufficient spatial coherence is that in typical scenarios the in-band energy excess caused by random wind bursts, incoherent from station to station within an infrasonic sensor array site, is far more frequent than the presence of sufficient spatial coherence (caused by microbaroms, etc.). If energy excess were used to populate the PDS there would be far more episodes of searching for coincidence detection that would occur with a spatial coherence detection statistic. In addition, traditional beamforming and energy detector sensitivity to vector slowness for a typical IMS scale infrasonic sensor array site does not lend itself to particularly well-defined estimates of arrival angle or

---

[14] As with the recognition of distinct contacts within a given slowness plane, in the formation of distinct clusters the seed contacts of any two clusters must be further apart in vector slowness than the criterion set by the extent of the exclusion region used in contact recognition.

magnitude slowness. It would be necessary to search a fan of several narrower 'beams' in the neighborhood of the vector slowness estimate from a traditional energy detector. Consequently, a PDS populated with a spatial coherence detection statistic is significantly more efficient than one populated with an excess energy detection statistic.

The PDS consists of a six-dimensional space of three external dimensions and three internal dimensions. The external dimensions are: spectral band B, sensor subgroup G, and epoch E. That is to say, the PDS can be thought of as a collection of cellular volumes in this external three-dimensional space. Because we do not know *a priori* where precisely in the spectrum and at what time a particular airborne infrasonic arrival may arrive at a given sensor site, it is necessary to seek detection over a number of frequency-overlapped spectral bands for an endless succession of time-overlapped processing epochs. Depending on the spatial diversity of the sensor site, it may be necessary to use particular subgroups of sensors for specific spectral bands. The endless succession of processing epochs would ultimately lead to memory problems in any processor with finite memory, given the current implementation of the automatic pipeline signal processing in the paradigm of the Detection and Feature Extraction algorithm (DFX). Consequently, a succession of time-overlapped processing intervals is used to achieve complete and on-going time coverage. Each time interval consists of dozens to thousands, but typically several hundred, time-overlapped processing epochs. To ensure complete coverage, several of the processing epochs at the start and end of each time interval (typically a few 10 s to 100 s of minutes of data) are repeats of the 'end' epochs in the previous interval or are going to be repeated as 'start' epochs in the next interval.

Within each 'cell' of the PDS's three external dimensions a statistically significant level of spatial coherence is sought. The mechanism for seeking this coherence is the slowness plane 'populated' via a suitable detection statistic. At the outset of the detection process we seek a signal whose spectral band, time of arrival, arrival angle, and magnitude phase velocity are all unknown. Thus it is necessary to search over all physically realizable combinations of spectral band, arrival time, azimuth and magnitude phase velocity. Since these are all continuum parameters to avoid an infinite search problem, the parameters are searched in quantized steps sufficiently large to reduce processing to a feasible burden, yet sufficiently small so as not to "step over" a detection which would occur only for a very well-matched and narrow range of one or more of these parameters. The search is conducted by producing an entire 'volume' of a three-dimensional subspace whose dimensions are: spectral band × sensor subgroup × processing epoch (for the arrival time parameter). Within each cell of this 3-D subspace, a three-dimensional subspace is generated whose dimensions are the numerical value of the detection statistic, arrival azimuth, and magnitude phase velocity. In this fashion, all unknown parameters of the signal which is to be detected, which parameters may influence the magnitude of the detection statistic, are searched. Within the PDS cells we have three internal dimensions; magnitude detection statistic, Γ, and two components of vector slowness s (in a plane). Thus the PDS is a six-dimensional space: $\Gamma = \Gamma\,(s\colon B, G, E)$. A number of alternative detection statistics can be formulated for estimating the degree of spatial coherence in a plane wave arrival, over a spatially diverse group of omni-directional sensors which are deployed in essentially a planar array. The detection statistic of interest to this narrative is the arithmetic mean of the aligned normalized cross-correlation coefficients $c_{ij}(l_i - l_j)$ for all independent sensor pair combinations which can be formed given the J sensors in sensor subgroup G. The waveform data used to generate the normalized cross-correlations functions is, of course, bandpass filtered to the spectral band defined by B, and consists of the time-domain samples required by processing the epoch E. The alignment by time lags l_i puts the waveform data for sensor i into the correct time-register for the hypothesized vector slowness s:

$$\Gamma(S) = \left[\frac{2}{J(J-1)}\right] \cdot \sum_{j=1}^{J-1}\left[\sum_{i=j+1}^{J} c_{ij}(l_i - l_j)\right]. \tag{15}$$

The factor $[J(J-1)/2]$ is simply the number of independent pair-wise combinations which can be developed from J sensor-channels; J ways to select the first sensor-channel, $(J-1)$ ways to select the second sensor-channel, divide by 2 to account for the fact that sensor-channel m correlated against sensor-channel n yields the same information as sensor-channel n correlated against sensor-channel m.

2.4 Slowness Plane Discretization

The PDS is populated with slowness planes which are themselves populated with a spatial coherence detection statistic. This population of each slowness plane is accomplished by time aligning the normalized cross-correlation functions generated from the bandpass-filtered waveform data from spectral band B from the sensor-pairs which can be formed for the sensors in subgroup G during the processing window corresponding to epoch E. Given a vector slowness value, s, and a known sensor geometry for subgroup G, each of the sensors in this group will have a unique time-delay τ_i relative to the reference coordinates of the sensor array site. These individual delays will form a self-consistent set of pairwise delays for all the independent sensor-channel pairs which can be formed for the sensors in subgroup G. The detection statistic is simply the arithmetic mean of the normalized cross correlations for each independent sensor-channel pair evaluated at the pairwise delay called-for by vector slowness value, s. Since the pairwise delays will only rarely equal the discrete delays at which the c_{ij} (Δl_{ij}) are generated, cubic interpolation is applied among the four available closest samples of c_{ij} (Δl_{ij}). $\Gamma(s)$ is readily calculated given the appropriate set of c_{ij} (Δl_{ij}). The question arises, how should the vector slowness space be sampled? The DFX infrasonic approach is to sample the physically realizable vector slowness space at a sufficiently fine grid of slowness values that the ideal array response function is sampled at no worse than a

1 dB loss for an ideal plane wave arrival. This necessitates a slowness sampling strategy which is somewhat dependent on the overall spectral band and bandwidth of the processing.

In practice, evaluating the PDS for evidence of infrasonic plane wave arrivals with statistically significant levels of spatial coherence requires a refinement algorithm. That is, if a given 'pixel' of the sampled slowness plane in a given 'cell' of the PDS exceeds its local threshold, then a finer gridwork of slowness pixels ($11 \times 11 = 121$ values in number) is generated surrounding the given pixel. These pixels are evaluated to find a local maximum, if any, in excess of the given pixel's maximum. Fitting and interpolation are performed in this fine gridwork to estimate the arriving plane wave's azimuth, magnitude slowness, level of spatial coherence, and Cramer-Rao estimates of the variances in the vector slowness components. This refinement procedure is often applied to several pixels in each slowness plane should there be more than a single pixel which exceeds its local threshold. The pixels which are refined in addition to being above their local thresholds must be local maxima and must be sufficiently far apart (in vector slowness) from one another to be considered distinct 'contacts'. This last requirement is imposed to avoid declaring multiple 'contacts' from the above threshold fluctuation structure of a given single contact's slowness lobe.

2.5 Normalized Cross-correlation Function Generation and Bandpass-filtering

The normalized cross-correlation coefficient at delay, Δl_{ij}, is defined as

$$
c_{ij}(\Delta l_{ij}) = \frac{\sum\limits_{t'=t_0}^{t_0+T} x_i^*(t') \cdot x_j(t' + \Delta l_{ij})}{\sqrt{\left[\sum\limits_{t'=t_0}^{t_0+T} |x_i(t')|^2\right] \cdot \left[\sum\limits_{t'=t_0}^{t_0+T} |x_j(t' + \Delta l_{ij})|^2\right]}} , \tag{16}
$$

where $x_i(t)$ is the bandpass-filtered time-domain waveform data from the ith sensor-channel for the time sample corresponding to epoch time t, t_0 is the start time of the processing window, T is the time duration of the processing window, and Δl_{ij} is the time delay in the same units as t. This time delay Δl_{ij} is of polarity such that an arrival which is later at sensor-channel j than at sensor-channel i will require $\Delta l_{ij} > 0$ to achieve waveform alignment. Given the physically accessible domain of delays, L_{ij}, (defined by the quotient of sensor baseline separation distance and minimum magnitude phase velocity of interest) the collection of the $c_{ij}(\Delta l_{ij})$ when arranged in order of ascending Δl_{ij} values, i.e., Δl_{ij} in the domain defined by $-L_{ij}/2 < \Delta l_{ij} < +L_{ij}/2$, can be treated as a normalized cross-correlation *function*.

It should be noted that to achieve exactly correct normalized cross-correlation functions, the sensor-channel i waveform data must be of fixed sample length corresponding to T, whereas a larger quantity of sensor-channel j waveform data must be available; $T + L_{ij}$, where L_{ij} is the full extent of the delay search space. For example, if $T = 30$ seconds, and the geometry of the sensor site can cause delays up to ± 5 seconds (over the longest baseline separation among the sensor-channels), then $L_{ij} = (+5) - (-5) = 10$ seconds. Channel i will require 30 seconds of waveform data but sensor-channel j will require 40 seconds of data. It is essential that the '$x^*_i x_j$' cross-product in the numerator of Equation (16) contain cross-products from only T seconds of waveform data in the correct alignment. Thus, if the Δl_{ij} is -3 seconds, sensor-channel i waveform data will span t_0 through $t_0 + 30$ and sensor-channel j waveform data will span $t_0 - 3$ through $t_0 + 27$, whereas if Δl_{ij} is $+2$ seconds, sensor-channel i waveform data will span t_0 through $t_0 + 30$ and sensor-channel j waveform data will span $t_0 + 2$ through $t_0 + 32$. Furthermore, the denominator term for a given sensor-channel must span exactly the same waveform data samples as the numerator term for that same sensor-channel, regardless of delay. Consequently, the sensor-channel i denominator term is a scalar $P_i(t_0, T)$, independent of Δl_{ij}. On the other hand, the sensor-channel j denominator term is an array $P_j(\Delta l_{ij}; t_0, T)$ of values L_{ij} in length. Approximations which replace $P_j(\Delta l_{ij}; t_0, T)$ by a scalar, e.g., the value it has at delay 0: $P_j(0; t_0, T)$, can cause spurious detection results.

In the name of computational speed, deficient algorithms for generating $c_{ij}(\Delta l_{ij})$ which allow the waveform data to 'wrap-around' are sometimes used. Where there is 'wrap-around' a Δl_{ij} value of -3 seconds will cause sensor-channel i waveform data span $t_0 + 3$ through $t_0 + 30$ followed by t_0 through $t_0 + 3$ and sensor-channel j waveform data span t_0 through $t_0 + 30$. Thus the last three seconds of sensor-channel j data are 'correlated' against the first three seconds of sensor-channel i data for this three second segment, yielding an actual delay of $+27$ seconds. Alternatively, when Δl_{ij} is $+2$ seconds, sensor-channel i waveform data span t_0 through $t_0 + 30$ and sensor-channel j waveform data span $t_0 + 2$ through $t_0 + 30$ followed by t_0 through $t_0 + 2$. Thus, the first two seconds of sensor-channel j data are 'correlated' against the last two seconds of sensor-channel i data for this two second segment, yielding an actual delay of -28 seconds. In either case, the sum in the numerator has invalid contributions from a significant fraction of its terms, degrading the detection sensitivity in a lag-dependent fashion and likely biasing the estimation of parameters.

The implementation of $c_{ij}(\Delta l_{ij})$ employed in the DFX infrasonic detector uses exact delay-dependent normalization and it also eliminates all 'wrap-around' effects. It is computationally efficient by using Fourier-based circular convolution in a novel and exactly correct fashion (accomplished in the frequency-domain). The bandpass-filtered waveform data, $x_m(t)$, which are input to the cross-correlation algorithm are filtered by means of acausal Finite Impulse Response (FIR) filters which are characterized by: (i) highly attenuated stop-band response (down > 50 dB from bandpass), (ii) steep transition regions ($< 0.3\%$ of bandpass), (iii) nearly flat in-band ripple (± 0.2 dB or smaller), and (iv) a constant time-delay independent of frequency. These attributes ensure firstly, the ambient out-of-band energy is actually not present at a detectable level in the bandpass-filtered waveform data, and 'red' spectral character of infrasonic ambient does not contaminate low frequency processing bands. Secondly, we are assured the waveform shape of the in-band process is indeed the waveform shape with which it became 'embedded' in the in-band ambient. This latter point is important where waveform shape, and/or spectral attributes, and/or statistical attributes are to be recovered from the received waveform data. The implementation of the FIR bandpass filtering is described by the formula below:

$$x_m(t) = \sum_{t'=0}^{\Delta T} h_B(t') \cdot y_m\left(t + t' - \left(\frac{\Delta T}{2}\right)\right) \tag{17}$$

where the time-domain filter kernel has the following attribute:

Unity Gain: $\sum_{t'=0}^{\Delta T} h_B(t') = 1$,

and the frequency-domain Fourier transform of the kernel has the following two attributes:

i) very low stop band response: $H_B(f) < \frac{\sqrt{10}}{1000}$, where $f < (f_1 - \delta f)$ or $f > (f_2 + \delta f)$; δf is the frequency width of the transition region (at each band edge), f_1 and f_2 are low-and high-frequency limits of the bandpass, respectively. Typically, $\delta f < 0.0015\ SRT$, where SRT is the sampling rate in samples/second of the digital time-domain waveform data.

ii) flat in-band response (i.e., low ripple amplitude): $H_B(f) = 1 \pm 0.046$, where $f_1 < f < f_2$.

The bandpass-filtering algorithm applied is implemented as a Fourier-based circular convolution (accomplished in the frequency-domain). The FIR kernels, $h_B(t)$, are generated from a compact previously designed stored library, on a best match basis.

2.6 Estimation: Output to Database Tables from Detection and Feature Extraction

The output from the DFX detection and feature extraction (i.e., estimation) processing is written to four database tables and, in the case of one of the tables, to a corresponding set of binary files external to the database. These four tables have the following names:

(i)   *wfdisc* inventory of auxiliary 'beam' information, traditional detection beams, and raw wave-form data;

(ii) *detection* an entry for each declared detection which includes initial feature estimates;

(iii) *arrival* an entry for each declared detection which includes indicative information and current feature estimates;

(iv) *infra_features* an entry for each declared detection which includes feature estimates relevant only to infrasound signal detections, and is linked to corresponding entries in *arrival* and *detection* via the field '*arid.*'

For infrasound processing a set of three auxiliary 'beams' are generated for each automatically processed time interval and sensor array site. These auxiliary beams are suitable for display in parallel with channels of raw and/or filtered sensor waveform data via the Analyst Review System. The most important auxiliary beam is the '*mx*' trace which is a time-series of the highest spatial coherence level in the PDS regardless of spectral band and/or sensor subgroup. The other two auxiliary beams are termed '*az*' and '*sl*' and are time-series of the azimuth and magnitude apparent phase velocity for which the corresponding '*mx*' value was realized. The entries which are relevant to infrasound processing in each of these three tables are:

Table 2

Detection table entries

Field	Description
arid	arrival identification (a unique integer)
jdate	Julian date corresponding to '*time*'
time	time of maximum energy in this arrival's energy packet (epoch time)
sta	station code (i.e., site name)
chan	channel code (i.e., '*sd*,' '*md*,' etc.)
cfreq	center frequency [in Hz] of the spectral band $(f_2 + f_1)/2$
seaz	initial estimate of azimuth of the arrival [in degrees] (i.e., back azimuth)
delaz	standard deviation in '*seaz*'
slow	initial estimate of magnitude slowness of the arrival [in s/km]
delslo	standard deviation in '*slow*'
snr	signal-to-noise ratio (magnitude of energy peak to rms magnitude of ambient background)
stav	short-term average of magnitude of energy peak
fstat	*F*-Statistic value of this detection
deltim	standard deviation in '*time*'
bandw	bandwidth [in Hz] of the spectral band $(f_2 - f_1)$
fkqual	an integer estimate of the *fk*-spectrum peak quality (1 good ... 4 bad)
commid	comment identification
lddate	date when this database entry was made

Table 3

Arrival table entries

Field	Description
sta	station code (i.e., site name)
time	time of maximum energy in this arrival's energy packet (epoch time)
arid	arrival identification (a unique integer)
jdate	Julian date corresponding to '*time*'
chanid	instrument identification
chan	channel code (i.e., '*sd*,' '*md*,' etc.)

Table 3

continued

Field	Description
iphase	reported phase
deltim	standard deviation in '*time*'
azimuth	observed azimuth of the arrival [in degrees] (i.e., back azimuth)
delaz	standard deviation in '*azimuth*'
slow	observed magnitude slowness of the arrival [in seconds/degree (of distance on the earth's surface)]
delslo	standard deviation in '*slow*'
amp*	amplitude (instrument corrected)
snr	signal-to-noise ratio (magnitude of energy peak to rms magnitude of ambient background)
auth	source/originator
commid	comment identification
lddate	date when this database entry was made

*not currently estimated.

Table 4

Infra_features table entries

Field	Description
arid	arrival identification (a unique integer)
eng_time	initial epoch time of this energy packet
eng_dur	time duration of this energy packet
eng_deldur	standard deviation in '*eng_dur*'
coh_time	initial epoch time at which this arrival's spatial coherence exceeds threshold
coh_dur	time duration of this interval of spatial coherence in excess of threshold
coh_deldur	standard deviation in '*coh_dur*'
coinc_time	initial epoch time at which this arrival's spatial coherence and energy both exceed their respective thresholds
coinc_dur	time duration of this interval of both spatial coherence and energy in excess of their respective thresholds
coinc_deldur	standard deviation in '*coinc_dur*'
ford	order of the filtering in performing
zrcr_freq	zero-crossing estimate of signal frequency [in Hz] (estimated in the time-domain)
zrcr_delfreq	standard deviation in '*zrcr_freq*'
crnr_freq	corner frequency estimate of signal frequency [in Hz] (estimated in the frequency-domain)
crnr_delfreq	standard deviation in '*crnr_freq*'
coh_per	coherent period [in seconds] (estimated in the frequency-domain)
coh_snr	ratio of inband coherent signal power to inband incoherent power on an equal channel basis (i.e., all sensor-channels have the same snr).
total_energy	total energy in this arrival's energy packet
auth	source/originator
commid	comment identification
lddate	date when this database entry was made

Most of the entries in these tables are self-explanatory; a few warrant further explanation. In the *detection* and *arrival* tables the azimuth entries (*seaz* and *azimuth*, respectively) and the magnitude of vector slowness entries (*slow*), as well as their standard deviations (*delaz* and *delslo*), are estimated from the vernier *fk*-spectrum output (adjusted for the nominal pre-alignment vector slowness value). In the case of *delaz* and *delslo* in both *detection* and *arrival* tables, these include both measurement and site-dependent modeling error. The parameter *time* in both the *detection* and *arrival* tables is the time at which the short-term average of bandpass-filtered beamformed[15] magnitudes (or squares if an *L2* norm is in use) is maximal for the given energy packet. This time ought to be negligibly different from the peak magnitude of any of the individual sensor waveform data series, allowing for beamforming delay, and is therefore a parameter which can be evaluated by inspection by an analyst. The entry *deltim* is a Cramer-Rao estimate of the measurement error associated with the parameter *time*. For the *detection* table *cfreq* and *bandw* are set in accordance with the spectral band used in forming the beam pair, $B_s(t)$ and $C_s(t)$, upon which coincidence detection was performed to obtain this particular detection. Neither *cfreq* nor *bandw* is an estimate; both are simply set from the bandpass-filter parameters, f_1 and f_2. The *F*-Statistic *fstat* is the value emitted by the internally invoked vernier *fk*-spectrum for this arrival. The integer-valued parameter *fkqual* is a measure of spatial coherence quality and is related to the sharpness and uniqueness of the global maxima in the *Fk*-spectrum (for the given vector slowness coverage Δs). In order to obtain a realistic value for this parameter, prior to each vernier *fk*-spectrum (where Δs is small and δs is very fine) a coarse *fk*-spectrum is calculated. This is also done on the pre-aligned bandpass-filtered waveform data in order to center the *Fk*-spectrum on the arrival of interest, wherein both Δs and δs are large in order to properly estimate the shape of the spatial coherence peak. The *fkqual* value emitted from the coarse *fk*-spectrum is then attributed to the results emitted by the vernier *fk*-spectrum.

In both the *detection* and *arrival* tables, the *snr* parameter is a measure of the ratio of the peak of the normalized short-term average (STA) of the bandpass-filtered beamformed magnitudes to the time-lagged[16] normalized[17] long-term average (LTA) of the bandpass-filtered beamformed magnitudes. The *stav* parameter in the detection table is the value of the normalized STA of the bandpass-filtered beamformed magnitudes (or squares if an *L2* norm is in use). The normalized

---

[15] The traditional beam $B_s(t)$.

[16] The LTA window is positioned in time such that the most recent datum in this window is 'gap' duration earlier in time than the oldest datum in the STA window. Thus, the LTA window is time-lagged relative to the STA window.

[17] Normalized in a sense that the sum of the data in the window is divided by the number of samples contributing. Thus, the measures from STA and LTA windows are comparable when applied to arrival-free White Gaussian Noise, their ratio ought to be a fluctuating value centered on unity.

LTA *ltav* can be determined by dividing the *stav* value by the *snr* value. The *iphase* value in the *arrival* table is currently one of three string parameters: 'I,' 'Ix,' or 'N.' These are phase designations and are discussed in Section three. Currently, the *amp* field of the *arrival* table is not set. It shall be set in a future release of this algorithm to the peak-to-peak amplitude in physical units (e.g., microPascals) of the peak value of each energy packet.

In the *infra_features* table there are several groupings of parameters which may serve to elucidate the nature of an infrasonic arrival. Since the heart of this detection algorithm is coincidence detection, one group of parameters seeks to define the time history of the energy excess for this arrival: *eng_time*, *eng_dur*, and *eng_deldur*; another group of parameters seeks to define the time history of the excess level of spatial coherence: *coh_time*, *coh_dur*, and *coh_deldur*; and the third group defines the time region of coincidence: *coinc_time*, *coinc_dur*, and *coinc_deldur*.

The parameters *zrcr_freq*, *crnr_freq*, and *coh_per* all seek to estimate the spectral content of this arrival's energy packet. The parameter *zrcr_freq* estimates the frequency by analyzing the interpolated time values of the zero-crossings of the bandpass-filtered beamformed waveform data of the beam $B_s(t)$. This list of zero-crossing intervals can be converted into a list of equivalent 'instantaneous' frequency values. Their mean is *zrcr_freq*, and their standard deviation is *zrcr_delfreq*. This list of 'instantaneous' frequency values may be amenable to further analysis for time-evolutions and trends. On the basis of a model in which the spectrum of the arrival's energy packet is flat out to a corner frequency then a decline of x dB/octave beyond that corner frequency, *crnr_freq* is the corner frequency value which yields the least sum of squared errors in a fit to this model. The *crnr_delfreq* parameter is then an estimate of the standard deviation in *crnr_freq*. (For the current implementation, x is a user-specified parameter with a default value of 6.) The estimate of the period corresponding to the dominant spectral frequency in this arrival's energy packet is given by *coh_per* and is estimated by inverting the interpolated frequency of the peak in a windowed Fourier-based power spectrum of $B_s(t)$ taken in the time-window of this arrival's energy packet. The parameter, *total_energy* is an estimate of the total power in the time window of this arrival's energy packet using as input the beam $B_s(t)$. Finally, *coh_snr* is the ratio of the in-band coherent power to the in-band incoherent power where the ratio is computed on an equal channel basis. It is directly related to the average of the aligned normalized cross-correlation coefficients and to the *F*-Statistic.

3. Phase Identification

Phase identification is an important prerequisite to source location. The basic premise here is that, regardless of the waveform technology, one assigns a phase for a given arrival based on the hypothesized propagation path, whether it be through the earth, ocean, or atmosphere. This hypothesized phase is then used to select an appropriate propagation model from which travel-time information can be extracted. For the propagation of infrasound through the atmosphere, a number of propagation paths have been identified. In particular, the refraction of sound in the atmospheric temperature gradient, and the formation and propagation of Lamb's acoustic-gravity wave have been well documented.[18]

The major difference between the seismic phase identification problem and that of infrasound is the variable nature of the propagation medium in the case of the latter. Recent work by GARCES *et al.* (1998) clearly indicates that the ambient wind has a strong influence on the determination of infrasound travel-time information in a temperature stratified atmosphere.

To help simplify the phase identification problem, we will classify infrasonic arrivals as being either non-seasonal deterministic, seasonal, or non-seasonal nondeterministic returns. Examples of the first type are the high-level thermospheric returns refracting in layers high in the thermosphere. While these signals may tend to be modified slightly by winds and solar effects, their presence will be largely independent of seasonal effects. Seasonal arrivals, such as the stratospheric returns, will depend substantially on the seasonal nature of the winds in the stratosphere.

The gross influence of the seasonal winds can be clearly observed in Figure 3 which shows the results of a ray-tracing study using a standard seasonal model for both ambient temperature and wind (HEDIN, 1991; HEDIN *et al.*, 1996).[19]

Here, approximately one half-million ray paths with varying vertical launch angles are calculated with all rays emanating from a common source located in the center at 0 km. The computational domain is then divided into numerous cells and the number of rays passing through each cell noted and assigned a unique color. The results for this calculation are for a source located at 40 North in January with no initial latitudinal component to the ray propagation. Plots of effective sound speed, $C_s^{(\text{eff})} = \sqrt{\gamma RT} + \boldsymbol{n} \cdot \boldsymbol{u}$, where γ is the ratio of the specific heats, R is the gas constant, T is the absolute Kelvin temperature, $\boldsymbol{n}$ is a unit vector in the direction of propagation, and $\boldsymbol{u}$ is the wind vector, have been shown for both against-wind (left) and with-wind (right) propagation cases.

An asymmetry between the 'against-' and the 'with-wind' propagation modes clearly exists. Both stratospheric and thermospheric reflections are a feature of the 'with-wind' propagation mode, whereas only thermospheric reflections are a feature

---

[18] Specific aspects of these propagation modes are discussed widely in the scientific literature and the individual references are too numerous to mention here. However, two publications of a more general nature which discuss this material and provide a good precis of the scientific literature are GOSSARD and HOOKE (1975), and MCKISIC (1997).

[19] These results have been predicted by a 3-D ray-tracing algorithm developed by one of the authors [DJB] and is based around the description of such a model given by GEORGES (1972). A more complete set of ray-tracing results are being presented in a later publication.

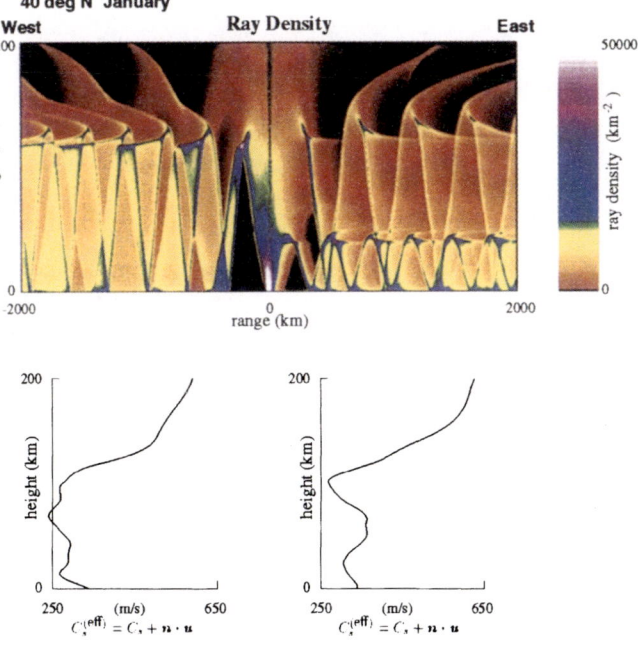

Figure 3

Theoretical ray-density structure for a ground-level source (shown here as the white region located at 0 km) located at 40°N in January. Trajectories for a half-million rays with varying vertical launch angle were determined using an Eikonal ray-tracing algorithm for acoustic waves in an atmosphere with seasonal variability in wind and temperature. The representative effective sound speed variation as a function of height is also displayed for both 'against-wind' (left) and 'with-wind' (right) propagation.

of the 'against-wind' propagation mode. The propagation direction for the stratospheric mode reverses in the complementary season.

Although one may reasonably expect a definite seasonal character to be present in the arrivals at a given infrasonic array over a sufficiently long-time period, we must also anticipate that a reasonable percentage of all arrivals in this same time period will be non-seasonal and nondeterministic. Whilst refraction off the atmospheric temperature gradient is in the main responsible for the thermospheric and stratospheric returns, tropospheric winds are generally responsible for the tropo-spheric-level arrivals (Dr. T. Armstrong, *private communication*). The appearance of these signals depends strongly on the nature of the local weather below the tropopause. Furthermore, recent studies by KULICHKOV (1998) suggest that scattering of infrasound energy off inhomogeneities in the atmosphere may be an important mechanism for the deviation from the theoretical ray-path geometry. Indeed, this work suggests that the scattered wave may be a nontrivial component of the signals recorded at any given microbarographic station. In addition, KULICHKOV (1998) indicates that ray-focusing due to scattering may be instrumental in generating apparent sound speeds for infrasonic arrivals which are less than acoustic velocities

Table 5

Phase naming nomenclature for infrasonic arrivals

Phase name	Arrival
Iw	Tropospheric level wind generated arrival
Is	Stratospheric arrival
It	Thermospheric arrival
N	Noise
ILm	Lamb wave arrival
Iwx	Tropospheric level wind generated coda phase arrival
Isx	Stratospheric coda phase arrival
Itx	Thermospheric coda phase arrival

at the ground. These are two examples of the non-seasonal nondeterministic infrasonic arrivals that are likely to occur. In practice, one will have considerable difficulty assigning travel-time information to these arrivals. Acoustic ray-tracing is conventionally used to map out the propagation paths for the first two arrival types (non-seasonal deterministic, and seasonal) however it becomes difficult or impossible to apply the same treatment to these latter arrivals.

The nomenclature we have developed for the purpose of infrasonic phase identification at the PIDC has been formulated with the following issues in mind:
1. The philosophy used by the other waveform technologies of describing an arrival in terms of the propagation path of the signal should be adopted for infrasonic arrivals.
2. The PIDC does not classify the generation mechanism of detected sources.
3. Refraction from major thermal boundaries (stratospheric, thermospheric) must be differentiated. This does not distinguish the two possible thermospheric returns that generally seem to be possible, but this would probably be a very difficult procedure in practice.
4. The wind-generated tropospheric returns must be represented.
5. Lamb wave arrivals must be represented.
6. The facility for assigning 'coda phases' to an automatic detection that is obviously part of the same arrival as a previous detection must be included.
7. Automatic noise detections must be accommodated.
With these requirements in mind, the set of nomenclature listed in Table 5 is proposed.

Referring to Table 5, the major distinguishing feature between the phases *Iw*, *Is* and *It* is the apparent velocity *v* recorded at the array. This is due to the generally increasing angle of incidence with the ground for the arrivals with higher turning points in the atmosphere. Phase identification therefore depends on the ability to decide arrival type based on apparent velocity. Temperature will be recorded at each IMS infrasound station. This will allow the accurate determination of the sound speed, C_s, through the relationship $C_s = \sqrt{\gamma R T}$, and hence the observed angle of incidence with the ground α through the relationship $\alpha = a\cos(C_s/v)$. This will allow one to infer the propagation path followed or, equivalently, assign a preliminary phase name to the arrival. However, without accurate seasonal tables indicating the velocity ranges for the different arrivals on a site-by-site basis, phase identification based on observed velocity will be a difficult proposition.

The infrasonic phase identification component of the system is still in the early stage of development and the nomenclature listed in Table 5 for identifying phases is not yet part of automatic processing. Currently, all infrasonic detections are labelled as either N (Noise), I or Ix arrivals; this classification separates out arrivals with observed speeds ranging from 270 m/s to 660 m/s (for both I and Ix arrivals), to those outside this range (N arrivals). The coda Ix phase is used here in a slightly different context to that in seismology. Here it is intended to indicate any arrival subsequent to an already declared detection in a group of arrivals that can all be considered to originate from the same source. It is designed to prevent the inclusion of the latter arrival(s) in an associated event that does not also contain the original detection.[20]

In future versions of the system it is intended initially to identify the phase based on the observed speed across the array[21] and then refine the identification in the event association stage through a multiple hypothesis scheme which will label the phase to be that one (Iw, Is, or It) whose travel-time information yields the smallest error in the hypothesized source location.

4. Event Association and Source Location

The problem of event association and, ultimately, source location involves the consideration of several inter-related issues. For example, how does one assign a weighting to the various detection parameters such as azimuth, speed (slowness) and time for an arrival of a particular type whether it be seismic, hydroacoustic or infrasonic? More to the point, what values should these weights take when combining arrivals of different types in mixed events, i.e., events formed by signals from two or more of the seismo-acoustic technologies. Another issue concerns the formulation of event definition criteria; how many of each defining parameter should be required to form an event. These questions and others relevant to the infrasound source location procedure will be addressed here.

The question of assigning weight values to the various parameters of an infrasonic arrival is difficult. The Fk plot displayed in Figure 4 demonstrates that azimuthal and slowness information for infrasonic arrivals can often be determined quite accurately.

---

[20] To achieve this, the observed parameters, in particular azimuth and time of arrival of the two detections, must be within some specified window.

[21] These velocity ranges will be dependent on season and array location.

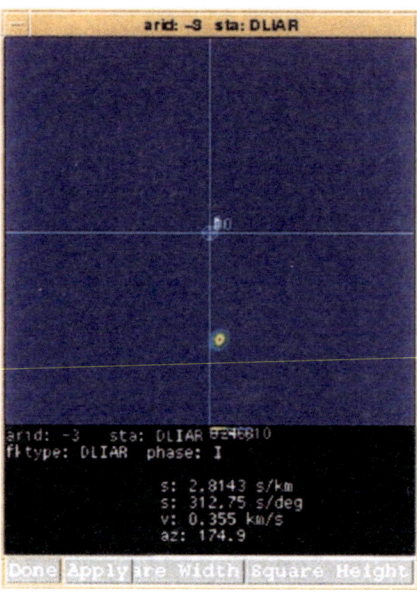

Figure 4
An *Fk* output showing the location of an infrasonic source in the slowness plane. The diagram indicates
zero slowness at the center and 5 s/km at the domain boundaries, and azimuth clockwise from the top.

The observed azimuth for this arrival is 174.9 degrees. However, ground-truth
information for this particular signal requires an azimuth, given by the great circle
through source and receiver, of 183.1 degrees, a deviation of some 8.2 degrees from
what was observed. In addition, two other arrivals from the same source exhibited
azimuthal deviations of 3.2 and 7.3 degrees, respectively. It is assumed that this
azimuthal deviation is attributable to the action of the wind along the propagation
path,[22] or scattering. Source location at the PIDC is based on an iterative nonlinear
least-squares inversion procedure that in the case of purely infrasonic arrivals weights
azimuthal arrival information more strongly than temporal information. Although
travel-time information is used to refine the source location, intersecting azimuths
becomes a primary requirement before a source location can be hypothesized. While
this procedure provides a rough, and perhaps satisfactory, idea of source location,
the azimuthal deviation due to the wind probably needs to be accommodated in the
accurate determination of the source location.

The infrasound propagation model currently in use at the PIDC is rudimentary.
The horizontal speed from source to receiver is assumed to be 320.0 m/s. Although
crude, this model does allow the formation of valid infrasound events; an example is

[22] Recently it has become evident that the deviation in observed azimuth and slowness due to the
relative height differences of the sensors in any infrasonic array could be significant (WANG, 1999).

the REB event 20252820 of December 9, 1998. The waveforms for this event are shown in Figure 5.

Signals recorded on three infrasonic arrays: SGAR, LSAR and PDIAR were used to estimate a source location. This hypothesized location is within the Nevada Test site. The semi-major axis of the 90% confidence ellipse is 330 km and the semi-minor axis is 74 km. These values are quite large and arise from the poor constraints provided by the atmospheric propagation model.

Another demonstration of the ability of infrasonic waves to determine azimuth with great accuracy is the result of a recent synergy experiment which combines infrasound with seismic data collected at the co-located TXAR seismic and TXIAR infrasonic arrays to locate two nearby mining explosion sources. This approach is part of the ongoing Multi-Sensor Data Fusion project at the PIDC to enhance our understanding of source location and characterization by the joint processing of seismic, hydroacoustic, and infrasonic data.

The waveforms for both the seismic and infrasonic channels can be viewed together in ARS and subsequently filtered while the *XfkDisplay* software performs the actual *Fk* computation and estimates azimuth, slowness, and *F*-statistic for a given arrival. A preliminary location and its associated error ellipse is determined using these measurements for only the seismic (P_n and L_g) arrivals and the standard travel-time tables derived from the *IASPEI* model (KENNETT, 1991). A final location is then computed using the two seismic and one infrasound arrival with an atmospheric sound speed of 320 m/s.

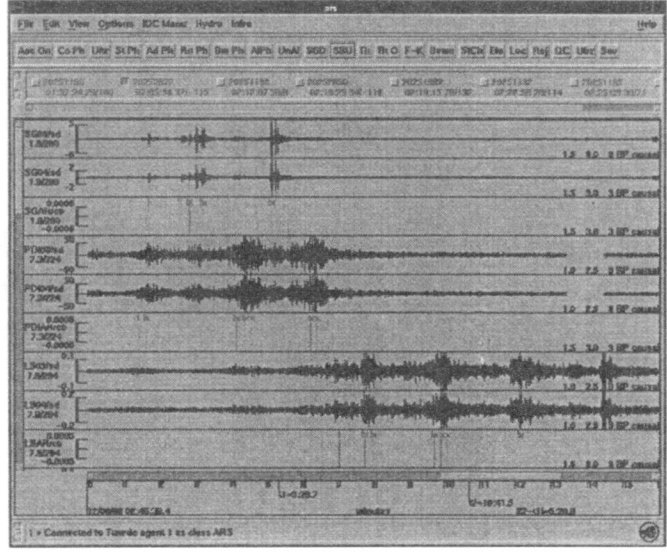

Figure 5
Infrasound waveforms contributing to REB event 20252820. Three-pole causal bandpass filtering is used with the bandpass indicated on each channel. Only two sensor channels per array are indicated.

For the first example, an event in the Micare coal district of northern Mexico, the seismic-only location has an azimuth of 98.2°, an error ellipse with axes 71 km by 49 km, and an estimated location approximately 20 km north of the mine. A single, clear infrasound arrival is detected, indicating an azimuth of 104.7°, and when combined with the seismic data produces a final location at 105.3° azimuth, with error ellipse axes 48 km by 25 km, which is in close agreement with the ground-truth azimuth of 105.9°. As expected, the addition of the *I* arrival improves the azimuth, reduces the error ellipse, and moves the estimated location farther to the south where it overlaps the mining area.

We note, however, that the reduction in the error ellipse is largely in the azimuth orientation while uncertainty in the distance to the event is not really improved. This is likely due to our oversimplified atmospheric model and not to the inferior ability of the infrasound arrivals to constrain the slowness and time parameters (hence, the *I* arrivals are azimuth defining and not slowness or time defining in the inversion as the P_n and L_g arrivals may be). Indeed, the slowness and arrival time of the incident infrasound wave across the array is measured quite accurately; we simply need a more realistic velocity model, perhaps including seasonal, wind, and temperature changes, with which to convert these measurements to more quantitative constraints on the source location estimation.

A second example, in the Tyrone copper mine of southwest New Mexico, showed similar results, although the nature and quality of the infrasound recordings required us to employ other means of computing the azimuth and slowness than the traditional *Fk* analysis. The preliminary location using only seismic information is at 305.5° azimuth with error ellipse axes of 97 km and 63 km and an estimated location approximately 30 km southwest of the mine. The infrasound recordings for this event emitted, instead of a single impulsive waveform, multiple arrivals similar to those shown in Figure 5 at LSAR, although of lower signal-to-noise. This multiple-arrival nature of the infrasound waves may indicate a complicated path through the atmosphere, possibly due to multiple reflections within atmospheric layers.

The emergent or packet-like appearance of the infrasound arrivals made them difficult to window for the *F-K* analysis, and we were not able to use the standard *XfkDisplay* to make an azimuth and slowness determination. In fact, our initial *F-K* analyses showed the effects of aliasing caused by the high frequency content of the signal combined with the large spacing of the infrasound sensors. As an alternative, we employ a technique based on envelope functions which appears to be more robust in extracting azimuth information from infrasound signals of poor quality (FLANAGAN *et al.*, 1999). Currently, this approach is being implemented as part of our ongoing efforts in data fusion at the PIDC to estimate an azimuth for hydroacoustic signals using widely spaced sensors. As the TXIAR array spacing is larger than many other infrasound arrays and similar to the proposed IMS standard, this approach is novel.

Using the envelope functions of the infrasound traces we obtain an azimuth of 315° and this is the value used in concert with the seismic parameters to compute the final location at 311.5° azimuth, with error ellipse axes of 59 km and 41 km. This agrees with the true azimuth to the source of 309.6°. Again, the combination of infrasound with seismic data improves the estimated location and moves it closer to the true location.

This work shows the promise of the envelope function approach to slowness-azimuth estimation by determining a more accurate azimuth, with a higher F-statistic for infrasound arrivals of low signal-to-noise. This suggests that envelope functions may be more reliable for characterizing infrasound data when a lack of signal coherency across the array prevents a classical Fk analysis. This approach may also be useful for infrasound wavefronts with higher phase velocities, indicating they are sampling higher velocities at greater altitudes in the atmosphere. This appears to be the case for the Micare event where the lone infrasound arrival time is more consistent with wave speeds in the higher altitude thermosphere (at 245 m/s) than the troposphere (at 320 m/s).

The PIDC is currently in the process of incorporating improved travel-time information into the source location procedure. This is being done in stages. The first stage utilizes seasonally and azimuthally varying travel times as a function of range for each infrasonic array. The travel-time information is being provided by an

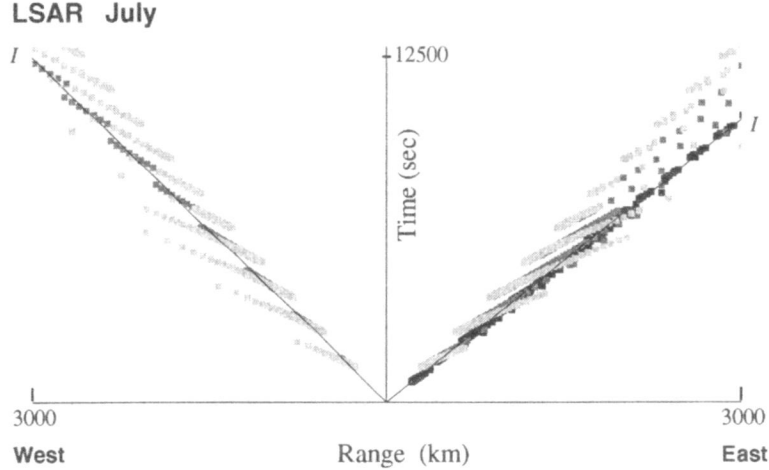

Figure 6
Theoretical bounce-point locations determined via an Eikonal ray-tracing model for acoustic waves propagating in an atmosphere with seasonal variation in wind and temperature, in this case the location is the LSAR array in July. Each individual bounce is recognized as a short sloping bar in which the component bounces merge together. The apparent velocity of the arrivals is given by the slope of the bar. A single line interpolating the different bounces is indicated. One such line per degree in azimuth will be used in a subsequent version of the system to provide travel-time information for each infrasonic array and for each season.

Eikonal ray-tracer for acoustic-gravity waves. A single phase is assumed with the travel time being given by the time of arrival of the stratospheric return if there is one, and the main thermospheric arrival otherwise. One travel-time curve for every degree in arrival azimuth is determined, and an example is shown in Figure 6.

In the second stage of development, travel-time information will be provided for each of the phases *Iw*, *Is* and *It*. In the event that a given phase type is not predicted by the model, travel-time information of a generic nature will be provided. An example is shown in Figure 7.

It is anticipated that the completed system will use the observed slowness to predict the most likely propagation path through the atmosphere, based on receiver location, and season. It would be useful to mimic the seismic community and, with accurately determined values of observed slowness, assign an individual travel-time curve to each arrival. One could then estimate single station source location using a number of clear arrivals recorded at a single array, assumed to be from the same source, by simply noting the difference in arrival time between the two arrivals and then applying the appropriate set of travel-time curves. However, this would require detailed knowledge of the atmosphere between source and receiver, as well as some method that would account for spurious arrivals. At the present time, there is no mechanism in place for making single-station source locations via infrasonic data processing alone, at the PIDC.

These ideas are reflected in the weighting assigned to the arrival parameters when forming events. Presently, azimuth for the infrasonic arrivals is given the greatest weighting, and predicted travel time a lower weighting. Slowness has zero weighting

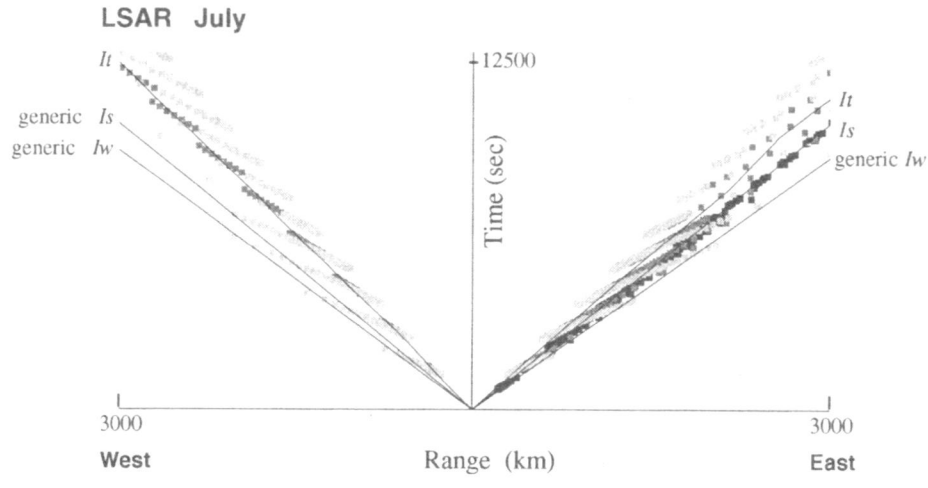

Figure 7
Same as Figure 6, however, each of the arrival types *Iw*, *Is* and *It* will have their own travel-time curve. In the event that various phase types are not predicted by the model, a generic curve will be used. This scheme will be incorporated in a later version of the system.

Table 6

Minimum number of defining phases for events of mixed data type

Mixed event	P time	P azimuth	H time	I time	I azimuth
	2	2	–	0	1
Seismic and	2	1	–	1	1
infrasonic	1	1	–	1	2
Hydroacoustic and	–	–	2	1	1
infrasonic	–	–	1	1	2
Seismic, infrasonic	1	0	1	1	1
and hydroacoustic	1	1	1	0	2

since it is not currently used in the event formation stage. Predicted travel time is made minimally defining in that it is used mainly to prevent causal violations when associating arrivals. If the 'slope' of the travel-time curves is ever used in the source location procedure, slowness will be made defining with a non-zero weighting. In the formation of mixed events, infrasound azimuth is given as much weight as P and H travel time. In arriving at a set of event definition criteria, it was decided that the 'minimal' infrasound event that should be allowed is one in which two well-defined azimuths intersect, and in which causality is not violated. For mixed events, the scheme shown in Table 6 is used.

A succinct summary of this diagram states that it takes at least three well-defined parameters (selected from P travel time, H travel time, I azimuth) to form an event. Less constraining parameters such as P azimuth and I travel time are to be introduced only after three well-defined parameters have been combined for a prospective event.

5. Interactive Processing

The final stage of data processing performed is interactive analysis. The routine interactive analysis of infrasonic and other waveform data ensures the integrity of the final event bulletins, as well as maintaining a check on the overall processing validity and stability of the system.

The analysts use a graphical user interface called the Analyst Review Station (ARS) to review infrasonic data. Various signal processing tools allow the analyst to appraise the signal, either on an individual sensor basis, or as a beamformed array. The distribution of signal power across the slowness plane is displayed using a tool called *XfkDisplay* and is demonstrated in Figure 4 for an arrival at the DLIAR array.

Due to high-frequency attenuation with distance, spectral content may be a major factor analysts will use to obtain a preliminary idea of the source to receiver distance, and will indeed be an important tool in discrimination studies. The tool used to extract single-sensor spectral information is known as *SpectraPlot*. With this tool, analysts can nominate a particular windowing whether it be box-car, squared-cosine, Hamming, Hanning, Blackman, Welch or Parzen tapering, and the degree of smoothing can be precisely controlled. An example *SpectraPlot* output taken from the waveforms shown in Figure 5 is shown in Figure 8.

Here, a sample of signal data is compared with a sample of signal-free background data. The signals recorded at the different arrays clearly show a progressive decrease in high frequency content and amplitude with distance.

A set of procedures has been established for the interactive processing of infrasonic data. However, due to the lack of IMS-type data during the major construction phase of this project, the preliminary nature of these procedures is emphasized. Data from the IMS infrasonic arrays will contain significantly more low frequency energy (<0.5 Hz) than most of the data presently available, and will also be generally less noisy since substantial effort is placed on seeking low-wind sites for each array location. The analyst procedures:

1. Provide a basic introduction to infrasound propagation physics.
2. Detail the observed characteristics (spectral content, signal duration, etc.) of events of interest (and those events generally not of interest) from a CTBT monitoring perspective.
3. Provide a basic set of filter bands that can be used when reviewing data. Certain caveats also must be provided with the use of these filter bands. These concern the nature of the signals recorded on the smaller aperture arrays (LSAR, PDIAR, SGAR) compared with those recorded on the larger IMS type arrays (DLIAR, TXIAR).
4. Detail the more common infrasound sources (microbaroms, open-cut mines, high-speed aircraft, etc.) and the features that make their signals unique.
5. Provide a set of suggested procedures to be followed when scanning for events missed by the automatic system.

6. *Evaluation of Operational Results*

The version of the infrasonic processing system currently in use at the PIDC has been operational since December 1998. The total number of detections recorded by each operational array during the period from December 1998 to July 1999 is indicated in Table 1. These detections are also displayed graphically in Figure 9.

The infrasonic arrays in the mainland US have been quite active during this period, recording approximately 100 detections per week per array.

Table 7 compares the automatically determined values of azimuth and slowness and their variances with the manually determined *XfkDisplay* values for all detections in a recent 100 hour segment of contiguous data.[23]

---

[23] The center frequency and bandpass interval were chosen to match those used in automatic processing.

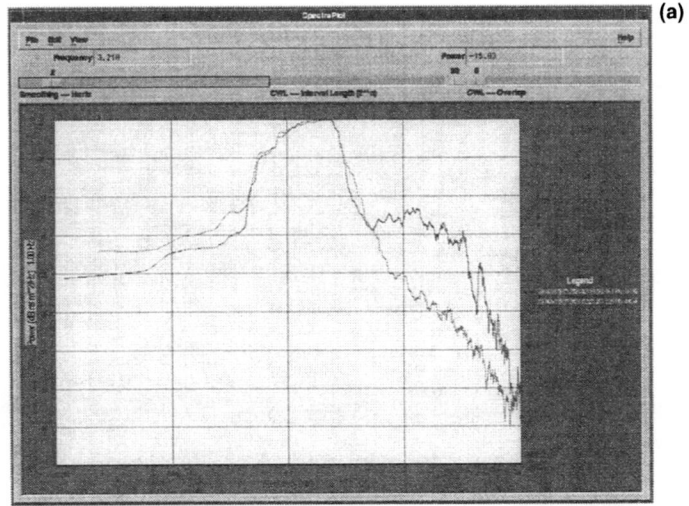

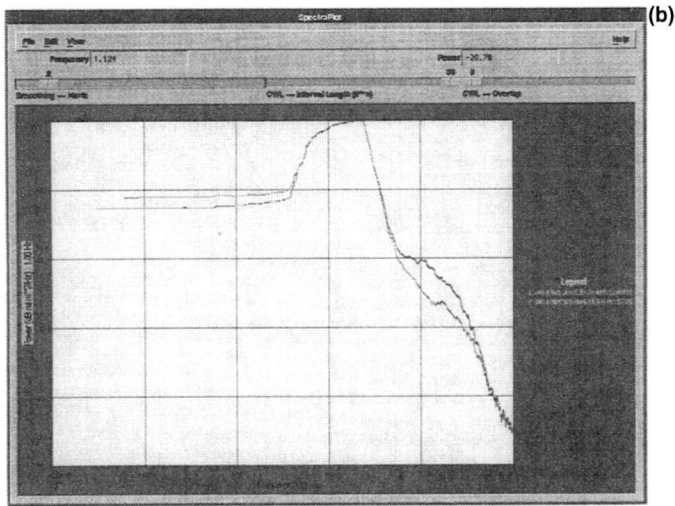

Figure 8
Power spectra for the signals indicated in Figure 5. Each figure shows a signal trace and a signal free trace.
(a) SGAR, (b) LSAR.

The automatically determined values for azimuth were on average 1.6 degrees away from the manually determined values, and the automatically determined values for speed were an average of 6.2 m/s away from the manually determined *XfkDisplay* values. This is quite reasonable, considering that the majority of

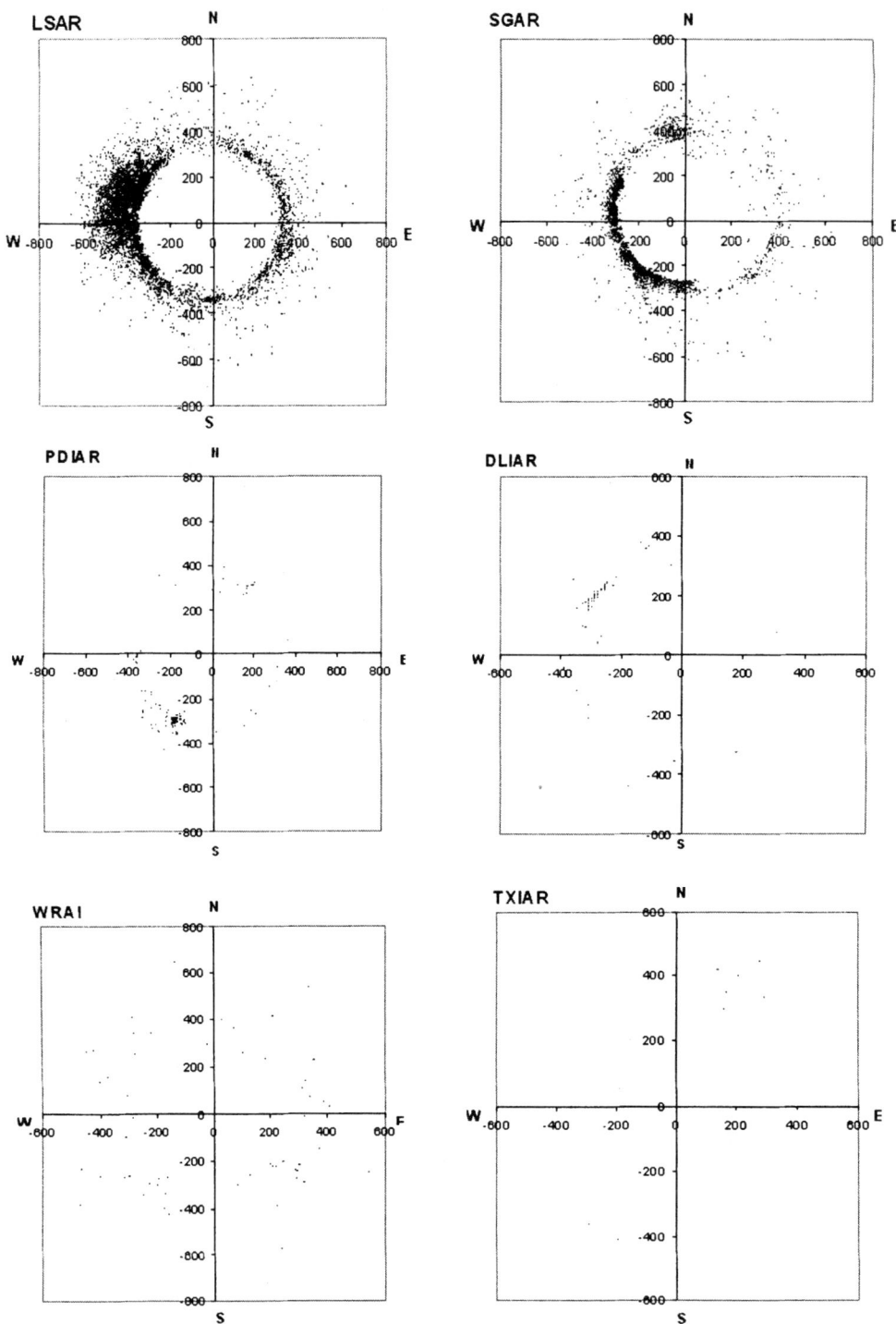

detections were made on the small aperture non-IMS standard arrays. One may anticipate that even more accurate results will be attainable using the larger aperture IMS arrays. It is of interest to note that the automatically determined values of the variance for both azimuth and speed appear to be on average 20–30% larger than the manually determined values. This discrepancy is currently under investigation. A total of 23 false alarms occurred. Nine of these were found to be due to microbaroms and 13 seemed to be due to triggering on side lobe energy of other significant (usually noise spikes) signals. There was one detection for which no apparent trigger could be found. Of concern are the eight failed detections. These failed detections are generally from the same direction as the microbaroms, and are likely due to the increased detection threshold enforced in these directions. This is an experimental feature that had been incorporated into the system to reduce microbarom detection and will have to be re-evaluated. Infrasonic associated events for this time period are summarized in Table 8 and Table 9.

Listed here are the statistics for pure infrasonic and mixed events. Table 8 details the number and makeup of all automatic events comprising an infrasonic arrival. The mixed events are indicated as being formed with either seismic (S) or Hydroacoustic (H) detections. In Table 9 an attempt has been made to evaluate the legitimacy of the infrasound only events. To maintain the integrity of the system while in the early development stage[24], only events consisting of arrivals on at least three distinct infrasonic stations were allowed to contribute to the REB during this period. SEL3 (Standard Event List) events are simply those automatically associated events that were formed in the third pipeline (which, due to the time delay must necessarily be where the infrasonic processing is performed) and have not been subjected to analyst scrutiny. The LEB (Late Event Bulletin) events are those in which analysts have reviewed the associations and have been either considered valid or erroneous. Table 9 indicates that possibly 11 of the 33 infrasound only events that were automatically formed were considered to be possibly legitimate. Of course without ground-truth information the true disposition of the events can never be known.

◀

Figure 9
Scatter diagram showing the azimuth and slowness of arrivals at the indicated arrays from Dec 04 1998 to July 01 1999. Units are in m/s.

---

[24] When regular interactive infrasonic processing commenced in December 1998, it coincided with a substantial military exercise in the Nevada area of the US. A significant number of infrasonic associations were formed as a result. To ensure that the sudden increase in infrasonic events did not impact other analyst processing, it was decided to process only three-station infrasonic events until analysts were more familiar with the technology.

Table 7

Automatically versus manually determined signal characteristics for a recent 100-hour segment of data

Detects	False alarm microbaroms side lobe triggers/ others	Genuine signal	Noise detects	Failed detections	Average azimuthal variance (deg.) auto/man	Average slowness variance (s/deg.) auto/man	Average difference between manually and automatically determined azimuth (deg.)	Average difference between manually and automatically determined speed (m/s)
118	23/9/13/1	65	30	8	4.6/3.8	25.3/21.8	1.6	6.2

Table 8

Pure and mixed infrasonic events formed during automatic processing, Dec. 04, 1998 to Jul. 01, 1999

Event type	Defining phases	SEL3 (Automatic processing)
Infrasound-only events	I I I	3
	I I	30
Mixed events	I I S*	7
	I I H**	4
	I H H	3
	I I H S S	1
	I I H S	1
	I H S S	1
	I I I H H S S	1
	I H H S	3
	I H H H	1
	I I I H H	1

*Associated seismic phase.
**H denotes any associated hydroacoustic phase. May be either *T* or *H*.
***Only infrasound phases legitimately associated.

Table 9

Evaluation of the automatically formed infrasound only events from Dec. 04, 1998 to Jul. 01, 1999

Event type	Defining phases	Possibly legitimate events	LEB	REB
* Infrasound-only events	I I I	2	2	1
	I I	9	0	0

7. Proposed System Enhancements

It will be necessary to assign an infrasonic magnitude, independent of equivalent nuclear yield, to infrasound associated events. This magnitude value, which would be accurate for all airborne sources, may then be included in event bulletins as a hypothesized 'size' for the source. The expression decided on (Infrasound source location experts meeting, San Diego, November 1998) is the logarithm of the pressure in Pascals measured a reference distance of 1 km from the hypothesized source location. The amplitude-distance relationship of WHITAKER (1995) will be used. This magnitude value can be written as

$$M_i = \frac{1}{N} \sum_{j=1}^{N} \log_{10}(10^{-0.019\nu} R^{1.36} P_j) \,, \tag{18}$$

where ν is the component of the stratospheric wind velocity in m/s in the direction of wavemotion, R is the source to receiver distance in kilometers, P_j is the observed pressure signal of the arrival at infrasonic array j in Pascals, and N is the total number of recording arrays.

8. Summary and Conclusions

An automatic and interactive data processing system is being established at the PIDC to analyze data from the 60-station IMS infrasonic network. The purpose of the system is to monitor compliance with the Comprehensive Test-Ban Treaty (CTBT). Although additional work is required, the system is completed to a stage where automatic events are being formed, submitted to an analyst for review, and written to the Reviewed Event Bulletin when considered legitimate.

Automatic signal detection is achieved via a coincidence detector which relies on two parameters achieving a predetermined threshold value simultaneously before a detection is declared. The first parameter is the normalized cross correlation, and the second is the STA/LTA ratio of a beam directed toward the region of the slowness plane with high correlation. Since December 4, 1998 the six infrasonic arrays listed in Table 1 are registering an average of 100 detections per week per array. The average deviation between automatically and manually determined azimuth is approximately 1.6 degrees, and the average deviation between automatically and manually measured speed is approximately 6.2 m/s for the 6 arrays listed in Table 1.

Presently the detections are classified broadly as being either I, Ix or N signals where I and Ix are legitimate signals (based on observed slowness), the Ix arrival considered to be from the same source as some previous I arrival; and the N designation referring to a detection which can be considered as noise (based on observed slowness). This classification scheme is being superseded by a phase identification system which attempts to classify the signals in terms of the hypothesized propagation path. The proposed classification scheme is given in Table 5.

Presently, a rudimentary propagation model is being used to provide travel-time information. This model is being superseded by a seasonal model which provides travel-time information via a ray-tracing algorithm for acoustic waves in a temperature stratified atmosphere with winds on a curved earth. It has been indicated in this paper that arrival azimuth for infrasonic signals can be measured quite accurately. It may therefore be more important for the source location procedure to have an accurate idea of the hypothesized arrival azimuth than hypothesized travel time. It may then be necessary to incorporate corrections to the observed azimuth due to meteorological effects, and due to the height differences of the recording sensors.

All automatically formed events are passed to analysts for review with a Reviewed Event Bulletin (REB) being the major product of the PIDC. It is therefore necessary to establish procedures for the interactive review of infrasonic data. A basic set of interactive infrasonic procedures is currently under trial. These procedures provide the analyst with a basic knowledge of infrasound propagation physics and the signal characteristics of the more common infrasonic signals (microbaroms, local mines, supersonic aircraft, etc.), and indeed the signals from events of interest as regards monitoring the CTBT.

Provisions are being made to allow for the determination of an Infrasonic Magnitude for all automatically formed events.

Acknowledgments

This work was Sponsored by U.S. Department of Defense, Defense Threat Reduction Agency, under Contract No. DTRA01-99-C-0025. Dr. Raymond Willeman played a major role in developing this project prior to his departure from the PIDC, and his insight and effort is recognized. The authors would like to thank researchers at the Los Alamos National Laboratory, in particular Drs. Rod Whitaker, Tom Armstrong and Doug Revelle, for providing guidance and expertise in the area of infrasonic monitoring. Discussions with colleagues at the IMS, in particular Drs. Doug Christie and Alberto Veloso, have been useful in creating an operational infrasonic detection and source location system. Dr. Rod Whitaker reviewed the manuscript and made a number of helpful suggestions. The authors would also like to acknowledge contributions made from participants at the infrasonic source location meeting in San Diego, November 1998.

References

BEDARD, A. J., GREENE, G. E., INTRIERI, J., and RODRIGUES, R. (1988), *On the Feasibility and Value of Detecting and Characterizing Avalanches Remotely by Monitoring Radiated Subaudible Atmospheric Sound at Long Distances*, Engineering Foundation Conference, Santa Barbara, CA.

BLANDFORD, R. R. (1974), *An Automatic Event Detector at the TONTO FORREST Seismic Observatory*, Geophys. *39*(5), 633–643.

BOWMAN, H. S., and BEDARD, A. J. (1971), *Observations of Infrasound and Subsonic Disturbances Related to Severe Weather*, Geophys. J.R. Astr. Soc. *20*, 215.

BROWN, R. F. (1963), *An Automatic Multichannel-correlator*, J. Res. Nat. Bur. Stand. *67C*, 33–38.

CHRISTIE, D. R. (1998), *Establishment of the IMS Infrasound Network-Scientific and Technical Issues*. In *Proceedings of the Informal workshop on Infrasounds*, Bruyeres-Le-Chatel, France.

DIGHE, K. A., WHITAKER, R. W., and ARMSTRONG, W. T. (1998), *Modelling Study of Infrasonic Detection of 1 kT Atmospheric Blast*. In *Proceedings, 20th Annual Seismic Research Symposium On Monitoring a Comprehensive Test-Ban Treaty*, Santa Fe NM, September 21–23, 571–578.

DOE-1994, (1994) *United States Nuclear Tests: July 1945 through September 1992*, Department of Energy Technical Report DOE/NV-209 (Rev. 14).

DONN, W. L., and EWING, M. (1962), *Atmospheric Waves from Nuclear Explosions*, J. Geophys. Res. *67*, 1855–1866.

DONN, W. L., and SHAW, D. M. (1967), *Exploring the Atmosphere with Nuclear Explosions*, Rev. Geophys. *5*(1), 53–82.

DONN, W. L., SHAW, D. M., and HUBBARD, A. C. (1963), *The Microbarographic Detection of Nuclear Explosions*, IEEE Trans. Nuc. Sci. *NS-10*, 285–296.

FLANAGAN, M. P., LEBRAS, R. J., HANSON, J., and JENKINS, R. (1999), *Analysis of Two Mining Explosions Recorded at TXAR Seismic and Infrasound Arrays*, Report No. SAIC-99/3002, Science Applications International Corporation, San Diego, CA.

GARCES, M. A., HANSEN, R. A., and LINDQUIST, K. G. (1998), *Traveltimes for Infrasonic Waves Propagating in a Stratified Atmosphere*, Geophys. J. Int. *135*, 255–263.

GEORGES, T. M. (1972), *3-D ray-tracing for acoustic gravity waves*. In *Proc. Conf. on Effects of Acoustic Gravity Waves on Electromagnetic Wave Propagation*, AGARD Proc. No. *115*: 2–1 to 2–8.

GEORGES, T. M. (1973), *Infrasound from Convective storms: Examining the evidence*, Rev. Geoph. and Space Phys., 11, 571–594.

GOSSARD, E. E., and HOOKE, W. H., *Waves in the Atmosphere* (Elsevier Scientific Publishing Co., New York. 1975).

GUTENBERG, B. (1939), *The Velocity of Sound Waves and the Temperature in the Stratosphere in Southern California*, Bull. Am. Met. Soc. *20*, 192–201.

HARKRIDER, D. G. (1964), *Theoretical and Observed Acoustic Gravity Waves from Explosive Sources in the Atmosphere*, J. Geophys. Res. *69*, 5295–5321.

HEDIN, A. E. (1991), *Extension of the MSIS Thermosphere Model in the Middle and Lower Atmosphere*, J. Geoph. Res. *96*, 1159–1172.

HEDIN, A. E., FLEMING, E. L., MANSON, A. H., SCHMIDLIN, F. J., AVERY, S. K., CLARK, R. R., FRANKE, S. J., FRASER, G. J., TSUDA, T., VIAL, F., and VINCENT, R. A. (1996), *Empirical Wind Model for the Upper, Middle and Lower Atmosphere*, J. Atmos. Terr. Phys. *58*, 1421—1444.

KATZ, CHARLES N. (1999), *Comparison of Infrasound Detectors Using Alternative Detection Statistics: Normalized Cross-Correlation Functions versus the F-Statistic*, SAIC Internal Technical Report, Monitoring Systems Operation.

KENNETT, B. L. N. (1991), *IASPEI 1991 Seismological Tables*, Research School of Earth Sciences, Australian National University, 167 pp.

KERR, A. U. (1971), *Digital Computer Programs for Recording and Processing Infrasonic Array Data*, Geophys. J. R. Astr. Soc. *26*, 21–40.

KULICHKOV, S. N. (1998), *On Problems of Infrasonic Monitoring of Small-energy Explosions*. In *Proceedings of the Informal Workshop on Infrasound*, Bruyeres-le-Chatel, France.

LINDEMAN, F. A., and DOBSON, G. M. B., (1923), *A Theory of Meteors, and the Density and Temperature of the Outer Atmosphere to which it Leads*, Proc. Roy. Soc. 102, 411–437.

MCKISIC, J. M. (1997), *Infrasound and the Infrasonic Monitoring of Atmospheric Nuclear Explosions*, Report No. PL-TR-97-2123, Phillips Laboratory, 310 pp.

MILNE, E. A. (1921), *Sound Waves in the Atmosphere*, Phil. Mag. *42*, 96–114.

PFEFFER, R., and ZARICHNY, J. (1962), *Acoustic Gravity Wave Propagation from Nuclear Explosions in the Earth's Atmosphere*, J. Atmos. Sci. *5*, 256–263.

PIERCE, A. D., and POSEY, J. W. (1970), *Theoretical Predictions of Acoustic-gravity Pressure Waveforms Generated by Large Explosions in the Atmosphere*. Prepared for Air Force Cambridge Research Laboratories, Office of Aerospace Research, USAF, Bedford, MA.

PRESS, F., and HARKRIDER, D. G. (1962), *Propagation of Acoustic Gravity Waves in the Atmosphere*, J. Geophys. Res. *67*, 3889–3908.

SMART, E., and FLINN, E. A. (1971), *Fast Frequency-Wavenumber Analysis and Fisher Signal Detection in Real-Time Infrasonic Array Data Processing*, Geophys. J. R. Astr. Soc. *26*, 279–284.

VELOSO, J. A. (1998), *Establishment of the IMS Infrasound Network*. In *Proceedings of the Informal Workshop on Infrasounds*, Bruyeres-Le-Chatel, France.

WANG, J. (1999), *Signal Detection and Estimation at the New IMS Array at Mina*. In *Proceedings of the 21st Annual Seismic Research Symposium*, Las Vegas, Nevada, September 21–24, 1999.

WANG, J., and WILLEMAN, R. (1997), *A Systematic Approach to Designing Detector Recipes for Infrasonic Arrays*, Center for Monitoring Research, Arlington, Virginia.

WHITAKER RODNEY W., PAUL MUTSCHLECNER, J., MASHA B. DAVIDSON, and SUSAN D. NOEL (1990), In *Proceedings of the Fourth International Symposium on Long Range Sound Propagation*, NASA Conference Publication 3101, compiled by William L. Willshire.

WHITAKER, R. W. (1995), *Infrasonic Monitoring*. In *Proceedings of the 17th Annual Seismic Research Symposium*, Scottsdale AZ, September 12, 1995, 996–1000.

(Received September 3, 1999, revised June 1, 2000, accepted June 8, 2000)

 To access this journal online:
http://www.birkhauser.ch

Pure appl. geophys. 159 (2002) 1127–1152
0033–4553/02/051127–26 $ 1.50 + 0.20/0

❙ Pure and Applied Geophysics

Surveying Infrasonic Noise on Oceanic Islands

MICHAEL A. H. HEDLIN,[1] JON BERGER[1], and FRANK L. VERNON[1]

Abstract — An essential step in the establishment of an International Monitoring System (IMS) infrasound station is the site survey. The survey seeks a location with relatively low infrasonic noise and the necessary logistical support. This paper reports results from our surveys of two of the oceanic sites in the IMS – the Azores and Cape Verde. Each survey sampled infrasonic noise, wind velocity, air temperature and humidity for ~3 weeks at 4 sites near the nominal IMS locations. The surveys were conducted on Sao Miguel (the main island in the Azores) and Maio (Cape Verde). Infrasonic noise was measured using the French MB2000 microbarometer.

During our 3-week experiment in January the trade winds at Cape Verde varied little from an azimuth of 63°. Because of the unvarying wind azimuth, the experiment gave us an opportunity to examine the effectiveness of a forest at reducing both wind speed and infrasonic noise. We find that the thick Acacia forest on Maio reduces wind speeds at a 2 m elevation by more than 50% but does not reduce infrasonic noise at frequencies below 0.25 Hz. This forest serves as a high-frequency filter and clearly does not reduce long-period noise levels which are due to large-scale turbulence in the atmospheric boundary layer above the forest. This is consistent with our observations in the Azores where the relationship between infrasonic noise and wind speed is more complex due to frequent changes in wind azimuth.

In Cape Verde, wind speed and infrasonic noise are relatively constant. The diurnal variations are clearly seen however the microbarom is only rarely sensed. In the Azores, during our 3-week experiment in November and December of 1998, wind speed and infrasonic noise change rapidly. At this location, daily noise level swings of 40 to 50 dB at 0.1 Hz are not uncommon in the early winter and are due to changes in wind speed and atmospheric turbulence. The effectiveness of an infrasound station in the Azores will be strongly dependent on time during the winter season.

The two surveys illustrate some of the difficulties inherent in the selection of sites for 1 to 3 km aperture arrays on oceanic islands. Due to elevated noise levels at these sites, 8 element, 2 km aperture arrays are strongly preferred.

Key words: Atmospheric turbulence, infrasonic noise, noise reduction.

1. Introduction

1.1. The CTBT and Infrasound

A natural consequence of any global nuclear-test ban treaty is the need for global monitoring. In the case of the recent Comprehensive Nuclear-Test-Ban-Treaty

[1] IGPP-University of California, San Diego, La Jolla, CA, 92093-0225, U.S.A.
E-mail: mhedlin@eos.ucsd.edu

(CTBT), which bans nuclear explosions of any yield, four monitoring networks will be used – seismic, hydroacoustic, air acoustic (or infrasound) and radionuclide. This study is concerned with infrasound, the inaudible acoustic energy from 0.01 to 20 Hz. Although there are myriad infrasound sources (including natural ones such as atmospheric turbulence, bolides, aurora, earthquakes, volcanic eruptions, ocean waves, avalanches, convective storms; GEORGES and YOUNG, 1972; KULICHKOV, 1992; MUTSCHLECNER and WHITAKER, 1994; WILSON et al., 1995; BONNER et al., 1998; and man-made ones such as supersonic aircraft or rockets, industrial explosions; YAMAMOTO, 1956; FEHR, 1967; LISZKA, 1974; GOSSARD and HOOKE, 1975; KULICHKOV, 1972; BONNER et al., 1998) the source of greatest interest, the kind that must be separated from the rest, are nuclear explosions. Since the era of atmospheric testing, it has been known that the shock front from a nuclear explosion will evolve into infrasonic energy that propagates efficiently through the lower atmosphere (LANDAU and LIFSHITZ, 1959; BERANEK, 1960). Buried nuclear tests cause a piston-like vertical ground motion which also produces infrasonic pressure waves (BLANC, 1985; CALAIS et al., 1998). Much of the acoustic energy from an infrasound source is refracted back to the Earth's surface at the thermocline (at 100 to 150 km altitude) or at lower elevations due to wind shear (SIMONS, 1995). A Lamb surface wave exists principally below ~30 km altitude (LAMB, 1932; FRANCIS, 1973) and will be sensed within 500 km of the source (BROCHE, 1977). The propagation of refracted and surface infrasonic energy through the atmosphere is dependent on the acoustic velocity and thus on wind, temperature, humidity and viscosity (e.g., STOKES, 1857; REYNOLDS, 1874; HAURWITZ, 1941; INGARD, 1953). Propagation of acoustic energy through the atmosphere thus depends strongly on time. The frequency content of infrasonic energy is strongly range-dependent due to attenuation (BLANC, 1984) and because of nonlinear stretching which favors the lower frequencies as propagation distance increases (SIMONS, 1995). Most signals of interest to the monitoring community lie between 0.02 and 4.0 Hz (CHRISTIE, personal communication).

On paper, any ideal monitoring network is perfectly uniform. On the earth's surface, the best possible global network will include a large number of stations on oceanic islands (Fig. 1). Of the 60 stations that will be in the IMS infrasound network, 23 will be on islands. Most of these island sites will be less than ideal in certain aspects considered important for a good station. For example, most sites will be subject to strong winds (Fig. 1). Wind is the most significant source of infrasonic noise between 0.01 and 1 Hz (MCDONALD et al., 1971). Microbaroms are 5–8 second, intermittent, infrasonic waves that are produced at the air-sea interface by ocean swell (DONN and POSMENTIER, 1967). This noise is most evident on islands and at continental edges. Some islands will offer few suitable locations to deploy the standard infrasound station – a 1 to 3 km aperture triangular array of 4 to 8 sensors – a configuration that has been shown to be optimal (HAUBRICH, 1968). Most islands are rugged and thus there exists a significant potential for signal blockage and the generation of noise from turbulence due to wind flow over topography.

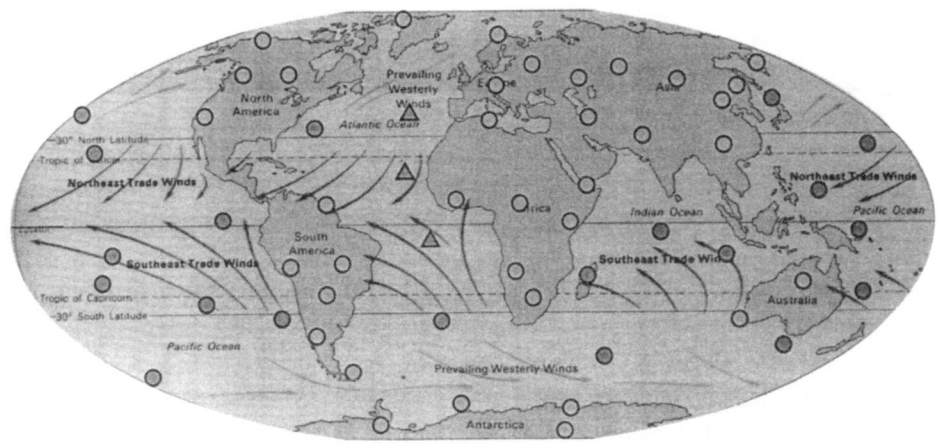

Figure 1

The planned global IMS infrasound network. Stations located on islands will, in general, be subject to strong winds and thus high infrasonic noise levels. Strong trade winds dominate in the tropics. In latitudes above 30°N and below 30°S the prevailing westerly winds dominate although local winds can be strongly influenced by local pressure cells. The three triangles in the Atlantic from north to south are the Azores, Cape Verde and Ascension. The background display of oceanic winds is adapted from the World Book Encyclopedia.

None of the oceanic or continental infrasound stations are to be selected without a site survey. The Provisional Technical Secretariat (PTS), which has issued a set of guidelines for site surveys, lists several factors as having distinct importance. A good site has low infrasonic noise between 0.02 to 4 Hz. The site is sheltered from winds by gentle topography and vegetation and has the necessary logistical support while being located as far as possible from cultural activity and other significant sources of noise – such as the ocean. Preference is given to sites located within 0.05° of the nominal station location and sites where a station from one of the other monitoring networks is located. In a standard site survey, infrasonic noise and meteorological data are collected at 4 favorable locations for a minimum of 2 weeks. A 2-week survey offers just a glimpse of the full range of meteorological phenomena that will occur through the year at an infrasound station, but should clearly indicate which site should be best.

1.2. Wind and Infrasonic Noise

The turbulent flow of wind is the most significant source of infrasonic noise in the band from 0.01 to 1 Hz (McDonald et al., 1971). The relationship between wind and infrasonic noise is given by Bernoulli's principle

$$p + (\rho v^2)/2 = C$$

where p is pressure, ρ is density, v is wind velocity and C is a constant. Differentiation of this formula shows that there is a simple linear scaling between variations in wind speed and pressure. Understanding the frequency dependence of infrasonic noise requires knowledge of the interaction of wind with topographic irregularities that cause the turbulence. In general terms, atmospheric turbulence is concentrated near the Earth's surface in the Atmospheric Boundary Layer (ABL; KAIMAL and FINNIGAN, 1994) – a region of high Reynolds-number flow. The upper extent of the ABL is limited by the capping inversion which is highest (at 1 to 2 km) during the day when the Earth's surface radiates heat and lowest (10's of meters) or nonexistent during the night (PANOFSKY and DUTTON, 1984; KAIMAL and FINNIGAN, 1994). The thickness of the boundary layer is also governed to some extent by topography as the turbulence of wind flow over the ground is increased by ground roughness. Large-scale atmospheric turbulence is maintained by thermal plumes and thus is most evident during the day. Wind shear introduces smaller eddies and "cascades" energy from the low to the higher frequencies (KAIMAL and FINNIGAN, 1994). There is no universal statistical characterization of atmospheric turbulence at large scales (WILSON *et al.*, 1999). The turbulence depends on the velocity and density profiles of the atmosphere as well as boundary effects such as topography and ground cover. The spatial structure of the atmospheric turbulence is dependent on the interaction of the wind with topography and can be modeled using one of a number of models including Gaussian, von Karman or Kolmogorov spectral models (WILSON *et al.*, 1999). Wind noise is known to be incoherent at spacing of 10's of meters (e.g., PRIESTLEY, 1966). There exists a direct scaling between the size of the wind eddies and the time-frequency of the noise they produce. At 0.25 Hz, the scale of the turbulence is ~25 m (GROVER, 1971).

1.3. Overview of this Paper

Our group has conducted two of the oceanic surveys: The Azores, Portugal and Cape Verde, Western Africa. In this paper we discuss the relationship between wind and infrasonic noise and the utility of vegetation and topography for reducing noise across the frequency band of interest to the monitoring community. We conclude with an assessment of the utility of these sites for nuclear test monitoring.

2. Infrasonic Noise Surveys in the Azores and Cape Verde

2.1. Field Equipment

For our surveys we used the MB2000 aneroid microbarometer fabricated by the French Département Analyse et Surveillance de l'Environnement (DASE). These sensors provide a filtered signal between 0.01 and 27 Hz (cut back to 9 Hz by an anti-aliasing filter) with an adjustable sensitivity. We deployed the low sensitivity (20 mV/Pa) version. The electronic noise of the sensor is 2 mPa rms, between 0.02 and 4 Hz, which is well below natural background noise (DASE Technical Manual, 1998).

Each sensor was placed in an insulating case and equipped with a microporous "front end" filter (DANIELS, 1959; BURRIDGE, 1971) for suppression of noise due to wind turbulence (COOK and BEDARD, 1971). The filter senses air pressure through microporosity ports distributed over a small area. Uncorrelated pressure variations, such as those due to wind turbulence, sum incoherently and are attenuated. Although noise and the signal of interest might have the same time-frequency, the noise, which is most commonly due to wind turbulence, is incoherent over shorter length scales. For example, at 0.25 Hz, wind noise is incoherent at spacing of 10's of meters while acoustic waves at the same frequency can be coherent at > 1 km (GROVER, 1971). Under windy conditions it has been demonstrated by numerous authors (incl. DANIELS, 1959; GROVER, 1971; BURRIDGE, 1971) that a spatial filter will suppress noise, however the suppression is highly dependent on frequency, hose configuration and wind speed (GROVER, 1971; NOEL and WHITAKER, 1991). For the surveys we deployed identical filters at all sites – a simple, 4 arm, 30-m aperture cross composed of microporous hose. Each sensor is deployed with ultrasonic wind velocity, air temperature and humidity sensors. The temperature and humidity sensors were located 1 m above the ground. The wind sensor was 2 m above the ground. The infrasound and meteorological signals were digitized at 20 and 1 sps respectively using a 24 bit Reftek 72A-08 data acquisition system. Each system was run on solar power.

The Atlantic site surveys were preceded by calibration tests in the field at the Pinon Flat Observatory (PFO) in southern California, and in the laboratory at IGPP in La Jolla, California. The tests were conducted to ensure all field systems were robust and yielded equal digitized signals for equal input.

2.2. Preliminary Site Selection

In the Azores and in Cape Verde, meteorological data have been collected for decades. These data together with ground cover, topography and infrastructure maps guided us in our selection of survey sites. In the Azores, meteorological data from the past 30 years indicates that winds on the main island, Sao Miguel, are relatively weak (Fig. 2). In this region, winds are strongly influenced by local pressure cells. In the summer a high pressure cell normally lies over the northern islands (Pedro Mata, meteorologist at Portugal's Instituto de Meteorologia). Clockwise air circulation brings winds to Sao Miguel from the northeast. For the rest of the year, western winds dominate. The faintness of the winds at the station in Ponta Delgada on Sao Miguel is at least in part due to shielding this large island provides from these winds. Sao Miguel is a 70×10 km island that has a population of 125,000. Topography on the island is dominated by three active stratovolcanoes, each of which has erupted at least once in the last 700 years (Panduronga Dessai, personal communication). The highest point on the island is the Agua de Pau volcanic center at 947 m. Eroded and vegetated mounds of soft welded tuffs or pumice dominate much of the fine-scale

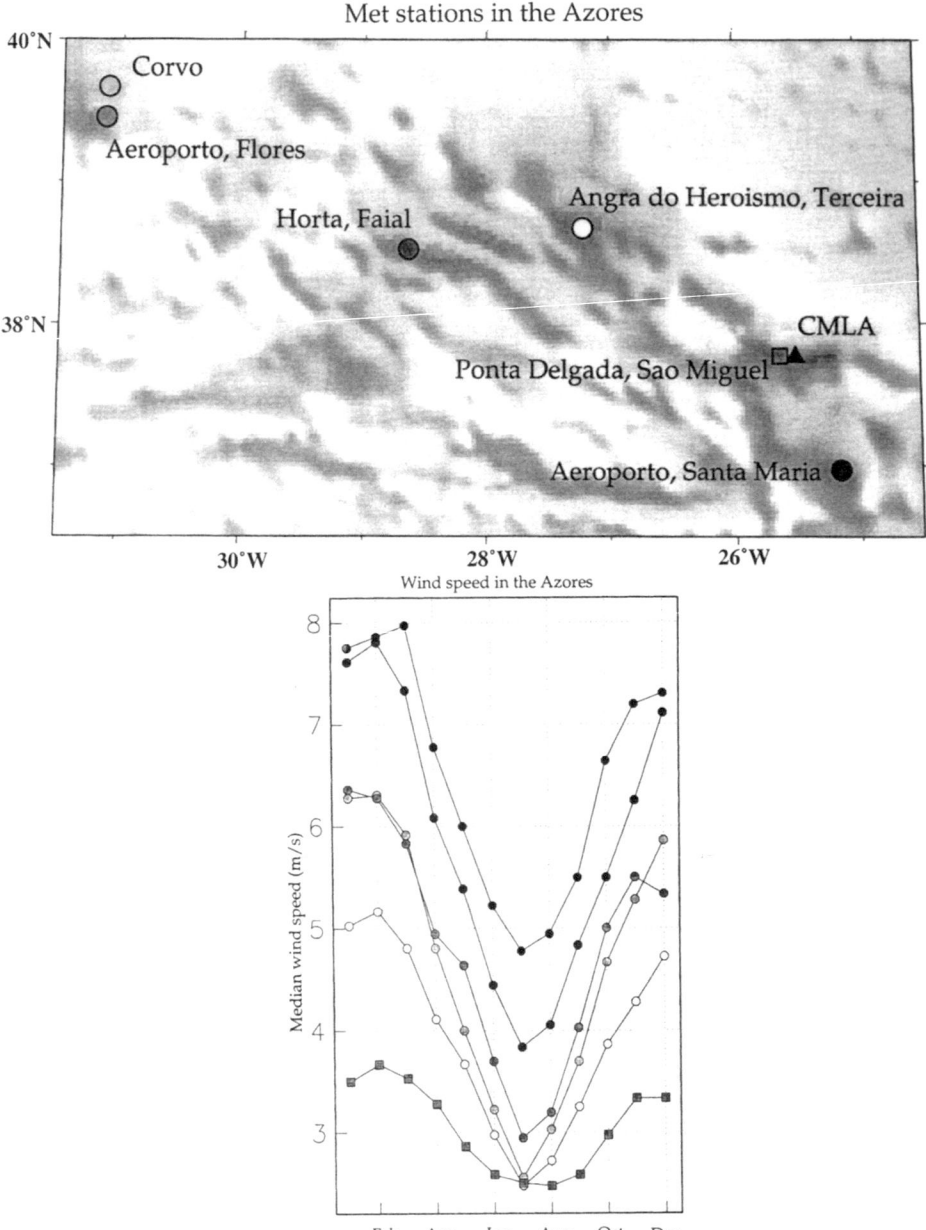

Figure 2
Meteorological data have been collected in the Azores for decades at the sites indicated above. As shown to the right, 30 year averaged wind speeds are strongly dependent on time and location. The square symbols represent wind speeds at Praia on Sao Miguel, the main island. The Met stations on the other islands are represented by the shaded circles. The shading is used to indicate the linkage between the curves on the right and the stations on the left. These data were collected between 1951 and 1980 and were published in "O Clima de Portugal", Instituto Nacional de Meteorologia e Geofisica, ISSN 0870-4767, 1991.

topography on this island. There are naturally occurring fumaroles in the east-central part of the island. The central, youngest, part of the island has no significant topography but is subject to relatively strong winds due to funneling around significant topography on both sides. All the islands in the Azores were once densely forested however they were settled in the 15th century and most of the trees have been cleared for agricultural or cultural use. Most of the flat areas on the island are a patchwork of stone walled pastures. Some dense forests exist in the more rugged areas. Most flat areas are a patchwork of pastures separated either by trees or rock fences.

The four survey points were selected to sample diverse means of reducing wind noise. A large, 5-km diameter, calderra exists at the west end of the island (Fig. 3). An eruption last occurred there in 1287 AD (BOOTH et al., 1978). The small population within the calderra is concentrated in Sete Cidades. The calderra was chosen as a survey point as it is large enough to accommodate an infrasound array. The walls of the calderra reach 275 m high. With the survey we planned to quantify the noise reduction, if any, that results from shielding provided by the walls. Two other sites, Pinhal da Paz and Cha do Macela, were located in thick Criptomeria forests in rolling topography. The latter site was co-located with the GSN station CMLA and lies closest to the nominal infrasound station location. The exact nominal location was not surveyed as this area is densely populated (Fig. 3) and is the site of considerable hydrothermal activity. The fourth site, Cha das Mulas, was

Figure 3
Sao Miguel, Azores and the four survey locations. The western site is located in a calderra. The two central sites are located in dense forests. Cha das Mulas is located on a plateau which has a rich low-level ground cover. Most of the original forest has been cleared. Most trees are in long rows and are used to decrease wind speed. The nominal station is located near the second largest city on the island, Ribeira Grande, in an active hydrothermal area. The recommended IMS infrasound array is at Cha das Mulas. One possible configuration is indicated by the triangle.

situated on a broad plateau near the east end of the island. The plateau is ~580 m above sea level and has large pastures separated by rows of Criptomeria trees.

The Cape Verde archipelago comprises 9 major islands (Fig. 4). All but one is unsuitable for infrasound. The western islands are relatively young and rugged. The extreme example is Fogo, a 2829 m high active volcano. The eastern, relatively flat, islands have little vegetation and are subject to high trade winds from the northeast. The exception is Maio which is sparsely populated, has a substantial forest and a line of site to the largest island, Santiago, where telecommunications infrastructure is located. All four survey sites were located on this island in the forest in a 3-km aperture centered equilateral triangle (Fig. 5).

Forests reduce wind speeds and thus will attenuate atmospheric noise. The thickness of the forest on Maio is variable. We have chosen sites at which the forest comprises mature, closely spaced, Acacia trees (MAO2) and sites where the trees are either broadly spaced (MAO1 and MAO3) or very young (MAO4). There is essentially no ground cover at any of these sites. A comparison of measurements made at these dissimilar locations will allow us to gauge the utility of Acacia forests for reducing wind speeds and noise in the infrasonic band between 0.02 and 5 Hz.

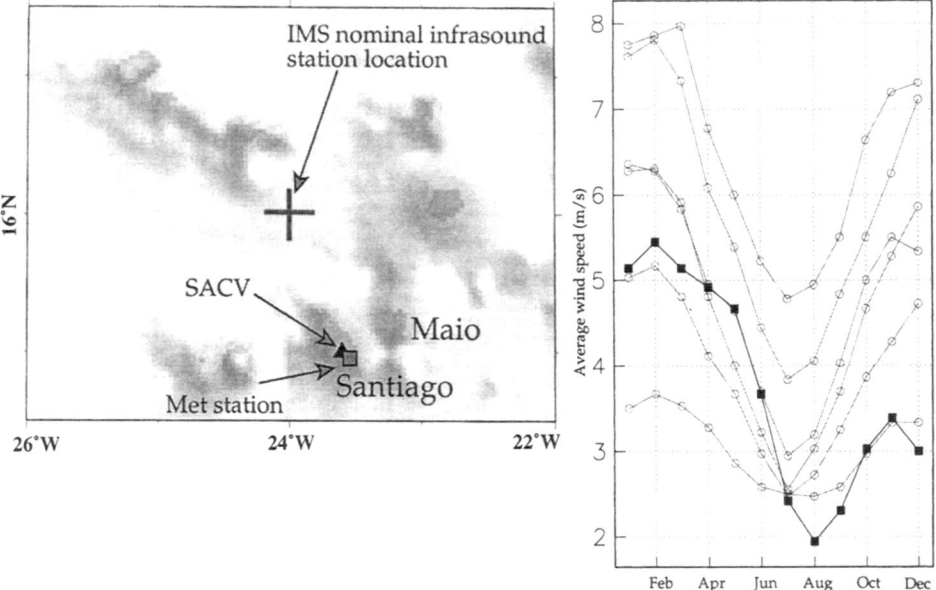

Figure 4
Meteorological observations have been made at Praia for decades. Seasonal trends in Cape Verde are similar to those in the Azores (right) however the wind in Cape Verde is dominated by the NE trades. Our survey occurred in January, a time of relatively strong winds. Meteorological data were taken from a 30 year study (Ó Clima de Portugal, 1931–1960).

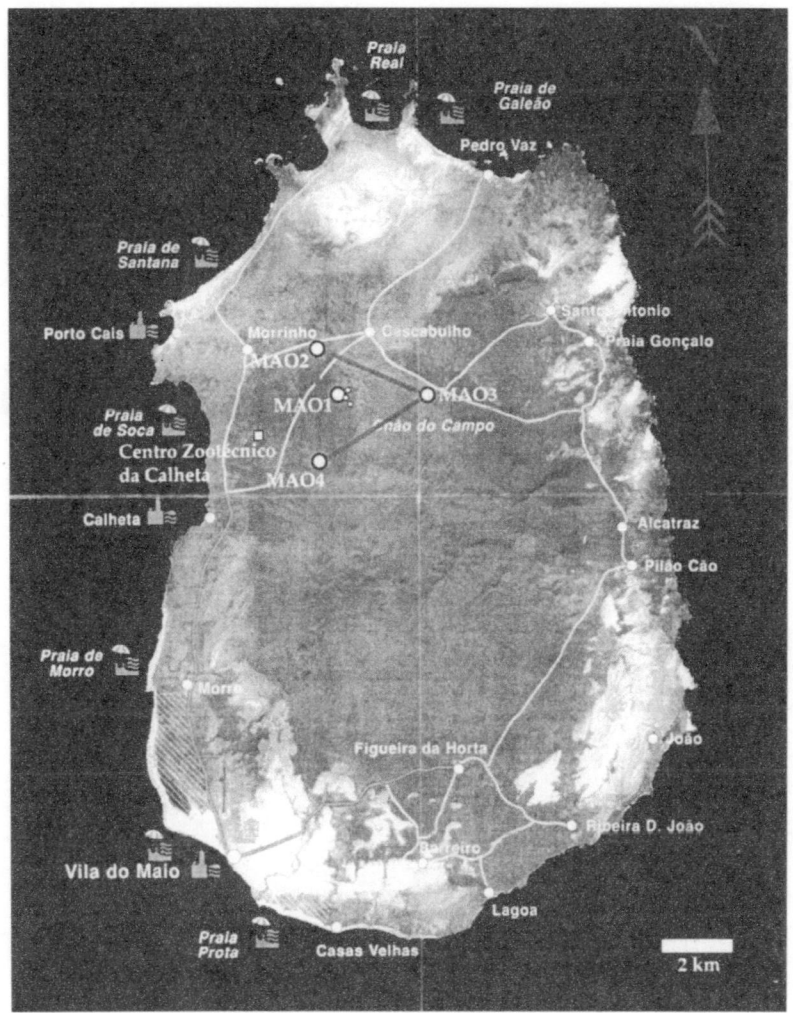

Figure 5

The island of Maio, Cape Verde has little vegetation except in an Acacia tree forest in the north. The island covers ~310 square km, roughly four times the size of Ascension Island, the site of another nominal IMS infrasound station. The four survey points are in a forest of Acacia trees.

2.3. Preliminary Results from the Surveys

The four survey points on Sao Miguel, Azores were occupied in November, 1998. After 19 days of simultaneous recording at all sites, the equipment was moved to Cape Verde (Figs. 1 and 4) for an additional three weeks of recording in January, 1999. Fifteen minutes of pressure and meteorological data from the Azores experiment are shown in Figure 6. The unfiltered pressure time series from the station at Cha das Mulas exhibits 30 s period fluctuations superimposed on

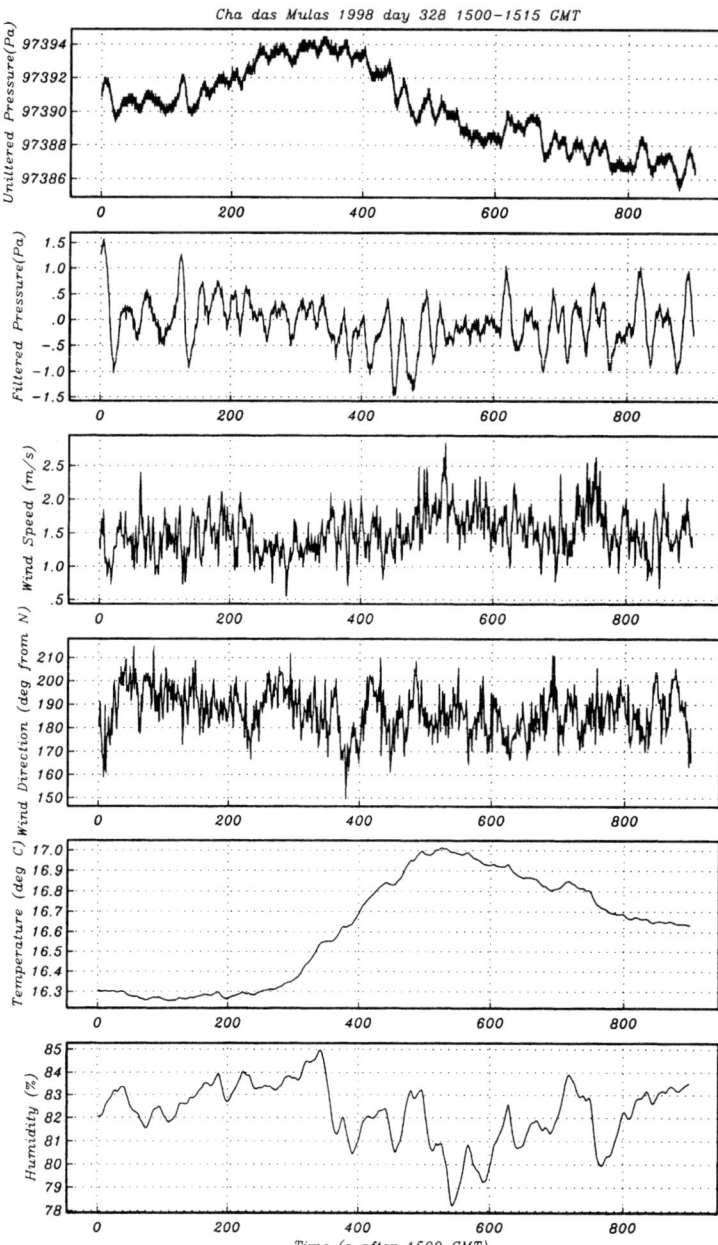

Figure 6

Atmospheric pressure and meteorological data from a 15-min interval starting at 15:00 GMT at Cha das Mulas. Raw and filtered pressure data are shown in the upper two panels. Wind speed, direction, temperature and humidity are shown below. As shown in the next figure, this is a time of relatively low infrasonic noise.

substantially longer period energy. The shorter period variations dominate the filtered record (second panel). In this time period, minor wind velocity fluctuations are constant. The temperature and humidity vary relatively slowly. The detailed structure of the filtered pressure record results from the superposed contributions of myriad atmospheric phenomena. This detail is impossible to fully understand or predict. Our study is concerned primarily with the dependence of the noise on frequency and the exact location of the sensor, and thus we will focus on the spectral properties of the noise.

To examine the spectral content of the filtered data we used Welch's method (WELCH, 1967) which yields a single power spectral estimate from an average of several taken at regular intervals in the time range of interest. The spectral estimates we have used in this study were derived from an average of four estimates, each taken from consecutive 204.8 s intervals. The power spectral density taken from the filtered record (Fig. 7) shows decay in power levels from the peak at 30 s to the corner of the anti-aliasing filter at 9 Hz. A strong microbarom peak is centered at 0.2 Hz. The peak at 30 s is somewhat misleading as it does not reflect an intrinsic lack of noise

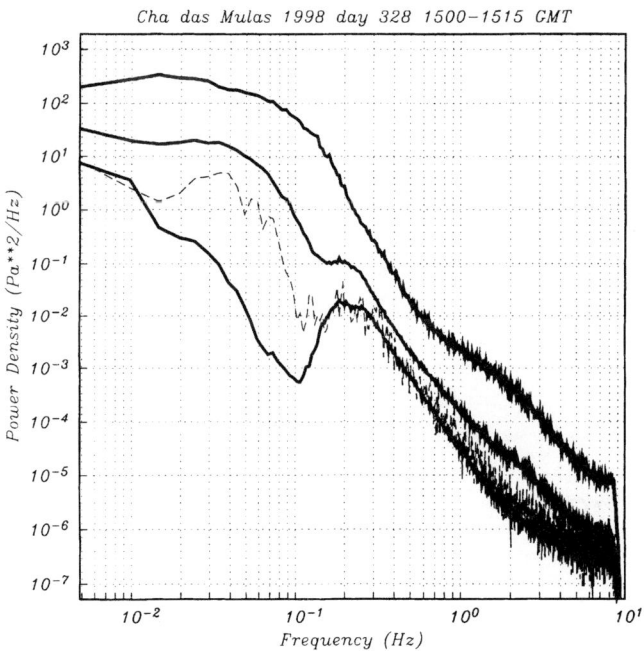

Figure 7

Noise power density at Cha das Mulas. In black is shown a single power density estimate taken from the filtered pressure data displayed in Figure 6. The thick curves are the 10th, 50th and 90th percentile noise levels from the entire experiment. These curves are based on 454 noise level estimates taken at 1 hour intervals. Each estimate is taken from 15 minutes of data. The 5 second microbarom is not observed at times of high noise.

power in the atmosphere at longer periods but is what remains after the sensor filter which bandpasses the raw signal to exclude energy below 0.01 Hz. The overall spectral shape is due to a well-known phenomena in the atmosphere which receives a significant input of energy at ∼0.1 to 1 mHz. Energy from large scale eddies cascades into smaller eddies, and thus into higher frequencies, as the large eddies are fragmented (KAIMAL and FINNIGAN, 1994).

From the full three weeks of recording at all four sites we calculated power spectral density at 454 time intervals. Each interval began at the turn of the hour and lasted 15 minutes. Tenth, 50th and 90 percentile noise levels at all frequencies from 0.005 Hz to 10.0 Hz are shown in Figure 7. At times of relatively low noise the microbarom peak is obvious at Cha das Mulas. This figure shows that the interval displayed in Figure 6 was a time of relative quiescence. From this figure it is evident that the longer periods depend more strongly on time. For example, the spread between the 10th and 90th percentiles is 50 dB at 0.1 Hz and just 20 dB at 1.0 Hz.

The same spectral character is seen at all four sites (Fig. 8). Median noise levels between 0.03 and 0.1 Hz are 5 to 10 dB lower in the calderra (thick dashed curve)

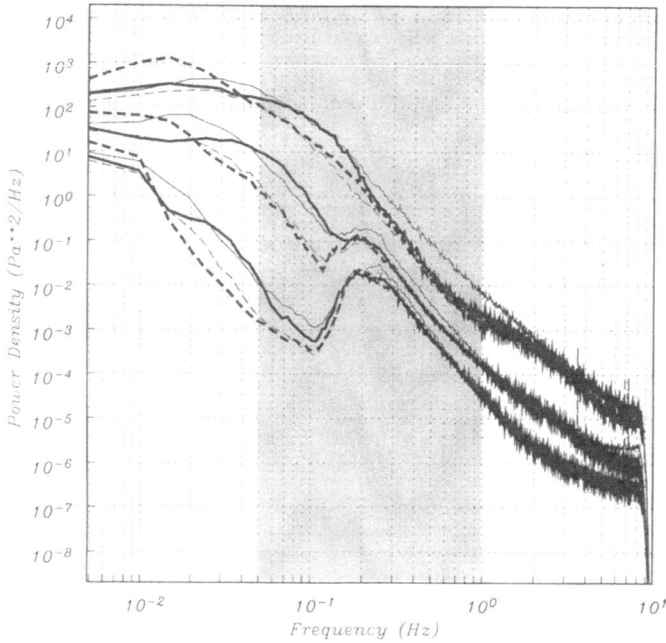

Figure 8

Tenth, 50th and 90th percentile noise curves at the four sites surveyed in the Azores. Noise levels in the calderra, at Pinhal da Paz, Cha da Macela and Cha das Mulas are represented by the thick dashed, thin dashed, thin solid and thick solid curves respectively. Between 0.02 and 0.2 Hz, noise at Cha das Mulas and Cha do Macela is 5 to 10 dB higher than in the calderra and in the forest at Pinhal da Paz. At higher frequencies, median noise levels are comparable at all sites. Most signals of interest to the monitoring community lie in the shaded region.

and in the forest at Pinhal da Paz (thin dashed curve) than on the plateau at Cha das Mulas (thick solid curve) and at Cha do Macela (thin solid curve). At other frequencies the noise levels are comparable. The noise spikes at 3.5 and 7.0 Hz at Pinhal da Paz are likely due to infrequent industrial activity. That survey point is near several quarries.

As expected, the noise power in the band between 0.005 Hz and 10 Hz is highly dependent on wind speed. In Figure 9 we display the power spectral density at several frequencies between 0.02 and 5 Hz as a function of wind speed. The wind speeds depend strongly on the site. In the calderra, the winds at an elevation of 2 m reach ~2.5 m/s. The winds are strongest in the forest at Cha da Macela where they reach 6.8 m/s. The winds are slightly weaker on the plateau at Cha das Mulas (max.

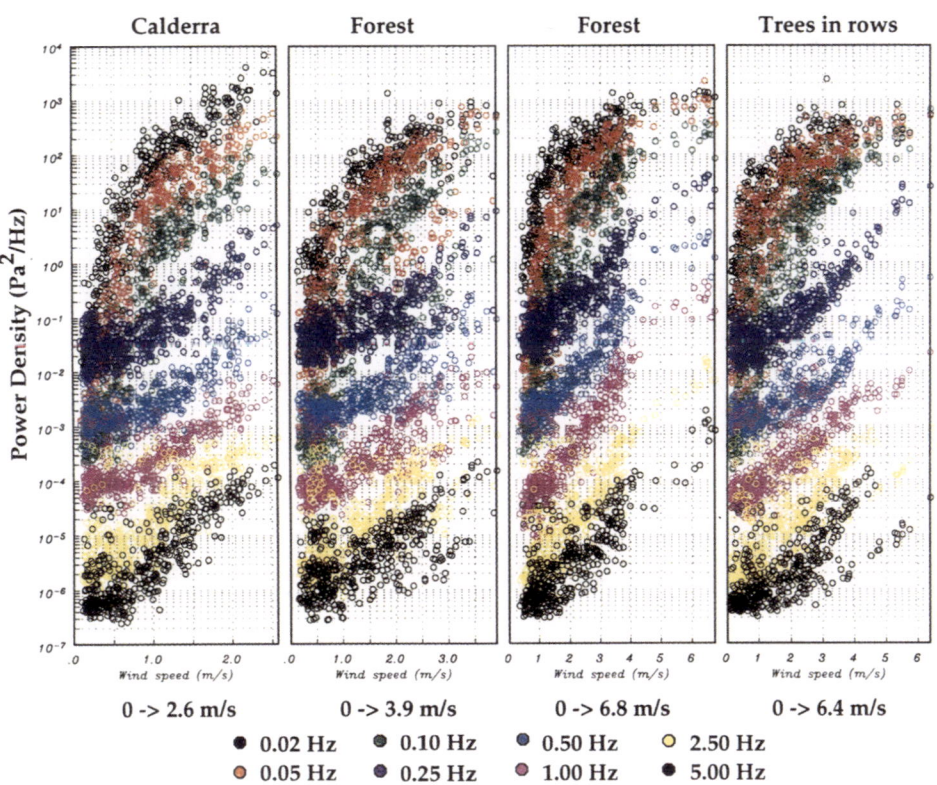

Figure 9

The four panels above show the strong dependence of infrasonic noise on wind speed and frequency at the four sites in the Azores. From top to bottom in each panel is noise power at 0.02 Hz (black); 0.05 Hz (red); 0.1 Hz (green); 0.25 Hz (dark blue); 0.5 Hz (light blue); 1.0 Hz (pink); 2.5 Hz (yellow); and 5.0 Hz (black). The wind speeds are relatively slight in the calderra however this doesn't appear to have yielded much of a noise reduction. Long-period noise levels are elevated at this location.

6.4 m/s) and are somewhat weaker in the forest at Pinhal da Paz where they reach a maximum of 3.9 m/s. Winds in the calderra are sharply reduced by the walls of the calderra which are 200 to 300 m above the central floor. The forest at Cha da Macela is thick however this location is adjacent to the topographically low, central, part of the islands where winds appear to be relatively strong. Figure 9 shows that the infrasonic noise at all frequencies between 0.02 and 5 Hz depends strongly on wind speed however the dependence is complicated by the highly variable wind azimuth (Fig. 10; left panels). In the Azores, mobile pressure cells dominantly influence local wind and thus the wind direction is highly variable. This variability is strongly dependent on receiver location. As a result, at each site there is a time-variant interaction of wind flow and local topographic features and thus the more complex scaling of wind power with wind speed (Fig. 9). The winds are weakest in the calderra, however at this location the noise power at long periods increases most rapidly with wind speed. The noise power increases relatively slowly at Cha das Mulas. This is likely because of the remoteness of this site from any rugged topography.

In Cape Verde the spectral character of the infrasonic noise is markedly different. The noise percentiles, taken from 400-hourly segments of data, are shown in Figure 11. In Cape Verde the noise levels are comparable to those in the Azores however the microbarom is not often sensed . It is not seen at any of the stations on the tenth percentile curves or above. This difference is unlikely due to proximity to the surf zone. The stations on Maio were between 3 and 6 km from the coast whereas the sites on Sao Miguel were located 4 to 8 km from the coast. The loss of the microbarom is due to the winds at Cape Verde which rarely fall below 1 m/s. As shown in Figure 11, at the times of lowest winds the microbarom is seen clearly. Considering all four sites together, this occurred less than 2% of the time and only when the 15 min average wind speed dropped below 0.3 to 0.4 m/s. The winds reached this level most often (~3% of the time) at MAO2 in the thickest part of the forest.

At Cape Verde the wind azimuth is relatively constant (Fig. 10; right panels) and there exists a simple dependence of infrasonic noise levels on wind speed (Fig. 12). As wind speeds at Cape Verde are relatively constant, the most significant source of variability in infrasonic noise levels is the diurnal effect (Fig. 13; right panel) which is predominantly due to day-night variations in the speed of the trade winds. Infrasonic noise levels are highly correlated with this effect and thus at any time of day, noise levels are quite easily predicted. Our survey indicates that significant wind speed and noise level variations are common in the Azores in the early winter. The experiment in November and December occurred at a time of advancing inclement weather. Toward the end of the experiment wind speed at Cha das Mulas could increase from 1 to 6 m/s in a few hours. Commensurate noise power increases of 40 to 50 dB occurred. Noise levels vary far more strongly with time than location. Periodically, any infrasound station on Sao Miguel will be "deafened" by wind noise. At times of high wind, the microbarom noise is obscured. Rapid changes in wind speed and noise

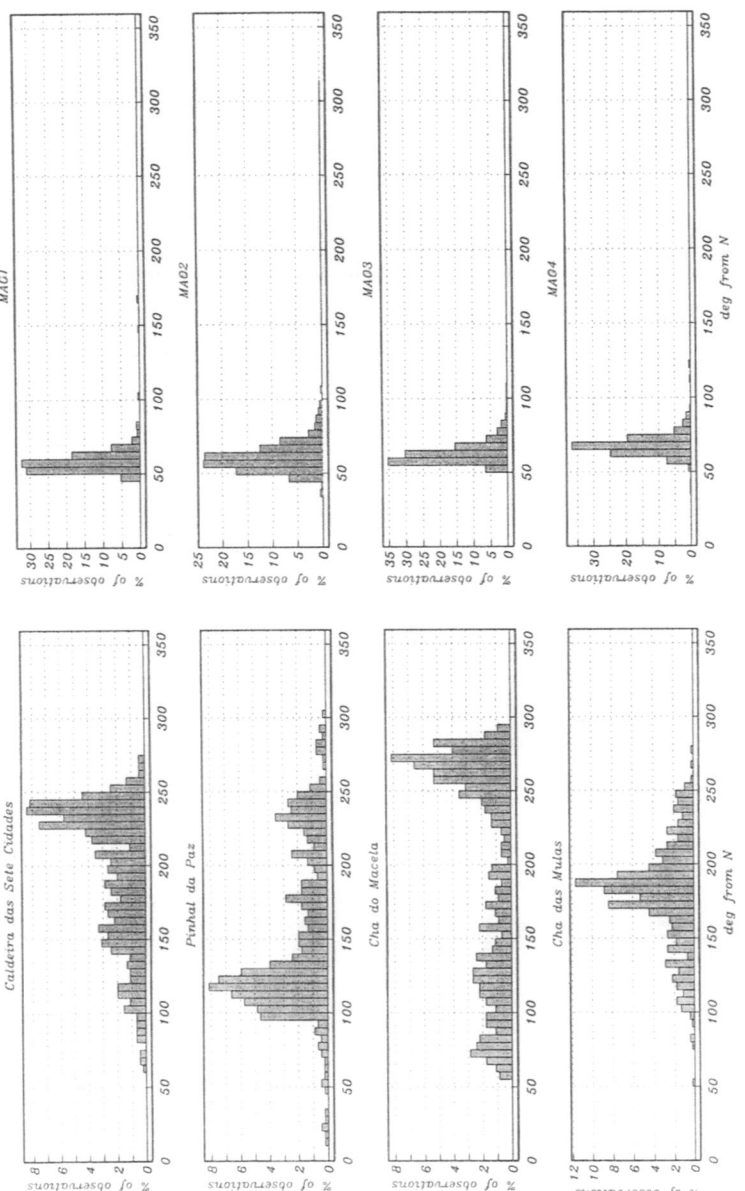

Figure 10

Wind direction at the 8 sites in the Azores and Cape Verde. In Cape Verde, the northeast trade winds dominate whereas in the Azores, wind is influenced by moving pressure cells.

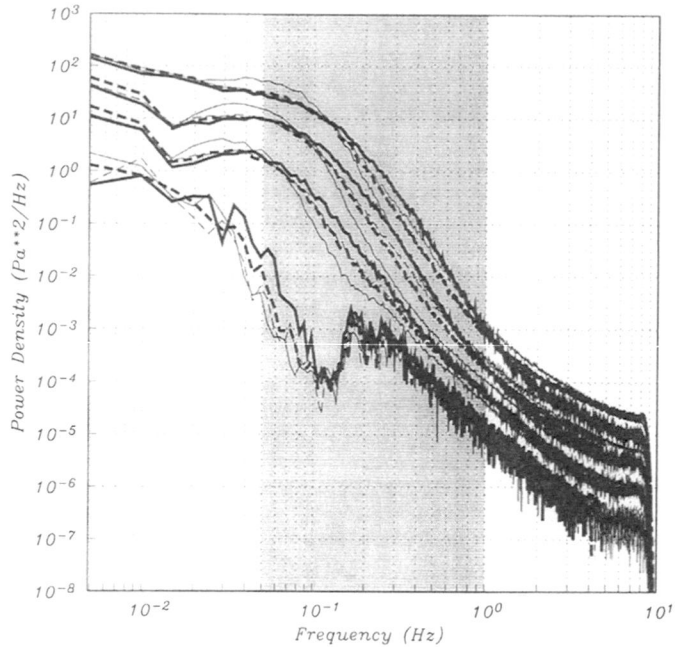

Figure 11

Minimum, 10th, 50th and 90th percentile noise curves from the Cape Verde experiment. The noise levels at MAO1, MAO2, MAO3 and MAO4 are indicated by the thick dashed, thin solid, thick solid and thin dashed curves, respectively. The four sites were located within 3 km of each other. MAO2 was located in the thickest part of the forest. MAO3 was located closest to the coast (Fig. 5) however MAO2 was just slightly further inland. The microbarom is seen just at times of minimal noise. Most signals of interest to the monitoring community lie in the shaded region.

levels, and thus in the effectiveness of any infrasonic station to sense signals from distant events, are not uncommon, at least in the winter in the Azores.

3. Discussion

3.1. Forests and Infrasonic Noise Reduction

Cape Verde is an excellent, natural, laboratory for the study of the generation and suppression of infrasonic noise. The archipelago is located at 16°N where northeast trade winds are dominant. During the 3-week experiment in January, the mean wind azimuth was 63° with a standard deviation of 9.6° (Fig. 10). The wind-topography interaction is relatively time invariant when the wind azimuth remains constant. As we see in Figure 12, these "wind tunnel" conditions produce a simple scaling between wind speed and wind noise at all frequencies from 0.02 to 5 Hz. Infrasonic noise levels are easily predicted (Figs. 12 and 13).

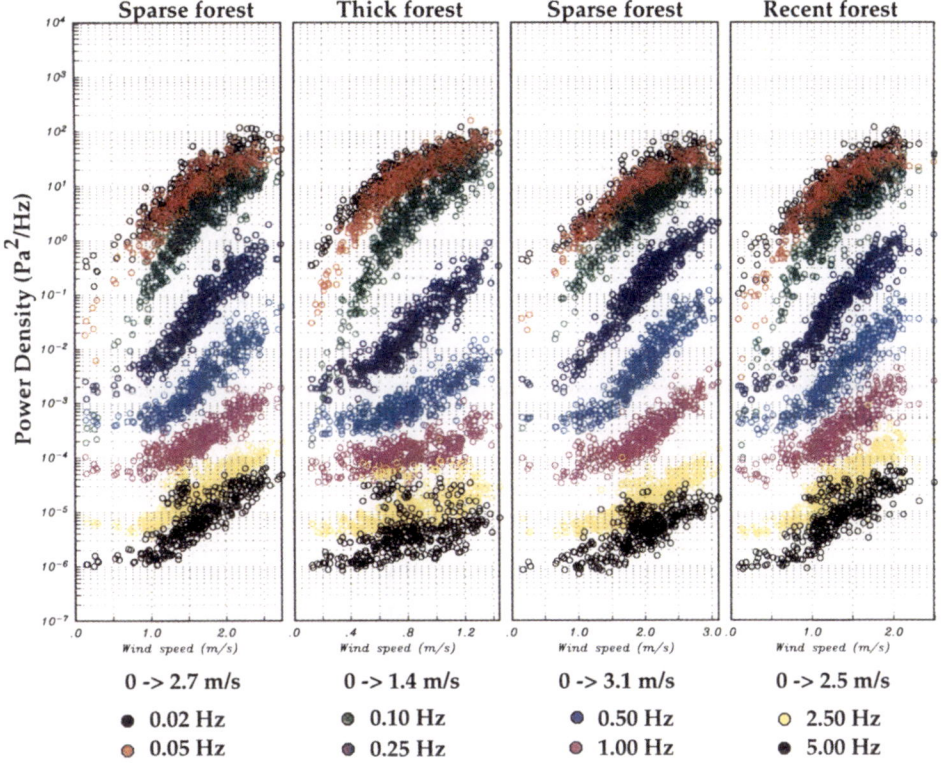

Figure 12
Noise power as a function of wind speed and frequency at the four locations on Maio. Wind speeds in the
thickest forest on Maio, which at MAO2 are lower than elsewhere, however noise power is not reduced at
frequencies lower than 0.25 Hz. Winds are weakest in the thickest part of the forest at MAO2. Noise power
has a relatively simple dependence on wind speed.

With azimuthal considerations effectively nullified it is possible to address the
issue of the effectiveness of a forest for reducing infrasonic noise in the frequency
band between 0.02 and 5 Hz. The Acacia forest on Maio is thickest at MAO2. At this
location the trees were planted in the 1950s and are ∼10 to 15 m high. The forest is
less mature, the trees are smaller (5 to 10 m) and more broadly spaced, at MAO1 and
MAO3. MAO4 was located in a grid of recently planted ∼5 m tall Acacia trees. As
we see in Figure 12, wind speeds (at an elevation of 2 m) at MAO2 are sharply
reduced. The maximum wind speed at this location is ∼1.4 m/s whereas at MAO1
and MAO4 the wind speed reached 2.7 m/s and 2.5 m/s, respectively and at MAO3
the wind speed reached 3.1 m/s. As we see in Figure 12, at frequencies below 0.25 Hz
there is no concomitant reduction in noise levels. For example, at all locations
infrasonic noise at a period of 20 s ranges from 0.01 Pa^2/Hz when there is no wind to
50 Pa^2/Hz when wind speeds are highest. It is apparent that infrasonic noise at

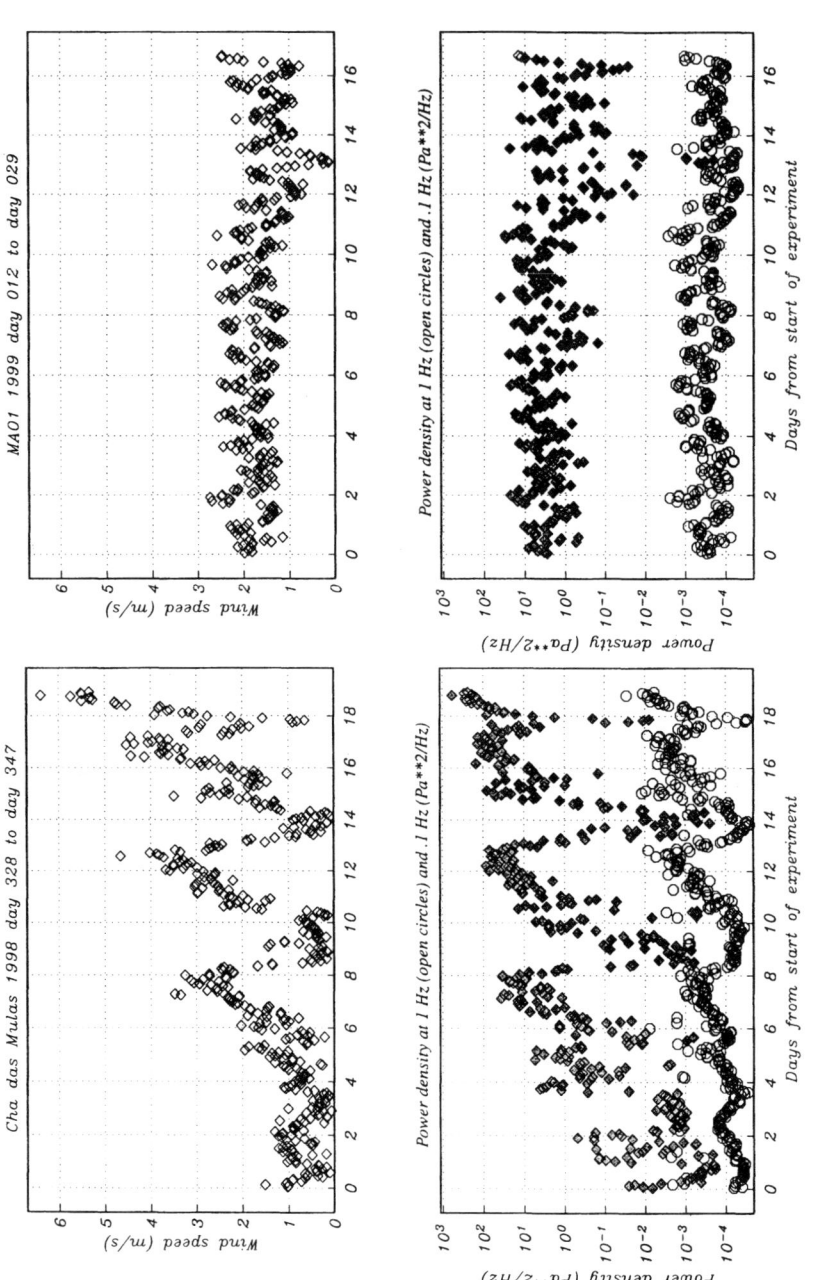

Figure 13

Noise power and wind speed as a function of time at one station in the Azores (left) and in Cape Verde. In the Azores, wind noise at 0.1 Hz (upper sequence of points in the lower panel) and at 1.0 Hz track closely the wind speed. Severe winds during a storm near the conclusion of the Azores experiment lead to wild swings in the noise level. At times, any infrasound station on the island will be deafened by the wind. In Cape Verde, diurnal variations are relatively slight.

frequencies below 0.25 Hz is not dependent on wind speed at 2 m elevation within the forest. Noise at these long periods depends on turbulence that is common to all sites – that which exists in the boundary layer above the trees. The thick forest at MAO2 clearly reduces noise levels at frequencies between 0.25 and 5.0 Hz. The noise estimates at MAO2 become progressively lower than at the other sites as the wind speeds increase. At times when the wind speed is near the maximum, noise levels in this frequency band at this site are 5 to 10 dB lower than at the other sites. Although the forest is reducing high-frequency infrasonic noise levels, the reduction is not complete. A comparison of noise levels at the maximum speed of 1.4 m/s at MAO2 with noise levels at the other sites at the same wind speed indicates that at frequencies at and above 1 Hz, noise levels in the thick forest are 5 to 15 dB higher than elsewhere. When wind speeds are low, the noise suppression within the thick forest extends to 0.1 Hz (Figs. 11 and 12).

The "wind tunnel" experiment in Cape Verde indicates that wind near the ground is not the cause of infrasonic noise at periods greater than 4 s. The most likely source of this noise is turbulence in the atmospheric boundary layer above the forest.

In the Azores, the dependence of infrasonic noise on wind speed is complicated by the azimuthal variations of the wind (Fig. 10). As shown in Figure 9 there is more scatter of noise estimates about a general trend to increased power at higher wind speeds. It is clear, however, that wind noise scales with wind speed as it does in Cape Verde. In the thickest forest at Pinhal da Paz, the wind speeds at an elevation of 2 m are reduced sharply (relative to Cha do Macela and Cha das Mulas) however the infrasonic noise levels at frequencies below 0.1 Hz are not also sharply reduced. Figures 8 and 9 indicate that the thick Criptomeria forest at Pinhal da Paz reduces noise levels between 0.05 and ∼0.1 Hz by 5 to 10 dB when winds are above average. Most trees at this location are mature and reach ∼25 m.

The forest is effective at reducing wind speeds. For example, at Pinhal da Paz, the maximum wind speed was 3.9 m/s and at Cha das Mulas the maximum wind speed was 6.4 m/s. At the peak wind speed, noise levels at Pinhal da Paz between 0.05 and 0.25 Hz were ∼5 dB lower than at Cha das Mulas. However it is clear that the forests are not dissipating all the energy at these frequencies. The noise power in this band at Cha das Mulas when wind speeds were ∼3.9 m/s was 5 to 15 dB lower than the peak.

3.2. Noise Levels Within a Calderra

Wind flow over a two-dimensional ridge is laminar upwind of the ridge crest and becomes turbulent on the lee side (KAIMAL and FINNIGAN, 1994). The turbulence results when the wind flow, which is compressed and accelerated at the ridge crest, decompresses on the lee side. A separation bubble forms at the location where some air flows in the opposite direction, back to the ridge, and is followed by a turbulent wake. The site in the calderra was chosen for our survey to determine if this turbulence would contribute noise in the band between 0.05 and 1.0 Hz or if, overall, the walls of the calderra would deflect the wind away from the site and reduce noise levels. Figure 9 indicates that noise levels are clearly increased at 0.02 Hz and suggests the presence of unusually energetic large-scale turbulence in this area. Figure 8 displays essentially the same result as it indicates that this noise enhancement at ~0.02 Hz is most pronounced on the 50th and 90th percentiles, i.e., when noise and wind speeds are elevated. It is difficult to make definitive conclusions about noise levels at higher frequencies because of the scatter in the noise estimates. However it is apparent that although the calderra effectively reduces wind speeds, noise levels at 0.05 Hz and above are not clearly reduced when wind speeds are high. When the wind is near-average, the calderra noise levels between 0.05 and ~0.15 Hz are comparable to those in the forest at Pinhal da Paz and 5 to 10 dB lower than at the other two sites (Fig. 8). As discussed in the previous section, at times of elevated winds, the calderra walls reduce wind speed although no notable reduction in infrasonic noise levels results (Fig. 9).

3.3. Final Site Selection

Choosing a site for a 1 to 3 km aperture infrasound array on an oceanic island is far from trivial. Infrasonic noise levels can vary sharply across a topographically rugged island. Aside from purely technical considerations, there are a number of practical issues such as land availability. Each array element is to be equipped with a noise reduction system that might span an area as much as 70 m across. Some nearby infrastructure (such as power lines, airports, etc.) should be in the vicinity of the array but not too close so as to influence noise levels. The two site surveys in this paper perhaps shed light on some of the difficulties inherent in the selection process.

3.3a Azores

From every perspective other than the most important one, noise, the plateau is clearly the best site and it will serve as the standard against which all other sites in the Azores will be judged. In addition to ample space to deploy array elements at the same elevation, there are unobstructed views to the horizon from most points. There is minor cultural activity and the site is relatively far from the ocean. A lush grass covers most of the plateau. Long rows of trees interrupt wind flow.

Median noise levels in the calderra on Sao Miguel are comparable to those at Cha das Mulas at all frequencies except between 0.03 and 0.1 Hz where the median noise levels in the calderra are 5 to 10 dB lower (Fig. 8). Most signals of interest lie between 0.05 and 1.0 Hz (PTS summary) and therefore this discrepancy could be important. Because of the short duration of the survey (~3 weeks) it is not clear if this noise spread is significant and indicates relative noise levels throughout the year. The calderra is large enough to accommodate an IMS infrasound array and could be an excellent site however it offers few options for array configurations. Two large lakes and a village lie within the calderra. Just one array configuration is possible – 4 elements in an isosceles triangle 2 1/2 km tall and 1 1/2 km at the base. As shown in Figure 12, noise levels at long periods are elevated, probably due to wind flow across the rim of the calderra. For these reasons, and because of the potential for signal blockage at some azimuths due to the walls of the calderra, this location is not preferred over Cha das Mulas.

The second site, Pinhal da Paz, lies in a dense Criptomeria forest on the axis of the island. This site is near several quarries but has median noise levels that are on par with those in the calderra (Fig. 8). This site lies in rolling hills and, like the calderra, offers few options for array design. Elements in any array deployed at this location would be separated vertically by ~50 m. If full scale (70 m aperture) Daniel's filters are required, it would be difficult to keep all ports at the same elevation. This site is also not preferred over Cha das Mulas for these reasons and because there is more cultural activity in this area. The third site, Cha da Macela, also lies in a dense forest but is subject to high winds directed through the central, low, part of the island (Figs. 3 and 9). Although median noise levels are comparable to those at Cha das Mulas, 90th percentile noise levels are high. Despite the proximity to the GSN station CMLA and the nominal infrasound coordinate (Fig. 3) this site is also not preferred.

Our survey indicates that the best site in the Azores is at Cha das Mulas. This broad plateau is large enough to accommodate any array configuration. Our brief survey indicates that this area is subject to substantial noise level swings (Fig. 13) however comparable noise swings occurred at all sites on Sao Miguel (and presumably at all other islands in the Azores). Any infrasound monitoring station located in the Azores will offer a highly variable performance, particularly during winter when storms are not uncommon.

3.3b Cape Verde

Due to limited options, the Cape Verde survey points were all located together on Maio. Because of topography, cultural activity or a lack of ground cover, the other islands are clearly unsuitable for an effective infrasound deployment. This "micro" deployment was in the configuration of a standard IMS array (a 3-km aperture centered equilateral triangle) in the event that all sites were shown to be adequate and all could be chosen to be sites for the permanent installation. All sites appear to be adequate however this configuration lies at the upper limit of the allowable range for IMS infrasound arrays and might not be appropriate for this rather windy site. Our survey revealed that thick forest reduces wind speed, however just as in the calderra in the Azores, the wind noise was not substantially reduced at long periods (Fig. 11). The best site for an infrasound array on this island is in the oldest, thickest, part of the Acacia forest near the site MAO2.

3.4. Recommended Arrays for these Oceanic Sites

There is an ongoing discussion regarding optimal designs for an IMS infrasound array. Given the constraint which will hold at most sites that no more than four elements are to be used, it is generally accepted that the best array is an equilateral triangle with one element at the center. At locations where infrasonic noise levels are high, the constraint on the number of elements will likely be loosened and as many as 8 will be allowed with the extra 4 sensors placed with the fourth at the center in a micro-array 200–300 m across. The aperture of the array is still an open issue. The PTS site survey summary (CHRISTIE, 1998) puts bounds on the acceptable array aperture at 1 and 3 km. Larger arrays will permit more accurate estimates of the source azimuth but only if the signal is coherent between the stations. At some sites where signals of interest will typically lie above noise and if the peak signal amplitudes lie at periods significantly longer than 1 s, a 3-km aperture might be optimal. It is expected that due to the frequency of the signals of interest and because of high-noise levels, infrasound arrays at most sites will have to be smaller than 3 km. A lower limit on the aperture of an effective infrasound array exists because of the loss of resolution at the lower apertures and owing to the microbaroms. An analysis of coherence (MACK and FLINN, 1971; BLANDFORD, 1997) indicates a 1-km aperture array gives the best overall balance between signal and microbarom coherence. In areas where the coherence drop-off of energetic microbaroms is slow, a 2-km array will be preferred. At sites where land is scarce (e.g., oceanic islands) nontechnical issues, such as land usage permits and topography, might render the scientific debate pointless.

On the Azores and in Cape Verde the array designs are still not finalized. Due to the strength of the microbarom and the presence of high noise due to winds, the best array designs will likely have an aperture of $\sim$2 km with extra 3 sites in a microarray at the center.

3.5. The Utility of Infrasound Site Surveys

Because of limited resources and manpower, the standard infrasound site survey lasts just 2 to 3 weeks. This provides just a glimpse of the full range of meteorological conditions that will occur in a typical year. Is it fair to assume that the brief period sampled is representative of the entire year and the survey will lead the surveyor to recommend the best site, or is a 2 to 3 week survey too brief to be useful? For a brief survey to be ineffective, climatic conditions at the time of the survey would have to be out of the ordinary and tilt the balance of infrasonic noise so that the relative noise levels at the sites are not representative of the yearly average. Intuitively, this seems rather unlikely and easy to check. The chief source of infrasonic noise is turbulence due to wind flow over topography. If wind direction changes in such a way that the wind-topography interaction at the time of the survey is unusual and, as a result, anomalously little turbulence is generated at certain sites, or excessive turbulence is generated at others, the brief survey taken at that time would be misleading. It should be possible to rule out this scenario by comparing wind velocity measurements made during the survey with those from long-term meteorological observations. A significant discrepancy in wind speed or direction combined with azimuthally dependent topography would be cause for concern. The winds observed during the Azores and Cape Verde surveys appear to be in line with long-term observations. Both surveys occurred at times of elevated winds, nonetheless these winds were as expected from historical data. There is no reason to believe that these winter surveys are misleading.

Another argument in favor of brief surveys, other than cost, is that a survey is not used to decide whether or not to locate a station near the nominal coordinates but simply to decide where. It is not necessary to survey for an entire year to be able to judge overall performance of the location, just relative performance of candidate sites.

3.6. The Utility of Oceanic Sites

There are numerous requirements for a good infrasound station (CHRISTIE, 1998). The exceptional island might meet all of them. Noise is the most important consideration. There are a plethora of noise sources – only a subset are of interest to us. Other sources such as earthquakes, rocket launches, volcanoes are easily identified using other means and are infrequent. Wind noise is the paramount concern. As we have seen, wind in the Azores, in the early winter at least, is capricious. This is not a trait of just oceanic sites although it is reasonable to expect such conditions to be more common on islands. Changes in wind speed of 5 to 6 m/s in a few hours appear to be common in the early winter in the Azores. No infrasound station deployed in such a setting can be expected to deliver invariant performance through the year. At times of high wind, such a station will be rendered ineffective for sensing all but the most energetic signals. There are several options for improving signal to noise levels. It is well known that spatial filters reduce noise and the effectiveness of these filters is strongly dependent on the design. It might be necessary to design the filters to suit local needs at each of the windy sites. The standard IMS infrasound array has four elements. Additional elements, located in a microarray at the center, will improve the performance of the overall array. In some areas flat ground will be unavailable and the best noise suppression scheme, beyond vegetation, will likely be spatially limited wind fences (e.g., REVELLE and WHITAKER, 1999).

The oceanic sites will make an important contribution to global monitoring but it is clear that the effectiveness of a typical oceanic, or continental, site will be highly dependent on local meteorological conditions. Global detection thresholds will be strongly dependent on time at all locations (TROST, 1997). Recent research by CLAUTER and BLANDFORD (1997) indicates that detection 90% confidence level two station detection levels will be between 0.1 kt and 0.3 kt over most continental areas and between 0.3 and 0.7 kt over most oceanic regions. Advances in empirical atmospheric wind models (e.g., DROB, 1999) and improvements in modeling the propagation of acoustic waves through a non-stationary medium (e.g., COLLINS, 1993; COLLINS *et al.*, 1995; NORRIS and GIBSON, 1999) should lead to improvements in estimating global monitoring thresholds as a function of time.

4. Conclusions

A three-week noise survey gives just a glimpse of the full range of meteorological conditions that will occur in a typical year. The survey should, however, reveal any serious discrepancy in noise levels that will render one site inferior to all others. Our survey found no significant noise level discrepancies between any site in the Azores and in Cape Verde.

In the Azores, at least in the early winter when our survey was conducted, noise level swings of 40 to 50 dB at 0.1 Hz are not uncommon. Any permanent infrasound array in this region will be periodically rendered ineffective at all frequencies of interest to the monitoring community. In Cape Verde, noise levels are relatively constant to the dominance of the NE trade winds over local air pressure systems. The most significant factor causing change in infrasonic noise levels is diurnal.

In the Azores, the preferred site is Cha das Mulas, the site that offers the most options for siting array elements. If array modifications are necessary, this area presents no limitations to what could be considered. The preferred location in Cape Verde, on the island of Maio, offers a similar degree of flexibility. In both the Azores and Cape Verde, it is clear that winds will be higher, on average, than at most continental sites. For this reason, it is also clear that these locations will require more array elements. For this reason 2-km aperture, 8-element arrays are recommended for both locations.

One goal of these site surveys was to quantify the infrasonic noise reduction possible in forests and in a large calderra. Our surveys provided evidence that topography and ground cover will reduce wind speed but will not affect noise levels at long periods (above 4 to 10 s). Noise at longer periods clearly stems from air flow at an elevation beyond the reach of the topography and forests we have considered in our surveys. Noise reduction at these periods might be beyond the reach of any practical noise suppression scheme.

Acknowledgments

Funding for the noise surveys was provided by the U.S. Department of Defense under contract DSWA 01-97-C-0163. Panduronga Dessai and his colleagues at SIVISA in Ponta Delgada provided essential field support in the Azores. The Cape Verde experiment benefited greatly from assistance provided by António Joaquim R. M. Fernandes, Cape Verde's minister of Infrastructure and Housing, Ana

Martins, Flavio Silva and Manuela Ribeira, the President of the municipality of Maio. Further technical support was provided by the IGPP north lab staff. The authors would like to thank Doug Christie and Alberto Veloso, at the CTBTO and Anton Dainty (Defense Threat Reduction Agency) for their assistance.

REFERENCES

BERANEK, L. L. *Noise Reduction* (McGraw-Hill Book Co., New York, 1960).

BLANC, E. (1985), *Observations in the Upper Atmosphere of Infrasonic Waves from Natural or Artificial Sources: A Summary*, Ann. Geophys. *3*, 673–688.

BLANDFORD, R. (1997), *Infrasound array design with respect to detection and slowness-estimation*, In *Proc. Infrasound Workshop for CTBT Monitoring, Aug 25-28, Santa Fe, NM*, pp. 37–70.

BONNER, J. L., GOLDEN, P., and HERRIN, G. (1998), *Acoustic Signals from Bolides and Explosions*, Proc. 20th Annual Seismic Res. Symp., 584–587.

BOOTH, CROASDALE, and WALKER (1978), *A Quantitative Study of 5000 Years of Volcanism on Sao Migueo Island*, Phil. Trans. Roy. Soc Lon. *288*, 271–319.

BROCHE, P. (1977), *Propagation des ondes acoustico-gravitationnelles excitees par des explosions.* Ann. Geophys. *33*, 3,281–3,288.

BURRIDGE, R. (1971), *The Acoustics of Pipe Arrays*, Geophys. J. R. Astr. Soc. *26*, 53.

CALAIS, E., MINSTER, J. B., HOFTON, M. A., and HEDLIN, M. A. H. (1998), *Ionospheric Signature of Surface Mine Blasts from Global Positioning System Measurements*, Geophys. J. Int. *132*, 191–202.

CHRISTIE, D. (1998), *Requirements for IMS Infrasound Station Site Surveys*, issued by the Preparatory Commission for the Comprehensive Nuclear-Test-Ban Treaty Organization, CTBT/PC/III/WBG/PTS/INF.3

CLAUTER, D. and BLANDFORD, R. (1997), *Capability modeling of the proposed international monitoring system 60-station infrasound network*, Proc. Infrasound Workshop for CTBT Monitoring, Santa Fe, NM, Aug 25-28, 1997, pp. 227–236.

COLLINS, M. D. (1993), *The Adiabatic Mode Parabolic Equation*, J. Acoust. Soc. Am. *94*, 2269–2278.

COLLINS, M. D., MCDONALD, B. E., KUPERMAN, W. A., and SIEGMANN, W. L. (1995), *Jovian Acoustics and Comet Shoemaker-Levy 9*, J. Acoust. Soc. Am. *97*, 2147–2158.

COOK, R. K. and BEDARD, A. J. (1971), *On the Measurement of Infrasound*, Geophys. J.R. Astr. Soc, *26*, 5.

DANIELS, F. B. (1959), *Noise-Reducing Line Microphone for Frequencies Below cps*, J. Acoust. Soc. Am. *31*, 529.

DASE TECHNICAL MANUAL (1998), *Technical Manual for the Microbarometre MB2000*, Departement analyse et surveillance de l'environment.

DONN, W. L. and POSMENTIER, E. S. (1967), *Infrasonic Waves from the Marine Storm of April 7, 1966*, J. Geophys. Res. *72*, 2053–2061.

DROB, D. P. (1999), *The Atmosphere from 35-160 km: Moving from Climatology to Nowcast*, Informal Session on Infrasound at Spring AGU, Boston, June 2, 1999.

FEHR, U. (1967), *Measurements of Infrasound from Artificial and Natural Sources*, J. Geophys. Res. *72*, 2403.

FRANCIS, S. H. (1973), *Acoustic-gravity Modes and Large-scale Traveling Ionospheric Disturbances of a Realistic, Dissipative Atmosphere*, J. Geophys. Res. *78*(13), 2278–2301.

GEORGES, T. M. and YOUNG, J. M., *Passive Sensing of Natural Acoustic-Gravity Waves at the Earth's Surface.* Ch. 21 of *Remote Sensing of the Troposphere* (ed Derr V. E.) (U.S. Gov. Printing Office, 1972).

GOSSARD, E. E. and HOOKE, W. H., *Waves in the Atmosphere* (Elsevier Scientific Publishing Co., New York, 1975).

GROVER, F. H. (1971), *Experimental Noise Reducers for an Active Microbarograph Array*, Geophys. J. R. Astr. Soc. *26*, 41.

HAUBRICH, R. A. (1968), *Array Design*, Bull. Seismol. Soc. *58*, 977.

HAURWITZ, B. (1941), *The propagation of Sound Through the Atmosphere*, J. Aeron. Sci. *9*, 35.

INGARD, U. (1953), *A review of the Influence of Meteorological Conditions on Sound Propagation*, J. Acoust. Soc. Am. *25*, 405.

KAIMAL, J. C. and FINNIGAN, J. J., *Atmospheric Boundary Layer Flows: Their Structure and Measurement* (*Oxford University Press*, 1994).

KULICHKOV, S. N. (1972), *Long-range Propagation of Sound in the Atmosphere, A Review*, Izv. Atm. and Ocean Phys. *28*, 253.

LAMB, H., *Hydrodynamics*, 6th edn. (London, 1932).

LANDAU, L. D. and LIFSHITZ, E. M., *Fluid Mechanics* (Pergamon Press Ltd., Oxford, 1959).

LISZKA, L. (1974), *Long-distance Propagation of Infrasound from Artificial Sources*, J. Acoust. Soc. Am. *56*, 1383.

MACK, H. and FLINN, E. A. (1971), *Fast Frequency-wavenumber Analysis and Fisher Signal Detection in Real-time Infrasonic Array Data Processing*, Geophys. J. R. Astr. Soc. *26*, 255–270.

McDONALD, J. A., DOUZE, E. J., and HERRIN, E. (1971), *The Structure of Atmospheric Turbulence and its Application to the Design of Pipe Arrays*, Geophys. J. R. Astr. Soc. *26*, 99–109.

MUTSCHLECNER, J. P. and WHITAKER, R. W. (1994), *Infrasonic Observations of the Northridge California Earthquake, 6th Internatil. Symp. Long-range Sound Propagation* (*Ottawa, Canada*) *12–14 June*, published by the National Research Council of Canada.

NOEL, S. D. and WHITAKER, R. W. (1991), *Comparison of Noise Reduction Systems*, Los Alamos National Laboratory Report, LA-12008-MS, June.

NORRIS, D. E. and GIBSON, R. G. (1999), Diurnal Variability in the Atmosphere and its effect on ray paths, EOS, Transactions, American Geophysical Union 1999 Spring Meeting *80*, S71.

PANOFSKY, H. A. and DUTTON, J. A. *Atmospheric Turbulence: Models and Methods for Engineering Applications* (John Wiley and Sons Inc., 1984).

PRIESTLEY, J. T. (1966), *Correlation Studies of Pressure Fluctuatinos on the Ground Beneath a Turbulent Boundary Layer*, Washington D.C., National Bureau of Standards Report No. 8942, U.S. Dept. of Commerce, National Bureau of Standard.

REVELLE, D. O. and WHITAKER, R. W. (1999), *Infrasonic Noise Reduction Using Shelters/Windbreaks*, EOS, Trans. Am. Geophys. Union 1999 Spring Meeting, *80*, S68.

REYNOLDS, O. (1874), *On the Reflection of Sound by the Atmosphere*, Proc. Roy. Soc. *22*, 531.

SIMONS, D. J. (1995), *Atmospheric methods for nuclear test monitoring*, Monitoring a comprehensive test ban treaty, NATO ASI series E, volume 303, 135–141 (eds. Husebye, E. S. and Dainty, A. M.).

STOKES. (1857), British Association Report

TROST, L. C. (1997), *High Altitude Wind Effects on Infrasound Network Performance*, Proc. the Infrasound Workshop for CTBT Monitoring, Santa Fe, NM, Aug 25–28, 1997, pp. 271–284.

WELCH, P. D. (1967), *The Use of Fast Fourier Transforms for the Estimation of Power Spectra: A Method Based on Time Averaging Over Short Modified Periodograms*, Trans IEEE AU, 70–73.

WILSON, C. R., OLSON, J. V., and SPELL, B. D. (1995), *Natural Infrasonic Waves in the Atmosphere: their Characteristics, Morphology and Detection*, Univ. of Alaska Technical Report, ARS-95-039, Prepared for ENSCO Corp.

WILSON, D. K., BRASSEUR, J. G., and GILBERT, K. E. (1999), *Acoustic Scattering and the Spectrum of Atmospheric Turbulence*, J. Acoust. Soc. Am. *105*, 30–34.

YAMAMOTO, R. (1956), *The Microbarographic Oscillations Produced by the Explosions of Hydrogen Bombs in the Marshall Islands*, Bull. Am. Met. Soc. *37*, 406.

(Received July 2, 1999, revised August 10, 2000, accepted August 14, 2000)

To access this journal online:
http://www.birkhauser.ch

Pure appl. geophys. 159 (2002) 1153–1181
0033–4553/02/051153–29 $ 1.50 + 0.20/0

© Birkhäuser Verlag, Basel, 2002

❙ Pure and Applied Geophysics

Seismic Precursors to Space Shuttle Shock Fronts

GORDON SORRELLS,[1] JESSIE BONNER,[2] and EUGENE T. HERRIN[2]

Abstract—Seismic precursors to space shuttle re-entry shock fronts are detected at TXAR in Southwest Texas when the ground track of the orbiter vehicle passes within ~150–200 km of the observatory. These precursors have been termed "shuttle-quakes" because their seismograms superficially mimic the seismograms of small earthquakes from shallow sources. Analysis of the "shuttle-quake" seismograms, however, reveals one important difference. Unlike ordinary earthquakes, the propagation azimuths and horizontal phase velocities of the individual phases of the "shuttle-quakes" are functionally related. From a theoretical model developed to account for the origin of these precursors it is found that the seismic phases of "shuttle-quakes" are "bow" waves. A "bow" wave originates at the advancing tip of the shock front trace (i.e., intersection of the re-entry shock front with the surface of the earth) when the ground speed of the orbiter vehicle exceeds the horizontal phase velocity of a particular seismic phase. "Bow" waves are shown to differ in two important respects from the ordinary seismic phases. They vanish ahead of the advancing tip of the shock front trace and their propagation azimuths and horizontal phase velocities are functionally related. The ground speed of the orbiter vehicle exceeds the horizontal phase velocities of crustal seismic phase over much of the re-entry flight profile. As a result, $P, S,$ and R_g "bow" waves will be seen as precursors to the re-entry shock front at stations located within a few hundred km of its ground track.

Key words: Seismic, Infrasound, Shock Fronts, Space Shuttle.

Introduction

Since the inaugural flight of the *Columbia* orbiter in April 1981, space shuttle missions have become commonplace. More than ninety missions have now been completed and a new mission is undertaken every six to eight weeks. this paper documents certain novel seismic consequences of the re-entry of the orbiter vehicle into the earth's atmosphere, after the completions of these missions. The typical re-entry flights profile begins about 8100 kilometers up-range of the recovery facility. The altitude and air speed of the orbiter vehicle at this point are approximately 122 km and 7.5 km/s, respectively. It remains in supersonic flight until it is only 40 km from the recovery facility. Its altitude at this point is 15 km. Thus, the shock

[1] Seismic Diagnostics, Inc., 2714 Country Club Parkway, Garland, TX 75043, U.S.A. E-mail: ggsorrells@attbi.com
[2] Southern Methodist University, Department of Geological Sciences, P.O. Box 0395, Dallas, TX 75275, U.S.A.

front created by the space shuttle orbiter vehicle as it re-enters the atmosphere is sustained along a ground track that is more than 8000 km in length.

Seismic disturbances are created by the intersection of the re-entry shock front with the surface of the earth. KANOMORI *et al.* (1991) use seismic data generated by a re-entry shock front to map its movement across the Los Angeles Basin. KAMAR (1995) uses similar seismic data to track a re-entry shock front that crosses the Pacific Northwest Seismic Network. An *ad hoc* model originally proposed by GOFORTH and MC DONALD (1968) and summarized by COOK *et al.* (1972) is invoked in both of these cases to account for the seismic observations. According to this model, the dominant contribution to the observed earth motion is assumed to be a quasi-static elastic deformation that occurs in response to the local atmospheric pressure changes introduced by the shock front. Since the deformation is quasi-static, significant earth motion is seen only in the immediate vicinity of the intersection of the shock front with the surface of the earth. However, in addition to the observation of the quasi-static signal, KANAMORI *et al.* (1990) also report the occurrence of a distinct pulse that precedes the shock front at the Pasadena station and follows it at the University of Southern California site.

Observations at TXAR

We have also detected seismic precursors to space shuttle shock fronts with short-period seismograms recorded at TXAR, an IMS Seismic Observatory located in far West Texas. For example, on March 9, 1996, the space shuttle *Columbia* re-entered the earth's atmosphere to complete the STS-75 mission with a landing at the Kennedy Space center. As shown in the figure 1, its re-entry ground track is oriented in an easterly direction and passes approximately 150–175 km north of TXAR. Based upon generic mission profile data found on the NASA World Wide Web Site and specific results reported by KAMAR (1995), the altitude and the air speed of the orbiter vehicle as its shock front passes TXAR are estimated to be approximately 60 km and 7 km/s, respectively. Thus, the speed of the orbiter vehicle at this stage of the re-entry *exceeds* the seismic wave velocities in the earth's crust. It is shown in a later section that this factor plays a decisive role in the creation of the seismic precursors identified below.

The STS-75 re-entry front "N" wave recorded by two elements of the TXAR infrasound array is seen in the lower panel of Figure 2. The data recorded by the TXAR seismic array during the same time interval is seen in the upper panel. Notice that the arrival of the seismic phase designated by the letter "N" coincides with the arrival of the shock front. This phase is an example of the quasi-static signal assumed by the Goforth and McDonald model. The signal that precedes it merits special attention. It was originally identified as the signal from a small shallow earthquake located at a regional distance from TXAR. However, further investigation revealed

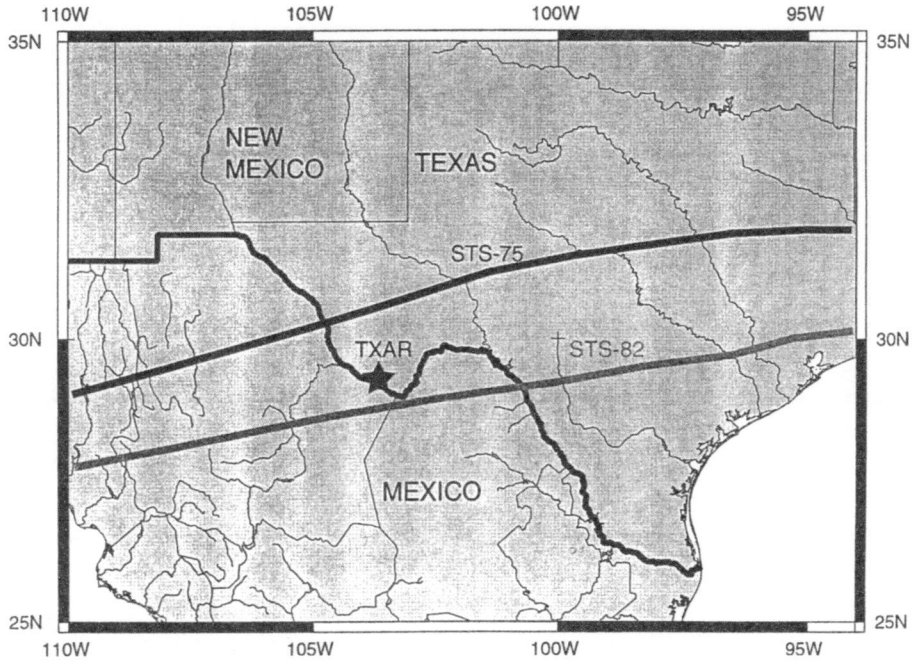

Figure 1
Re-entry ground tracks for shuttle missions STS-75 and STS-82. The missions were bearing east to the Kennedy Space Flight Center for landing. 'Shuttle-quakes' were observed at TXAR from both of these missions.

that similar signals preceded other re-entry shock fronts whenever space shuttle ground track passed within about 150–200 km of TXAR. These observations have become so common that we now use the term, "shuttle-quake" to identify this particular type of seismic signal.

The seismic and infrasound records of another "shuttle-quake" are shown in Figure 3. These data were acquired during the re-entry phase of the STS-82 mission on February 21 1997. They were chosen for presentation because the STS-82 re-entry ground track is considerably closer to TXAR than the STS-75 ground track, passing about 30–60 km south of the station. The STS-82 re-entry shock front is also produced an audible "boom" that was heard by the residents of Terlingua, a small village near TXAR. Microbarograms of the STS-82 shock front "N" wave are plotted in the lower panel of Figure 3. Its impulsive start and relatively large pressure amplitude (~ 100 μbars) account for the audible "boom." The seismic data recorded during the same time interval are plotted in the upper panel of Figure 3. It is seen that in this case the "shuttle-quake" signal now precedes the shock front arrival by only about 30 s and resembles the records of a small local earthquake caused by a shallow source.

Although the seismograms of "shuttle-quakes" mimic ordinary earthquake seismograms, the results of the FK analysis of the STS-75 and STS-82 shuttle-quake

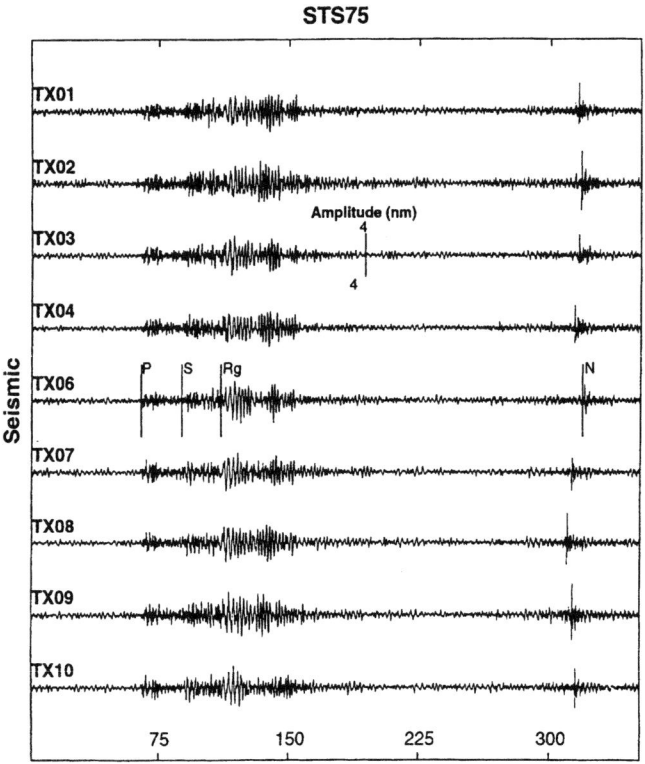

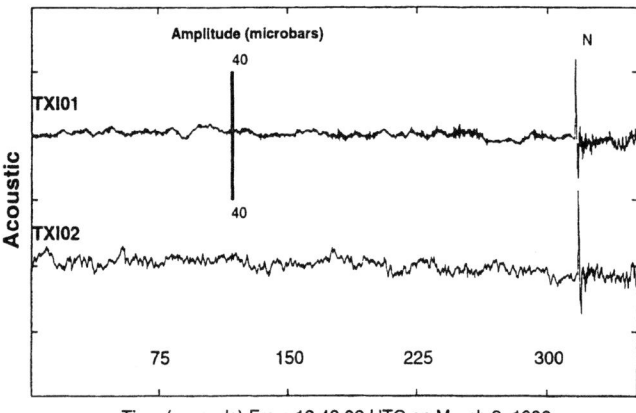

Figure 2

Shuttle-quakes and *N* waves from STS-75 shuttle mission. Distinct phases include *P, S* and *Rg* recorded on the vertical component of the seismometers of the TXAR array. The seismometers and infrasound sensors recorded the *N* wave approximately 225 seconds after the *P* arrival.

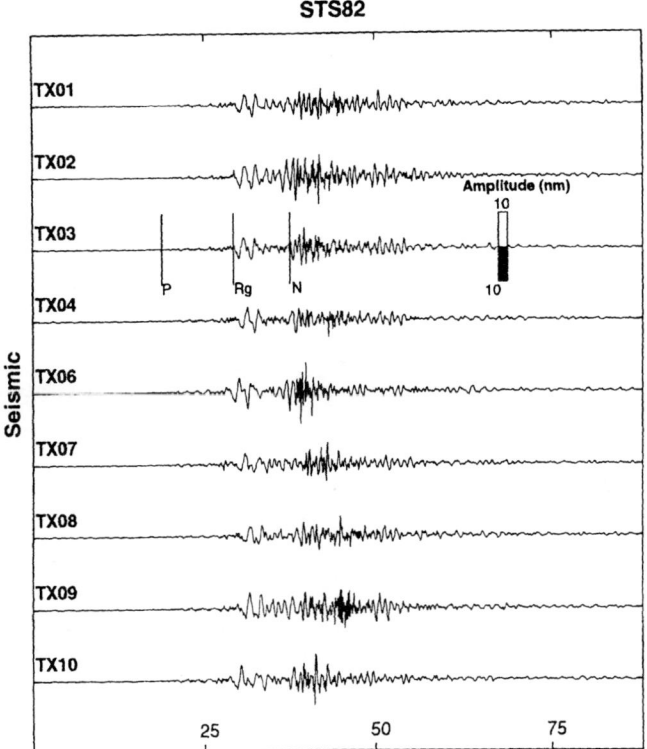

Time (seconds) From 08:37:06 UTC on February 21, 1997

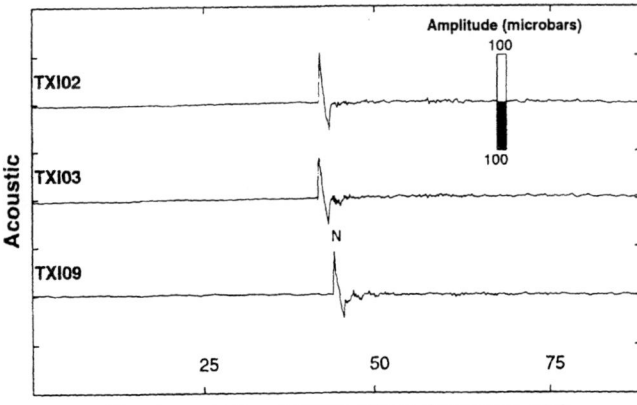

Time (seconds) From 08:37:06 UTC on February 21, 1997

Figure 3

Seismograms and barograms from the STS-82 shuttle mission. Distinct phases include Pg and short-period Rayleigh (Rg) recorded on the vertical component of the seismometers of the TXAR array. The infrasound sensors of the TXAR seismo-acoustic array recorded the classic "N wave" from the passage of the supersonic pressure wave. This wave coupled into the ground and can be seen as a packet of different frequencies on the seismic channels (labeled as N on the upper plot).

Table 1

Summary of the results of the FK analysis of the STS-75 and STS-82 "shuttle-quake" seismograms

Mission	Distance to ground track (km)	Year	Day	Phase	Arrival time	Horizontal phase velocity (km/sec)	Propagation azimuth (degrees)
STS-75	~150–175	1996	69	P	13:44:06.0	6.9	109
				S	13:44:32.7	4.1	139
				Rg	13:44:54.7	2.9	163
				N	13:48:27.3	0.356	170
STS-82	~30–60	1997	52	Pg	08:17:55.6	6.1	53
				Rg	08:18:04.3	2.5	22
				N	08:18:15.4	0.723	8

seismogram summarized in Table 1, illustrate that a different propagation azimuth characterizes each phase. It is also found that when the orbiter vehicle passes north of TXAR (STS-75) the propagation azimuths increase as the phase velocity decreases. However, when it passes south of TXAR (STS-82) they decrease with decreasing phase velocity. A theoretical model that accounts for the origin of "shuttle-quakes" and explains the observed functional relationship between the propagation azimuths and horizontal phase velocities of their individual phases is described in the following paragraphs.

Theory

A Simplified Model of the Re-entry Shock Front

The plane, $z = 0$, of a Cartesian coordinate system is chosen to represent the surface of the earth. It is assumed that during the time interval, $0 \leq t_s \leq \tau$, of the re-entry phase, the orbiter vehicle is in level flight above this plane. The origin of the coordinate system is chosen to be the image of the re-entry point on the plane, $z = 0$. The x axis of the coordinate system is chosen to coincide with the re-entry ground track of the orbiter vehicle and is positive in the direction of the flight. The constants, z_0, and v identify the altitude and air speed of the orbiter vehicle. It is also assumed that v is greater than the sound speed in the atmosphere, so that a shock front approximated by a conical surface as illustrated in Figure 4 is created. The dashed line seen in this figure delineates the intersection of the shock front with the surface of the earth. This line is referred to as the "shock front trace." Now, as shown in Figure 5, x_s, and y_s are the coordinates of an arbitrarily chosen point on the shock front trace. It then follows directly from the definitions above that if the shock front is approximated by a right circular cone, then the expression for the shock front trace is given by

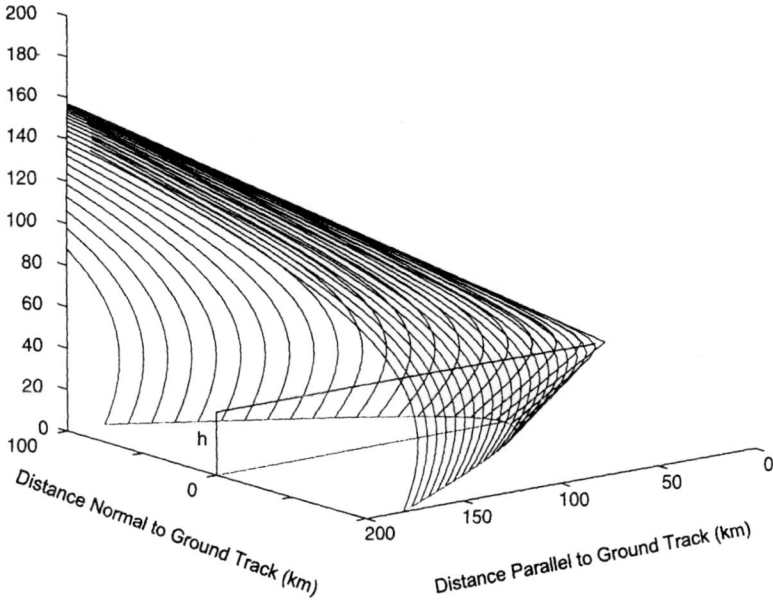

Figure 4

Illustration of an idealized re-entry shock front. It is represented by a right circular conical surface. The altitude of the orbiter vehicle is z_0. Its altitude, flight path and ground track are shown as solid lines. The intersection of the shock front with the surface of the earth is a hyperbola that is also shown is a solid line.

$$x_s(t_s, y_s) = vt_s - b\sqrt{y_s^2 + z_0^2} \ . \tag{1}$$

The constant, b, which appears on the right hand side (rhs) of eq. (1) is defined by

$$b = \cot \phi \tag{2}$$

where 2ϕ is the apex angle of the right circular cone.

Now let ω denote angular frequency, and let the spectrum of the overpressure distribution on the shock front trace be defined by $P(\omega, y_s)$. To simplify analysis, it is assumed that

$$P(\omega, -y_s) \ = \ P(\omega, y_s) \tag{3}$$

and that

$$\begin{aligned} |P(\omega, y_s)| &\geq 0 \quad 0 \leq y_s \leq Y \ , \\ |P(\omega, \lambda_s)| &= 0 \quad |y_s| > Y \ . \end{aligned} \tag{4}$$

These constraints are imposed to approximately account for the critical refraction of the shock front in the lower atmosphere. Y is defined to be the normal distance between the ground track and the two points on the shock front trace where critical acoustic refraction first occurs. These points are commonly referred to as the "lateral cut-off points" of the shock front.

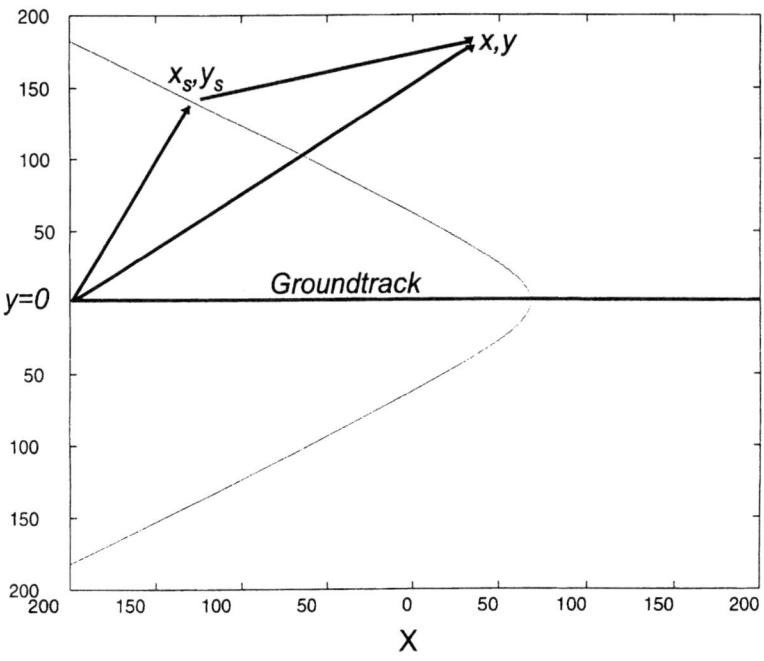

Figure 5

Illustration of a hypothetical shock front trace in the plane $z = 0$. The line, $y = 0$ corresponds to the orbiter vehicle ground track. Points on the shock front trace are identified by the coordinates $[x_s, y_s]$. Observation points are identified by the coordinates $[x, y]$.

The Spectrum of Rayleigh Waves Excited by Space Shuttle Shock Fronts

An examination of the spectrum of the Rayleigh waves excited by the application of the simplified source model described above, to the surface of an isotropic, homogeneous, elastic half-space, provides the basis for an explanation of the "shuttle-quakes" observed at TXAR. The upper surface of this half-space is chosen to coincide with the plane, $z = 0$, in the coordinate system defined above. The symbols, μ and c_R identify the shear modulus and the Rayleigh wave speed in the half-space. As shown in Figure 5, x and y are the coordinates of an observation point located on the surface of the half-space.

If δu_z is the spectrum of the vertical displacement component of the Rayleigh wave excited at the point $[x_s, y_s]$ and seen at the point $[x, y]$, it then follows from EWING *et al.* (1957, p. 45) that

$$\delta u_z = -\frac{i\kappa B}{2\mu} P(\omega, y_s)\delta A e^{-i\omega t_s} H_0^{(2)}(\kappa r). \tag{5}$$

The unidentified symbols that appear on the rhs of eq. (2) are identified as follows

$$\kappa = \frac{\omega}{c_R} \ . \tag{6}$$

and

$$r = \sqrt{(x - x_s)^2 + (y - y_s)^2} \ . \tag{7}$$

$H_0^{(2)}$ is a Hankel function of the second kind and order 0. B is a dimensionless constant whose value is functionally dependent upon c_R and the body-wave speeds in the half-space. The product, $P(\omega, y_s)\delta A e^{-i\omega t_s}$, that appears on the rhs of eq. (5), is recognized as the spectrum of a vertically directed point force applied to the surface area element, δA located at the point $[x_s, y_s]$ at time, t_s. Since the shock front trace propagates at speed, v, in the positive x direction, it is convenient to represent δA as;

$$\delta A = v \, dt_s \, dS \ , \tag{8}$$

where dS is a differential length element measured along the shock front trace.

Now let $t_s = 0$ denote the time that the orbiter vehicles re-enters the earth's atmosphere. Then if u_z is the spectrum of the vertical displacement component of the Rayleigh wave excited by the sum of all point forces acting on the surface, in the interval $0 \leq t_s \leq \tau$, it follows that

$$u_z = \int_A \delta u_z = -\frac{i\kappa v B}{2\mu} \int_S P(\omega, y_s) \int_0^\tau H_0^{(2)}(\kappa r) e^{-i\omega t_s} \, dt_s \, dS \ . \tag{9}$$

The line integral appearing on the rhs of eq. (9) is transformed to an integration path along the y-axis. with the substitution

$$dS = -\text{sgn}(y_s)\sqrt{1 + \left(\frac{dx_s}{dy_s}\right)^2} \, dy_s \ , \tag{10}$$

where, $\text{sgn}(y_s)$ is the signum function.

Now express $H_0^{(2)}$ as an integral sum of plane waves, as follows. If k is a continuous variable defined over the interval $-\infty \leq k \leq \infty$ and

$$\xi^2 = k^2 - \kappa^2 \ , \tag{11}$$

then from EWING et al. (1957, p. 28, eq. (2–44′))

$$H_0^{(2)}(\kappa r) = \frac{i}{\pi} \int_{-\infty}^{\infty} e^{-\xi|y-y_s|} e^{-ik(x-x_s)} \frac{dk}{\xi} \ . \tag{12}$$

In the following derivations, Real ξ is chosen to be greater than or equal to zero to insure that u_z remains finite as $y \to \pm\infty$. Now if x_s, $H_0^{(2)}$ and dS are replaced by their definitions (eqs. (1), (12), and (10)), the order of integration is exchanged between the t_s and k integrals, and the integration is performed over the t_s integral, it follows that

$$u_z = \int_{-Y}^{Y} \operatorname{sgn}(y_s) F(y_s) u_z^p(\omega, y - y_s, x - v\tau + b\rho(y_s)) dy_s \ , \tag{13}$$

where

$$u_z^p = \frac{\kappa B P(\omega, y_s)}{2\pi i \mu} \left[e^{i\omega\tau} \int_{-\infty}^{\infty} \frac{e^{-\xi|y-y_s|} e^{-ik(x-v\tau+b\rho(y_s))}}{\xi(k-k_v)} dk - \int_{-\infty}^{\infty} \frac{e^{-\xi|y-y_s|} e^{-ik(x+b\rho(y_s))}}{\xi(k-k_v)} dk \right] \ . \tag{14}$$

The unidentified symbols that appear on the rhs of eqs. (13) and (14) are defined as follows

$$F(y_s) = \sqrt{1 + \left(\frac{dx_s}{dy_s}\right)^2} = \frac{\sqrt{y_s^2 + (z_0 \sin \phi)^2}}{\rho(y_s) \sin \phi} \ , \tag{15}$$

$$\rho(y_s) = \sqrt{y_s^2 + z_0^2} \ , \tag{16}$$

$$k_v = \frac{\omega}{v} \ . \tag{17}$$

The function, u_z^p, that is defined by eq. (14) describes the spectrum of the Rayleigh wave generated by a moving, point pressure source in the convenient form of an integral sum of plane waves. The point source moves at speed, v, in the positive x direction and is located at the point, $[v\tau - b\rho(y_s), y_s]$ at time, τ.

Now for the sake of illustration assume that $y \geq 0$. Then, by taking advantage of the symmetry of $P(\omega, y_s)$ implied by eq. (4), it is readily shown that

$$u_z = \int_0^Y F(y_s) \hat{u}_z^p(\omega, y - y_s, x - v\tau + b\rho(y_s)) dy_s \ , \tag{18}$$

where if $y > y_s$

$$\hat{u}_z^p = \frac{\kappa B P(\omega, y_s)}{\pi i \mu} \left[e^{-i\omega\tau} \int_{-\infty}^{\infty} \frac{e^{-\xi y} \sinh(\xi y_s) e^{-ik(x-v\tau+b\rho(y_s))}}{\xi(k-k_v)} dk \right.$$

$$\left. - \int_{-\infty}^{\infty} \frac{e^{-\xi y} \sinh(\xi y_s) e^{-ik(x+b\rho(y_s))}}{\xi(k-k_v)} dk \right] \ . \tag{19}$$

and if $y \leq y_s$

$$\hat{u}_z^p = \frac{\kappa B P(\omega, y_s)}{\pi i \mu} \left[e^{-i\omega\tau} \int_{-\infty}^{\infty} \frac{e^{-\xi y_s} \sinh(\xi y) e^{-ik(x - v\tau + b\rho(y_s))}}{\xi(k - k_v)} \, dk \right.$$

$$\left. - \int_{-\infty}^{\infty} \frac{e^{-\xi y_s} \sinh(\xi y) e^{-ik(x + b\rho(y_s))}}{\xi(k - k_v)} \, dk \right] . \tag{20}$$

$\hat{u}_z^p$ is the spectrum of the sum of the Rayleigh waves generated by the pair of moving point pressure sources located at the positions whose coordinates are $[v\tau - b\rho(y_s), \pm y_s]$ at time, τ.

Equations (19) and (20) are evaluated in Appendix A by the methods of contour integration. It is shown there that the expressions for $\hat{u}_z^p$ are resolvable into contributions from a branch line integral, and a simple, real, pole. The pole contribution is of particular significance since it owes it existence solely to the movement of the source loads at a constant rate, v. It is referred to as the source pole in the following derivations.

Now, for the sake of illustration assume that $x > 0$. Then, if the source pole contribution is represented by $_s\hat{u}_z^p$ and the effects of anelastic attenuation are neglected, it is found from eqs. (A-24) through (A-30) that

$$_s\hat{u}_z^p = 0; \quad x > v\tau - b\rho(y_s) , \tag{21}$$

$$_s\hat{u}_z^p = -\frac{2BP(\omega, y_s)}{\mu\sqrt{\varepsilon^2 - 1}} \sinh\left(\kappa y_s \sqrt{\varepsilon^2 - 1}\right) e^{-\kappa y \sqrt{\varepsilon^2 - 1}} e^{-ik_v(x + b\rho(y_s))};$$

$$y > y_s; \quad 0 < x \leq v\tau - b\rho(y_s) , \tag{22}$$

$$_s\hat{u}_z^p = -\frac{2BP(\omega, y_s)}{\mu\sqrt{\varepsilon^2 - 1}} \sinh\left(\kappa y \sqrt{\varepsilon^2 - 1}\right) e^{-\kappa y_s \sqrt{\varepsilon^2 - 1}} e^{-ik_v(x + b\rho(y_s))};$$

$$0 \leq y \leq y_s; \quad 0 < x \leq v\tau - b\rho(y_s) , \tag{23}$$

where

$$\varepsilon = \frac{c_R}{v} . \tag{24}$$

Note that the signs of the radicals appearing on the rhs of eqs. (22) and (23) are chosen to be positive. These choices are dictated by the fact that, as shown in Appendix A, both $\mathrm{Re}\,\xi$ and $\mathrm{Im}\,\xi$ are positive on the real axis. The above equations illustrate that the source pole contribution possesses novel attributes that are not found in the expressions for Rayleigh waves from stationary sources. It is seen from eq. (21) that the contribution vanishes ahead of the pair of advancing point pressure loads, and it is seen from eqs. (22)–(24) that its propagation characteristics change dramatically, depending upon the value of ε.

For example, if $\varepsilon > 1$, the radicals appearing on the rhs of eqs. (22) and (23) are real and positive. Therefore, in this case $_s\hat{u}_z^p$ is seen to be the spectrum of a plane wave. At high frequencies, this wave will attenuate rapidly in directions normal to the ground track. In addition, the imaginary argument of the second exponential factor that appears on the rhs of eqs. (22) and (23) shows that it propagates in the horizontal plane, not at the speed of a Rayleigh wave but, at a speed equal to the speed of the point pressure loads. Its propagation direction is parallel to their ground track. Consequently, it follows that when $\varepsilon > 1$, the source pole contribution has the essential attributes of an *inhomogeneous* wave.

On the other hand, if $\varepsilon > 1$, the radicals appearing on the rhs of eqs. (22) and (23) are imaginary and positive. Now, let

$$\cos \theta_v = \varepsilon = \frac{c_R}{v}. \tag{25}$$

It then follows that

$$_s\hat{u}_z^p = -\frac{2BP(\omega, y_s)}{\mu \sin \theta_v} \sin(\kappa y_s \sin \theta_v) e^{-i\kappa[(x + b\rho(y_s)) \cos \theta_v + y \sin \theta_v]};$$

$$y > y_s; \quad 0 < x \leq v\tau - b\rho(y_s) , \tag{26}$$

$$_s\hat{u}_z^p = -\frac{2BP(\omega, y_s)}{\mu \sin \theta_v} \sin(\kappa y \sin \theta_v) e^{-i\kappa[(x + b\rho(y_s)) \cos \theta_v + y_s \sin \theta_v]};$$

$$0 \leq y \leq y_s; \quad 0 < x \leq v\tau - b\rho(y_s) . \tag{27}$$

Inspection of the exponential factors appearing on the rhs of eqs. (26) and (27) illustrates that $_p\hat{u}_z^p$ is still the spectrum of a plane wave. In this case, however, the wave propagates at speed c_R in the horizontal plane and in the direction, θ_v, as measured with respect to the ground track. Therefore, it may be legitimately classified as a Rayleigh wave. The form of $_p\hat{u}_z^p$ as described by eqs. (21) and (25) through (27) is referred to as the spectrum of the point source "bow" wave, since its propagation characteristics are similar to those of the waves created by a ship moving at high speed through water.

Now let $_s u_z$ represent the contribution of the source pole to the Rayleigh wave spectrum of the seismic disturbance excited by the overpressures distributed along the entire shock front trace. It then follows from eq. (18) that

$$_s u_z = \int_0^Y F(y_s) {}_s\hat{u}_z^p(\omega, y - y_s, x - vt + b\rho(y_s)) dy_s . \tag{28}$$

The homogeneous forms of $_s\hat{u}_z^p$ are used in eq. (28) when $c_R > v$ and the "bow" wave representations are used when $c_R < v$. From the description of the observations at TXAR, the speed of the orbiter vehicle exceeds the seismic wave velocities in the earth's crust. It is therefore appropriate to use the "bow" wave formulation of the point source spectrum for the purposes of this paper. An approximate solution for

$_s u_z$ when $c_R < v$ is derived in Appendix B. It is valid for large values of $\kappa \sin \theta_v$, and is expressed in terms of functions represented by two-dimensional asymptotic series. The expressions listed in eqs. (B-25) through (B-29) represent the approximate solution for the source pole contribution for all possible locations of the observation point in the first quadrant of the x-y plane. The terms described by these equations are the spectra of multiple, discrete seismic disturbances with distinctively different amplitude and phase characteristics. It is seen from eq. (B-25) that all terms of $_s u_z^p$ vanish ahead of the advancing tip of the shock front trace. It is also important to note that the spectral term represented by $U_z^{(1)}(\kappa \sin \theta_v, 0)$ exists at all points behind or adjacent to the advancing tip of the stock front trace. The existence of the other spectral terms depends upon the location of the observation point with respect to the shock front trace. The investigation of these latter terms is beyond the scope of this paper; however, the occurrence of $U_z^{(1)}(\kappa \sin \theta_v, 0)$ in the solution for the source pole contribution is of particular significance, since it provides the theoretical basis for the explanation of the "shuttle-quakes" observed at TXAR.

It is found from eq. (B-21) that

$$U_z^{(1)}(\kappa \sin \theta_v, 0) = \frac{2B}{\mu \sin \theta_v} e^{-i\kappa(x \cos \theta_v + y \sin \theta_v)} \sum_{m=0}^{\infty} q_{2m} I_m^{(1)}(\kappa \sin \theta_v, 0) \qquad (29)$$

where from eqs. (B-11) and (B-14)

$$q_{2m} = \frac{I}{(2m)!} \left(\frac{\partial}{\partial y_s} \right)^{2m} [P(\omega, y_s) F(y_s)]_{y_s=0} \qquad (30)$$

and from eq. (B-19)

$$I_m^{(1)}(\kappa \sin \theta_v, 0) = \frac{(-1)^m}{(\sin \theta_v)^{2m}} \left\{ y_s e^{-i\kappa b \rho(y_s) \cos \theta_v} \left(\frac{\partial}{\partial \kappa} \right)^{2m} n_0(\kappa y_s \sin \theta_v) \right.$$

$$\left. + \sin \theta_v \sum_{\ell=1} (\kappa y_s C)^{\ell+1} \frac{h_{\ell-1}^{(2)}(\kappa b \rho(y_s) \cos \theta_v)}{\rho(y_s)^{\ell-1}} \left(\frac{\partial}{\partial \kappa} \right)^{2m} \frac{n_\ell(\kappa y_s \sin \theta_v)}{\kappa^\ell} \right\}_{y_s=0} \qquad (31)$$

The symbol, n_ℓ seen on the rhs of eq. (31) identifies a spherical Neumann function of order, ℓ. Whereas, the symbol, $h_{\ell-1}^{(2)}$ identifies a spherical Hankel function of the second kind and order, $\ell - 1$, and the constant C is defined as

$$C = b \cot \theta_v . \qquad (32)$$

Now, we can show

$$[y_s^{\ell+1} n_\ell(\kappa y_s \sin \theta_v)]_{y_s=0} = -\frac{(2\ell)!}{2^\ell \ell! (\kappa \sin \theta_v)^{\ell+1}} . \qquad (33)$$

Then, with the use of this expression and noting that

$$\left(\frac{\partial}{\partial \kappa}\right)^{2m}\left(\frac{1}{\kappa}\right) = \frac{(2m)!}{\kappa^{2m+1}} \tag{34}$$

and

$$\left(\frac{\partial}{\partial \kappa}\right)^{2m}\left(\frac{1}{\kappa^{2\ell+1}}\right) = \frac{(2(\ell+m))!}{(2\ell)!\kappa^{2(\ell+m)+1}} \tag{35}$$

it follows that

$$I_m^{(1)}(\kappa\sin\theta_v, 0) = -\frac{(-1)^m(2m)!}{(\kappa\sin\theta_v)^{2m}}\left\{\frac{e^{-i\kappa bz_0\cos\theta_v}}{\kappa\sin\theta_v} + z_0\sum_{\ell=1}A_{\ell m}C^{\ell+1}\frac{h_{\ell-1}^{(2)}(\kappa bz_0\cos\theta)}{(z_0\kappa\sin\theta_v)^\ell}\right\}, \tag{36}$$

where

$$A_{\ell m} = \frac{(2(\ell+m))!}{2^\ell\ell!(2m)!} . \tag{37}$$

In the case of the TXAR observations it is safe to assume that if $\kappa\sin\theta_v \gg 1$, then $\kappa z_0 b\cos\theta_v \gg 1$. It therefore follows that

$$h_{\ell-1}^{(2)}(\kappa bz_0\cos\theta_v) \approx \frac{(i)^\ell e^{-i\kappa bz_0\cos\theta_v}}{\kappa bz_0\cos\theta_v} . \tag{38}$$

Thus

$$I_m^{(1)}(\kappa\sin\theta_v, 0) \approx -\frac{(-1)^m}{(\kappa\sin\theta_v)^{2m+1}}e^{-i\kappa z_0 b\cos\theta_v}\sum_{\ell=0}A_{\ell m}\left(\frac{iC}{z_0\kappa\sin\theta_v}\right)^\ell . \tag{39}$$

It is then seen from eqs. (29) and (30) that for the case of the TXAR observations

$$U_z^{(1)}(\kappa\sin\theta_v, 0) \approx -\frac{2Be^{-i\kappa[(x+bz_0)\cos\theta_v+y\sin\theta_v]}}{\mu\kappa(\sin\theta_v)^2}\sum_{m=0}^{\infty}\frac{(-1)^m}{(\kappa\sin\theta_v)^{2m}}\left(\frac{\partial}{\partial y_s}\right)^{2m}$$

$$\times [P(\omega, 0)F(0)]\sum_{\ell=0}A_{\ell m}\left(\frac{iC}{z_0\kappa\sin\theta_v}\right)^\ell . \tag{40}$$

and to a first order in $1/\kappa\sin\theta_v$

$$U_z^{(1)}(\kappa\sin\theta_v, 0) \approx -\frac{2BP(\omega, 0)e^{-i\kappa[(x+bz_0)\cos\theta_v+y\sin\theta_v]}}{\mu\kappa(\sin\theta_v)^2} . \tag{41}$$

The argument of the exponential appearing in this expression is referenced to a point on the shock front trace by noting that from eq. (1) at $t_s = \tau$ and $y_s = 0$

$$bz_0 = v\tau - x_s(0, \tau) . \tag{42}$$

Therefore

$$U_z^{(1)}(\kappa \sin \theta_v, 0) \approx -\frac{2BP(\omega, 0)e^{-i\omega\tau}e^{-i\kappa[(x-x_s(0,\tau))\cos\theta_v + y\sin\theta_v]}}{\mu\kappa(\sin\theta_v)^2} . \tag{43}$$

It is seen from this expression that at high frequencies and $\theta_v \neq 0$, $U_z^{(1)}(\kappa \sin \theta_v, 0)$ is approximated by the spectrum of a plane wave. This wave originates at $t_s = \tau$ and at the point $[x_s(0,\tau), 0]$. These coordinates identify the position of the tip of the shock front trace at $t_s = \tau$. τ may be interpreted as the duration of the time interval during which $v > c_R$. Therefore, this form of the exponential argument shows that the spatial relationship between $U_z^{(1)}(\kappa \sin \theta_v, 0)$ and the shock front trace is time-stationary as long as the ground speed of the orbiter vehicle exceeds the speed of Rayleigh waves in the half-space. It is also seen that the wave whose spectrum is represented by $U_z^{(1)}(\kappa \sin \theta_v, 0)$, propagates at speed, c_R in the horizontal plane, and in the direction, θ_v, as measured counter-clockwise with respect to the orbiter vehicle ground track, where it will be recalled from eq. (24) that

$$\theta_v = \cos^{-1}\left(\frac{c_R}{v}\right) . \tag{44}$$

In this regard it is seen that the high frequency propagation characteristics of $U_z^{(1)}(\kappa \sin \theta_v, 0)$ are analogous to those attributed to the point source "bow" wave, described earlier and, thus is identified as the spectrum of the Rayleigh "bow" wave generated at the shock front trace. Notice that, unlike Rayleigh waves generated by stationary sources, the Rayleigh "bow" wave amplitude spectrum is independent of the distance between its point of origination and the location of the observation point. It then follows that depending upon the intensity of the shock front overpressures and the effects of attenuation and scattering, it should be possible to detect this wave at stations remote from the orbiter vehicle ground track.

It is also important to recognize that the functional relationship between the propagation azimuth and the speed of an elastic wave in the horizontal plane as represented by eq. (44) is a unique property of "bow" waves, in general. This property may be used to discriminate between "bow" waves and signals generated by stationary sources.

The above derivations are based upon the assumption that $y \geq 0$. Because of the mirror image symmetry of $P(\omega, y_s)$ and the shock front trace, the solutions for the various terms of $_su_z^p$ for $y \geq 0$ and $y < 0$ are also mirror images. The wavefront diagram seen in Figure 6 illustrates the expected spatial relationship between the shock front trace and the Rayleigh "bow" wave. Notice that at a fixed observation point the Rayleigh "bow" wave will be seen as a precursor to the shock front "N" wave except at locations near the ground track. It should also be noted that the time interval between the arrival of the Rayleigh "bow" wave and the shock front "N" wave as seen at a fixed observation point will increase as its distance from the ground track increases. Finally, it is important to note that the propagation azimuths of the

Rayleigh "bow" wave as seen on opposite sides of the ground track are equal in magnitude, but opposite in sign when referenced to the ground track.

Comparison of Theory with Observations

The predicted attributes of Rayleigh "bow" waves described above correspond to the observed attributes of the R_g phases of the TXAR "shuttle-quakes." As shown in Figures 2 and 3 the R_g phase is a precursor to the shock front "N" wave in both cases. The data summarized in Table 1 show that when the distance between the orbiter vehicle ground track and TXAR is $\sim$30–60 km (STS-82 mission), the R_g phase precedes the "N" wave arrival by only about 11 seconds. When this distance is increased to $\sim$150–175 km (STS-75 mission) the arrival time difference between the R_g phase and "N" wave increases to 212 seconds. As shown in Table 2a, the observed R_g propagation azimuths also closely correspond to the predicted propagation azimuths of a Rayleigh "bow" waves with the same horizontal phase velocities. The ground track azimuths listed in this table were estimated from Figure 1. The predicted values of θ_y were obtained from eq. (44) by using the values of the R_g horizontal phase velocities listed in Table 1 as c_R and assuming that $v = 7$ km/sec. The sign of θ_y was chosen by noting that TXAR lies in the $y < 0$ half-plane with respect to the STS-75 ground track and in the $y > 0$ half-plane with respect to the STS-82 ground track. It is seen from this comparison that the differences between the observed R_g propagation azimuths and predicted propagation azimuths of a Rayleigh "bow" wave are insignificant. It follows then, that the R_g phases of the TXAR "shuttle-quakes" are Rayleigh "bow" waves generated by space-shuttle re-entry shock fronts.

The P and S phases of the TXAR "shuttle-quakes" are also "bow" waves. Recognition that the spectrum of the Rayleigh "bow" wave is a discrete component of the source pole contribution provides the theoretical basis for this claim. Recall that the existence of a source pole is dependent only on the constant of rate movement of the shock front. It is clear then, that it will appear, not only in the integral representation for the spectrum of the Rayleigh waves, but also in the analogous representations for the spectra of P and S waves. Evaluation of the source pole contribution to these spectra will, in the high frequency approximation, lead to expressions of the forms of eqs. (40) and (41) if the shock front speed exceeds both body wave horizontal phase velocities. This constraint is satisfied for seismic crustal phases as long as the speed of the orbiter vehicle is greater than about 6 km/sec. Therefore, it is to be expected that the Rayleigh "bow" wave will be accompanied by crustal P and S "bow" waves when the orbiter vehicle ground-track passes within 150–200 km of TXAR. If the P and S phases of the "shuttle-quake" are "bow" waves, then their propagation azimuths and horizontal phase velocities must be related by an expression of the form of eq. (44). The observed propagation azimuths of the "shuttle-quake" P and S phases are compared to the predicted propagation azimuths of "bow" waves with the same horizontal phase velocities in Table 2b. The procedure described above for the calculation of predicted R_g azimuths is also used to determine the predicted P and S azimuths. It is seen from the results summarized in this table that the difference between the observed and predicted P and S azimuths is small, thus confirming that the P and S phases of the "shuttle-quakes" are "bow" waves.

The identification of the "shuttle-quake" phases as shock front "bow" waves permits the seismic estimation of the normal distance, D, from TXAR to the orbiter vehicle ground track. If ΔT_{21} is the difference in arrival times for two shuttle-quake phases whose horizontal phase velocities are c_1 and c_2, it is a simple geometric exercise to show that

$$D = v \frac{\cot \theta_1 \cot \theta_2}{|\cot \theta_1 - \cot \theta_2|} \Delta T_{21} , \tag{45}$$

where

$$\theta_j = \cos^{-1}\left(\frac{c_j}{v}\right); \quad j = 1,2 . \tag{46}$$

Consequently from the information contained in Tables 1, 2a, and 2b it is found that D is 167 km for the STS-75 mission and 29 km for the STS-82 mission.

Conclusions

We conclude that the "shuttle-quake" signals seen at TXAR are "bow" waves. They originate at the tip of the advancing re-entry shock front trace and appear as

Table 2a

Comparison of observed and predicted propogation azimuths for shuttle-quake Rg phases

Mission	Ground track azimuth	θ_v	Predicted azimuth	Observed Rg azimuth	Residual
STS-75	~95°	−66°	161°	163°	2°
STS-82	~85°	+69°	16°	22°	6°

Table 2b

Comparison of observed and predicted propogation azimuths for shuttle-quake P and S phases

Mission	Ground track azimuth	Phase	θ_v	Predicted azimuth	Observed Rg azimuth	Residual
STS-75	~95°	P	−10°	105°	109°	4°
		S	−54°	149°	139°	−10°
STS-82	~85°	P	+29°	56°	23°	−3°

short-period precursors to re-entry shock fronts as represented by the "*N*" waves seen at TXAR. The results summarized in Tables 2a and 2b also demonstrate that the observed functional relationship between the propagation azimuths and horizontal phase velocities of the TXAR "shuttle-quake" phases is explained by an expression of the form of eq. (44). It is apparent that "shuttle-quake" will exist in essentially the same form as long as the orbiter vehicle ground speed exceeds the seismic wave velocities in the earth's crust. Since the orbiter vehicle air speed is 7.5 km/sec at the beginning of the re-entry phase of all missions, this condition will invariably be satisfied for every mission over some initial fraction of their re-entry ground track. The fact that the STS-75 and STS-82 "shuttle-quakes" were observed at TXAR at a distance of 6000–7000 km from the mission re-entry points implies that the magnitude of this initial fraction is at least 75–80% of the total length of the re-entry ground track. It follows then, that since the re-entry ground tracks frequently span the continental United States, "shuttle-quake" observations should not be unique to the TXAR site. In particular, similar signals should be seen at U.S. stations located west of TXAR and within a few hundred kilometers of the re-entry ground track when the Kennedy Space Center is used as the recovery facility for the orbiter vehicle.

They will not be seen during the final stage of the re-entry flight profile. During this stage the ground speed of the orbiter vehicle is systematically reduced in preparation for the landing at the recovery facility. As the ground speed is reduced, it may be inferred from eqs. (21)–(24) that the individual seismic phases of the "shuttle-quakes" will separately vanish. The extinction of the "bow" waves will be complete when the ground speed drops below ~ 3 km/sec. They are replaced by inhomogeneous waves. The inhomogeneous waves originate at the tip of the advancing shock front trace and propagate parallel to the ground track at the ground speed of the orbiter vehicle. They will appear as low frequency seismic precursors to the re-entry shock front but will only be detected at stations near the ground track. The low frequency pulses that were reported by KANAMORI *et al.* (1991) are believed to be inhomogeneous waves.

There are other potential sources for seismic "bow" waves. For example, the entry of meteorites into the earth's atmosphere will also generate shock fronts. Their ground speeds (~ 12–30 km/sec) will also generally exceed the re-entry ground speed of the space shuttle orbiter vehicle. Based upon the results summarized above, it then follows that seismic "bow" waves will be generated if their shock fronts intersect the surface of the earth before their flight trajectories are terminated. The seismic observations reported by ANGLIN and HADDON (1987) confirm this expectation for a visible meteorite whose ground track traversed the Yellow Knife Array (YKA) in Northwest Territory, Canada. Figure 2 of their report shows a weak high frequency *P* wave followed by a much stronger, high frequency R_g wave. It is clear from the seismograms seen in this figure that the propagation azimuths of these two phases across the array are significantly different, thus identifying them as "bow" waves. The remainder of the reported observations of seismic disturbances associated with the passage of a visible meteorite cannot be explained by the invocation of the "bow" wave model. These observations are commonly explained as being the result of the explosive disintegration of the meteorite in the atmosphere (see KAMAR, 1995 for example of this type of event and a reference list). Based upon the published evidence, it therefore appears that the shock fronts of visible meteorites rarely reach the surface of the earth before their flight paths are terminated.

Appendix A

Evaluation of the Expressions for the Rayleigh Wave Spectrum Excited by Moving, Point Pressure Source Pairs

It is seen by inspection of eqs. (19) and (20) in the text that the expressions for $\hat{u}_z^p$ involve integrals of the form

$$I(\alpha, \beta, \chi) = \int_{-\infty}^{\infty} \frac{e^{-\chi\alpha} \sinh(\xi\beta) e^{-ik_x}}{\xi(k - k_v)} dk; \quad \text{where } \alpha \geq \beta \ . \tag{A-1}$$

Terms of this form cannot be evaluated by direct integration and numerical evaluation is computationally intensive. Given this situation, it is useful to replace the variable of integration, k, with the complex variable, $\zeta = k + i\eta$, and apply the methods of contour integration in the ζ plane. Also, to aid the implementation of this approach, it is useful to include the effects of attenuation in the earth in the derivations. This is accomplished by replacing κ with;

$$\kappa = \kappa_R - i\kappa^* \ , \tag{A-2}$$

where

$$\kappa_R = \frac{|\omega|}{c_R} \ , \tag{A-3}$$

$$\kappa^* = \frac{|\omega|}{2c_R Q_R} \tag{A-4}$$

and Q_R is the spatial dissipation factor for Rayleigh waves in the half-space. It is seen fom eq. (A-1) that the movement of the shock front in the positive x direction at a constant speed, v, introduces a simple, real pole at the point $[k_v, 0]$. In addition, since the integrand of I is not an even function of ξ, there are branch points at $\pm\kappa$. The path of integration in ζ plane is chosen to satisfy the condition, $\text{Re}\xi \geq 0$. This constraint is imposed to insure that u_z remains finite as $y \to \pm\infty$. It then follows from eq. (11) in the accompanying text and eq. (A-2) that in the ζ plane

$$\xi^2 = \zeta^2 - \kappa^2 = (\zeta + \kappa)(\zeta - \kappa) = |\zeta^2 - \kappa^2| e^{i(\vartheta_1 + \vartheta_2)} , \tag{A-5}$$

where

$$\vartheta_1 = \tan^{-1}\left[\frac{\eta - \kappa^*}{k + \kappa_R}\right] , \tag{A-6}$$

$$\vartheta_2 = \tan^{-1}\left[\frac{\eta + \kappa^*}{k - \kappa_R}\right] . \tag{A-7}$$

Therefore

$$\xi = |\zeta^2 - \kappa^2|^{1/2} e^{i(\vartheta_1 + \vartheta_2)/2} . \tag{A-8}$$

It then follows that

$$\mathrm{Re}\,\xi = |\zeta^2 - \kappa^2|^{1/2} \cos\left(\frac{\vartheta_1 + \vartheta_2}{2}\right) , \tag{A-9}$$

$$\mathrm{Im}\,\xi = |\zeta^2 - \kappa^2|^{1/2} \sin\left(\frac{\vartheta_1 + \vartheta_2}{2}\right) . \tag{A-10}$$

If η is set to zero in eqs. (A-6) and (A-7), it is readily shown from eqs. (A-9) and (A-10) that on the real axis of the ζ plane, $\mathrm{Re}\,\xi \geq 0$ and $\mathrm{Im}\,\xi \geq 0$, and so it may be included as part of the integration path. Then, as shown in Figures (A-1a) and (A-1b), to insure convergence the integration path is closed by an infinite arc in the lower half plane if $\chi > 0$ or a similar arc in the upper half plane if $\chi < 0$. The source pole is excluded by deforming the path on the real axis to pass below it when $\chi > 0$ and above it when $\chi < 0$. The exclusion of the branch points requires branch cuts. The path of the branch cuts in the ζ plane is chosen to satisfy the condition, $\mathrm{Re}\,\xi = 0$. If $\mathrm{Re}\,\xi = 0$, then ξ^2 must be real and negative. If ξ^2 is real it follows from eq. (A-5) that

$$\sin(\vartheta_1 + \vartheta_2) = 0 . \tag{A-11}$$

With the aid of eqs. (A-6) and (A-7), it is found that eq. (A-11) is satisfied by

$$\eta = -\frac{\kappa_R \kappa^*}{k} . \tag{A-12}$$

Equation (A-12) constrains the branch cuts to the hyperbolic path shown as dashed lines in Figures A-1a and A-1b. Now, if ξ^2 is negative, it follows from eq. (A-5) that

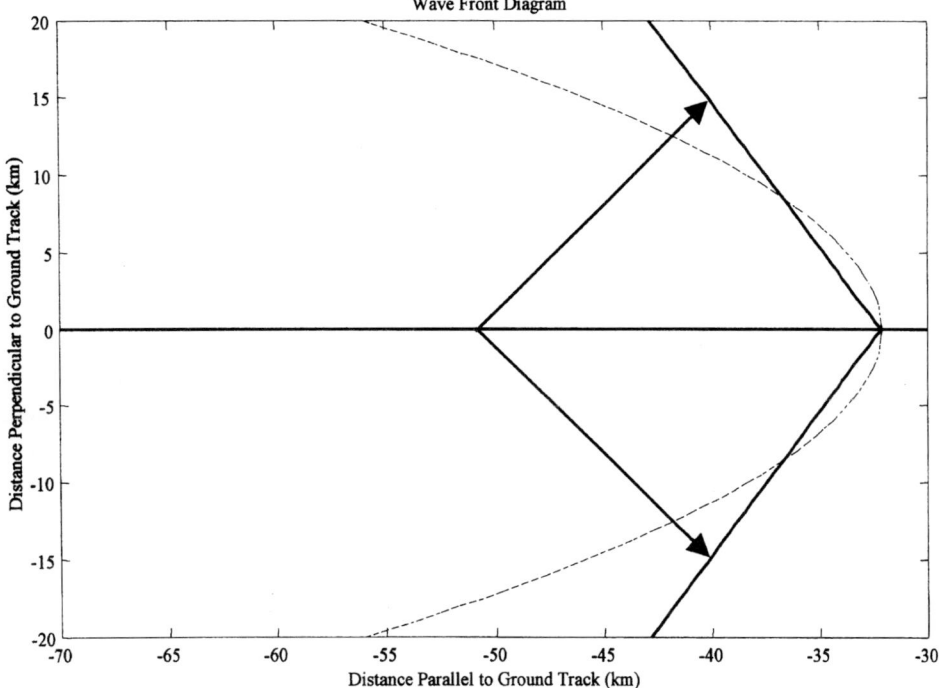

Figure 6

A wavefront diagram illustrating the relationship between the shock front trace and the Rayleigh "bow" wave. The dashed curve represents the shock front trace. The solid horizontal line identifies the space shuttle ground track. The two solid lines extending from the tip of the shock front trace identify the two wings of the Rayleigh "bow" wave. The two arrows are representative "bow" wave propagation vectors.

$$\cos(\vartheta_1 + \vartheta_2) < 0 \qquad (\text{A-13})$$

and with the aid of eqs. (A-6) and (A-7) it is found that eq. (A-13) implies

$$k^2 - \eta^2 \le \kappa_R^2 - \kappa^{*2} \;. \qquad (\text{A-14})$$

Equation (A-14) identifies the part of the hyperbola that is to be used as the branch cut. Evaluation of this inequality reveals that the branch cuts terminate at $\zeta = \pm\kappa$. Finally, with the application of arguments originally posed by EWING *et al.* (1957, pp. 44–47), it may be shown that the signs of Imζ on either side of the branch cuts are as illustrated in Figures (A-1a) and (A-1b). It then follows that

$$I(\alpha, \beta, \chi) = i\pi \, Res\,(\alpha, \beta, \chi) + L_k^+(\alpha, \beta, \chi); \quad \chi > 0 \qquad (\text{A-15})$$

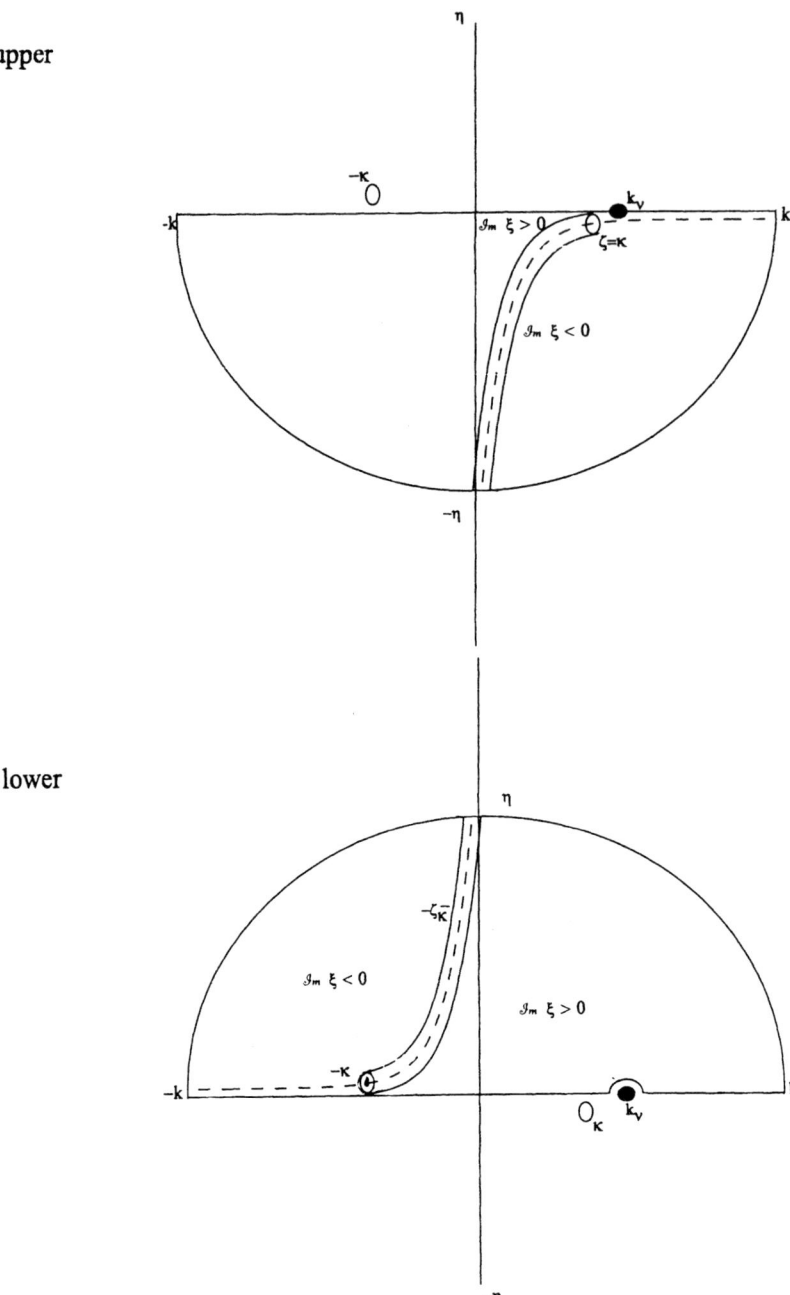

Figure A-1

The contour used for the evaluation of $I(\alpha, B, \chi)$ when $\chi > 0$ (upper) and when $\chi < 0$ (lower).

and

$$I(\alpha, \beta, \chi) = -i\pi \, Res \,(\alpha, \beta, \chi) - L_k^-(\alpha, \beta, \chi); \quad \chi < 0 \,, \tag{A-16}$$

where $Res \,(\alpha, \beta, \chi)$ is the residue at the source pole and L_k^+ and L_k^- are the branch line integrals in the fourth and second quadrants, respectively.

Now let

$$\chi(\tau) = x - v\tau + b\rho(y_s) \,. \tag{A-17}$$

It then follows from eqs. (19) and (20) in the accompanying text that

$$\hat{u}_z^p = \frac{\kappa BP(\omega, y_s)}{\pi i \mu} \left[e^{-i\omega\tau} I(y, y_s, \chi(\tau)) - I(y, y_s, \chi(0)) \right]; \quad y > y_s \tag{A-18}$$

and

$$\hat{u}_z^p = \frac{\kappa BP(\omega, y_s)}{\pi i \mu} \left[e^{-i\omega\tau} I(y_s, y, \chi(\tau)) - I(y_s, y, \chi(0)) \right]; \quad y \le y_s \,. \tag{A-19}$$

Now for the sake of illustration assume that $x > 0$. It then follows from eqs. (1) and (16) in the accompanying text that $\chi(0) > 0$ and that when $x > v\tau - b\rho(y_s)$, $\chi(\tau) > 0$. However, if $x < v\tau - b\rho(y_s)$ then $\chi(\tau) < 0$.

It is then seen from eqs. (A-15), (A-16) and (A-19) that if $x > v\tau - b\rho(y_s)$, and $y > y_s$;

$$\hat{u}_z^p = \frac{\kappa BP(\omega, y_s)}{\mu} \left\{ e^{-i\omega\tau} Res \,(y, y_s, \chi(\tau)) - Res \,(y, y_s, \chi(0)) \right.$$

$$\left. + \frac{1}{\pi i} \left[e^{-i\omega\tau} L_\kappa^+(y, y_s, \chi(\tau)) - L_\kappa^+(y, y_s, \chi(0)) \right] \right\} \,, \tag{A-20}$$

and if $x > v\tau - b\rho(y_s)$, and $y \le y_s$;

$$\hat{u}_z^p = \frac{\kappa BP(\omega, y_s)}{\mu} \left\{ e^{-i\omega\tau} Res \,(y_s, y, \chi(\tau)) - Res \,(y_s, y, \chi(0)) \right.$$

$$\left. + \frac{1}{\pi i} \left[e^{-i\omega\tau} L_\kappa^+(y, y_s, \chi(\tau)) - L_\kappa^+(y, y_s, \chi(0)) \right] \right\} \,. \tag{A-21}$$

Similarly, if $x < v\tau - b\rho(y_s)$ and $y > y_s$

$$\hat{u}_z^p = -\frac{\kappa BP(\omega, y_s)}{\mu} \left\{ e^{-i\omega\tau} Res \,(y, y_s, \chi(\tau)) + Res \,(y, y_s, \chi(0)) \right.$$

$$\left. + \frac{1}{\pi i} \left[e^{-i\omega\tau} L_\kappa^-(y, y_s, \chi(\tau)) + L_\kappa^+(y, y_s, \chi(0)) \right] \right\} \tag{A-22}$$

and if $x < v\tau - b\rho(y_s)$ and $y \leq y_s$

$$\hat{u}_z^p = -\frac{\kappa B P(\omega, y_s)}{\mu} \left\{ e^{-i\omega\tau} \, Res \, (y_s, y, \chi(\tau)) + Res \, (y_s, y, \chi(0)) \right.$$

$$\left. + \frac{1}{\pi i} \left[e^{-i\omega\tau} L_\kappa^- (y_s, y, \chi(\tau)) + L_\kappa^+ (y_s, y, \chi(0)) \right] \right\} . \qquad (A-23)$$

Equations (A-20) through (A-23) constitute the formal solution for the spectrum of the vertical component of the Rayleigh wave generated by a pair of point pressure loads moving at a constant speed, v, in the positive x direction. It may be shown that the branch line integrals appearing on the rhs of these equations do not contribute to the formation of the seismic disturbances with the properties of the "shuttle quakes" observed at TXAR. Consequently, the remainder of this appendix will be devoted to the examination of the contributions from the source pole. These contributions are represented by the residues that appear on the rhs of the equations referenced above.

It follows from eq. (A-1) that

$$Res \, (\alpha, \beta, \chi) = \frac{e^{-\xi_v \alpha} \sinh(\xi_v \beta) e^{-i k_v \chi}}{\xi_v}; \quad \alpha \geq \beta , \qquad (A-24)$$

where

$$\xi_v = \sqrt{k_v^2 - \kappa^2} . \qquad (A-25)$$

Then, with the aid of eqs. (17-a) it may be shown that

$$e^{-i\omega\tau} Res \, (y, y_s, \chi(\tau)) = Res \, (y, y_s, \chi(0)), \quad y > y_s \qquad (A-26)$$

and

$$e^{-i\omega\tau} Res \, (y_s, y, \chi(\tau)) = Res \, (y_s, y, \chi(0)); \quad y \leq y_s . \qquad (A-27)$$

Now let $_s\hat{u}_z^p$ represent the contribution from the source pole to the point source spectrum. Then, by retaining the residue terms seen on the rhs of eqs. (A-20) through (A-23), and with the aid of eqs. (A-26) and (A-27), it is found that

$$_s\hat{u}_z^p = 0; \quad x > v\tau - b\rho(y_s) . \qquad (A-28)$$

It also follows that if $x < v\tau - b\rho(y_s)$

$$_s\hat{u}_z^p = -\frac{2\kappa B P(\omega, y_s)}{\mu} Res \, (y, y_s, \chi(0)); \quad y > y_s \qquad (A-29)$$

and

$$_s\hat{u}_z^p = -\frac{2\kappa B P(\omega, y_s)}{\mu} Res \, (y_s, y, \chi(0)); \quad y > y_s . \qquad (A-30)$$

Appendix B

An Expression for the Source Pole Contribution to the Rayleigh Wave Spectrum Excited by a Space Shuttle Shock Front when $c_R < v$

Equations (21) and (25) through (27) in the accompanying text imply that the solution for the $_p u_z$ takes on several different forms, depending on the location of the observation point with respect to the advancing shock front. In particular, it is found that if $y > Y$, then the observation point lies outside the region swept by the shock front and it follows from eq. (28) in the accompanying text that

$$_s u_z = 0; \quad x > v\tau - bz_0 , \tag{B-1}$$

and if $x < v\tau - bz_0$

$$_s u_z = -\frac{2B}{\mu \sin \theta_v} e^{-i\kappa(x \cos \theta_v + y \sin \theta_v)} \int_0^{\tilde{y}} F(y_s) P(\omega, y_s) \sin(\kappa y_s \sin \theta_v) e^{-i\kappa b \rho(y_s) \cos \theta_v} \, dy_s , \tag{B-2}$$

where

$$\tilde{y} = \frac{1}{b} \sqrt{(v\tau - x)^2 - (bz_0)^2}; \quad v\tau - b\rho(Y) < x < v\tau - bz_0, \tag{B-3}$$

$$\tilde{y} = Y; \quad x \le v\tau - b\rho(Y) .$$

The situation is somewhat more complicated if $y \le Y$ for then the observation point lies within the region swept by the shock front. In this case, when x lies in the interval, $v\tau - bz_0 < x < v\tau - b\rho(y)$, it is found that

$$_s u_z = -\frac{2B}{\mu \sin \theta_v} e^{-i\kappa(x \cos \theta_v + y \sin \theta_v)} \int_0^{\tilde{y}} F(y_s) P(\omega, y_s) \sin(\kappa y_s \sin \theta_v) e^{-i\kappa b \rho(y_s) \cos \theta_v} \, dy_s \tag{B-4}$$

and, when x lies in the interval,

$$v\tau - b\rho(y) \le x \le v\tau - b\rho(Y) ; \tag{B-5}$$

$$_s u_z =_s u_z^{(1)} + _s u_z^{(2)} , \tag{B-6}$$

where

$$_s u_z^{(1)} = -\frac{2B}{\mu \sin \theta_v} e^{-i\kappa(x \cos \theta_v + y \sin \theta_v)} \int_0^y F(y_s) P(\omega, y_s) \sin(\kappa y_s \sin \theta_v) e^{-i\kappa b \rho(y_s) \cos \theta_v} \, dy_s \tag{B-7}$$

$$_s u_z^{(2)} = -\frac{2B}{\mu \sin \theta_v} e^{-i\kappa(x \cos \theta_v)} \sin(\kappa y \sin \theta_v) \int_y^{\tilde{y}} F(y_s) P(\omega, y_s) e^{-i\kappa(b\rho(y_s) \cos \theta_v + y_s \sin \theta_v)} \, dy_s . \tag{B-8}$$

It is seen by inspection of the rhs of eqs. (B-2) through (B-8) that the expression for $_s u_z$ involve integrals of the form

$$I^{(1)}(\kappa \sin \theta_v) = \int Q(\omega, y_s) e^{-i\kappa b \rho(y_s) \cos \theta_v} \sin(\kappa y_s \sin \theta_v) dy_s \qquad \text{(B-9)}$$

and

$$I^{(2)}(\kappa \sin \theta_v) = \int Q(\omega, y_s) e^{-i\kappa b \rho(y_s) \cos \theta_v} e^{-i\kappa y_s \sin \theta_v} dy_s \ , \qquad \text{(B-10)}$$

where

$$Q(\omega, y_s) = P(\omega, y_s) F(y_s) \ . \qquad \text{(B-11)}$$

It is seen by inspection of the rhs of eq. (B-9) that $I^{(1)}(\kappa \sin \theta_v)$ is the finite Fourier Sine Transform of the function, $Q(\omega, y_s) e^{-i\kappa b \rho}(y_s) \cos \theta_v$. Similarly, it is seen from eq. (B-10) that $I^{(2)}(\kappa \sin \theta_v)$ is the finite Fourier Transform of the same function. It then follows that the Fast Fourier Transform (FFT) algorithm may be used to numerically evaluate $I^{(1)}(\kappa \sin \theta_v)$ and $I^{(2)}(\kappa \sin \theta_v)$. Initial attempts to directly implement this approach yielded acceptable results although at a cost of considerable CPU time. Consequently, an alternative approach was adopted. This approach and the results of its implementation are summarized below.

It can be reasonably assumed that $Q(\omega, y_s)$ is a smoothly, varying function of y_s so that all derivatives with respect to this variable are continuous in the interval $0 \le y_s \le Y$. We therefore express $Q(\omega, y_s)$ as

$$Q(\omega, y_s) = \sum_{m=0}^{\infty} q_m y_s^m \ , \qquad \text{(B-12)}$$

where from eqs. (4) and (1) in the accompanying text

$$q_m = 0; \quad m \text{ odd} \qquad \text{(B-13)}$$

and

$$q_m = \frac{1}{m!} \left(\frac{\partial}{\partial y_s} \right)^m Q(\omega, y_s)|_{y_s=0}; \quad m \text{ even} \ . \qquad \text{(B-14)}$$

Then, assuming uniform convergence in the interval, $0 \le y_s \le Y$, it follows that

$$I^{(1)}(\kappa \sin \theta_v) = \sum_{m=0}^{\infty} q_{2m} I_m^{(1)}(\kappa \sin \theta_v) \qquad \text{(B-15)}$$

and

$$I^{(2)}(\kappa \sin \theta_v) = \sum_{m=0}^{\infty} q_{2m} I_m^{(2)}(\kappa \sin \theta_v) \ , \tag{B-16}$$

where

$$I_m^{(1)}(\kappa \sin \theta_v) = \int y_s^{2m} e^{-i\kappa b \rho(y_s) \cos \theta_v} \sin(\kappa y_s \sin \theta_v) dy_s \tag{B-17}$$

and

$$I_m^{(2)}(\kappa \sin \theta_v) = \int y_s^{2m} e^{-i\kappa b \rho(y_s) \cos \theta_v} e^{-i\kappa y_s \sin \theta_v} \, dy_s \ . \tag{B-18}$$

Approximate solutions to eqs. (B-17) and (B-18), in the form of asymptotic series, valid, for $\kappa \sin \theta_v \gg 1$, are then found by repeated application of the method of integration by parts. these solutions are

$$
\begin{aligned}
I_m^{(1)}(\kappa \sin \theta_v) = \frac{(-1)^m}{(\sin \theta_v)^{2m}} & \left\{ y_s e^{-i\kappa b \rho \cos \theta_v} \left(\frac{\partial}{\partial \kappa} \right)^{2m} n_0(\kappa y_s \sin \theta_v) \right. \\
& + \sin \theta_v \sum_{\ell=1} (\kappa y_s b \cot \theta_v)^{\ell+1} \frac{h_{\ell-1}^{(2)}(\kappa b \cos \theta_v)}{\rho^{\ell-1}} \\
& \left. \times \left(\frac{\partial}{\partial \kappa} \right)^{2m} \frac{n_1(\kappa y_s \sin \theta_v)}{\kappa^\ell} \right\} \ ,
\end{aligned}
\tag{B-19}
$$

$$
\begin{aligned}
I_m^{(2)}(\kappa \sin \theta_v) = \frac{(-1)^m}{(\sin \theta_v)^{2m}} & \left\{ y_s e^{i\kappa b \rho \cos \theta_v} \left(\frac{\partial}{\partial \kappa} \right)^{2m} h_0^{(?)}(\kappa y_s \sin \theta_v) \right. \\
& + \sin \theta_v \sum_{\ell=1} (\kappa y_s b \cot \theta_v)^{\ell+1} \frac{h_{\ell-1}^{(2)}(\kappa b \cos \theta_v)}{\rho^{\ell-1}} \left(\frac{\partial}{\partial \kappa} \right)^{2m} \\
& \left. \times \frac{h_\ell^{(2)}(\kappa y_s \sin \theta_v)}{\kappa^\ell} \right\} \ .
\end{aligned}
\tag{B-20}
$$

The symbol, n_ℓ seen on the rhs of eqs. (B-19) and (B-20) identifies a spherical Neumann function of order, ℓ. Whereas the symbol, $h_\ell^{(2)}$ identifies a spherical Hankel function of the second kind and order, ℓ.

Now, to simplify the notation let

$$U_z^{(1)}(\kappa \sin \theta_v, 0) = \frac{2B}{\mu \sin \theta_v} e^{-i\kappa(x \cos \theta_v + y \sin \theta_v)} \sum_{m=0}^{\infty} q_{2m} I_m^{(1)}(\kappa \sin \theta_v, 0) \ , \tag{B-21}$$

$$U_z^{(1)}(\kappa \sin \theta_v, \tilde{y}) = \frac{2B}{\mu \sin \theta_v} e^{-i\kappa(x \cos \theta_v + y \sin \theta_v)} \sum_{m=0}^{\infty} q_{2m} I_m^{(1)}(\kappa \sin \theta_v, \tilde{y}) \ , \tag{B-22}$$

$$U_z^{(2)}(\kappa \sin \theta_v, 0) = \frac{2B}{\mu \sin \theta_v} e^{-i\kappa x \cos \theta_v} \sum_{m=0}^{\infty} q_{2m} \left[\sin(\kappa y \sin \theta_v) I_m^{(2)}(\kappa \sin \theta_v, y) \right.$$

$$\left. - e^{-iky \sin \theta_v} I_m^{(1)}(\kappa \sin \theta_v, y) \right] , \qquad \text{(B-23)}$$

$$U_z^{(2)}(\kappa \sin \theta_v, \tilde{y}) = \frac{2B}{\mu \sin \theta_v} \sin(\kappa y \sin \theta_v) e^{-i\kappa x \cos \theta_v} \sum_{m=0}^{\infty} q_{2m} I_m^{(2)}(\kappa \sin \theta_v, \tilde{y}) , \qquad \text{(B-24)}$$

where $I_m^{(j)}(\kappa \sin \theta_v, \alpha)$; $j = 1, 2$ defines the value of $I_m^{(j)}$ at $y_s = \alpha$. It then follows from eqs. (B-1) when the observation point lies ahead of the advancing tip of the shock front trace.

$$_s u_z = 0 . \qquad \text{(B-25)}$$

If the observation point lies outside the region swept by the shock front trace and behind the tip of the shock front trace but ahead of its lateral cut-off points, it then follows from eq. (B-2) that

$$_s u_z = U_z^{(1)}(\kappa \sin \theta_v, 0) - U_z^{(1)}(\kappa \sin \theta_v, \tilde{y}) \qquad \text{(B-26)}$$

and as the lateral cut-off point passes the observation point, the expression for $_s u_z$ becomes

$$_s u_z = U_z^{(1)}(\kappa \sin \theta_v, 0) - U_z^{(1)}(\kappa \sin \theta_v, Y) . \qquad \text{(B-27)}$$

Equations (B-26) and (B-27) are also applicable inside the region swept by the shock front trace as long as the observation point lies ahead of its advancing wing. Then, as the shock front trace reaches, then passes the observation point, the following expression for $_s u_z$ is applicable:

$$_s u_z = U_z^{(1)}(\kappa \sin \theta_v, 0) + U_z^{(2)}(\kappa \sin \theta_v, y) - U_z^{(2)}(\kappa \sin \theta_v, \tilde{y}) \qquad \text{(B-28)}$$

and finally as the lateral cut-off points reach, then pass the observation point;

$$_s u_z = U_z^{(1)}(\kappa \sin \theta_v, 0) + U_z^{(2)}(\kappa \sin \theta_v, y) - U_z^{(2)}(\kappa \sin \theta_v, Y) . \qquad \text{(B-29)}$$

REFERENCES

ANGLIN, F. M. and HADDON, R. A. W. (1987), *Meteoroid Sonic Shock-wave-generated Seismic Signals Observed at a Seismic Array*, Nature *328*, 607–609.

COOK, J. C., GOFORTH, T., and COOK, R. K. (1972), *Seismic and Underwater Responses to Sonic Boom*, J. Acous. Soc. Am. *51*, 729–742.

EWING, M. W., JARDETZKY, W. S., and PRESS, F., *Elastic Waves in Layer Media* (McGraw-Hill Book Co., New York, 1957) 380 p.

GOFORTH, T. T. and MCDONALD, J. A., *Seismic Effects of Sonic Booms* (NASA CR-1137, 1968) 118 pp.

KAMAR, A. (1995), *Space Shuttle and Meteroid – Tracking Supersonic Objects in the Atmosphere with Seismographs*, Seis. Res. Lett *66*, 6–12.

KANAMORI, H., MORI, J., ANDERSON, D. L., and HEATON, T. H. (1991), *Seismic Excitation by the Space Shuttle, Columbia*, Nature *349*, 781–782.

(Received June 12, 2000, revised/accepted July 14, 2000)

 To access this journal online:
http://www.birkhauser.ch